Springer Proceedings in Earth and Environmental Sciences

The series Springer Proceedings in Earth and Environmental Sciences publishes proceedings from scholarly meetings and workshops on all topics related to Environmental and Earth Sciences and related sciences. This series constitutes a comprehensive up-to-date source of reference on a field or subfield of relevance in Earth and Environmental Sciences. In addition to an overall evaluation of the interest, scientific quality, and timeliness of each proposal at the hands of the publisher, individual contributions are all refereed to the high quality standards of leading journals in the field. Thus, this series provides the research community with well-edited, authoritative reports on developments in the most exciting areas of environmental sciences, earth sciences and related fields.

Kirtikumar Randive · Ashok Kumar Nandi ·
Pradeep Kumar Jain · Sanjeevani Jawadand
Editors

Current Trends in Mineral-Based Products and Utilization of Wastes: Recent Studies from India

Prospects and Challenges of Mineral Based Products and Utilization of Wastes for the 'Make in India' Initiative, Nagpur November 10–11, 2022

Springer

Editors
Kirtikumar Randive
Department of Geology
Rashtrasant Tukadoji Maharaj Nagpur
University
Nagpur, India

Ashok Kumar Nandi
International Bauxite, Alumina
and Aluminium Society (IBAAS)
Mineral Information and Development
Centre
Nagpur, India

Pradeep Kumar Jain
Indian Bureau of Mines (Retd.)
Nagpur, India

Sanjeevani Jawadand
Shri Mathuradas Mohota College
of Science
Nagpur, India

ISSN 2524-342X ISSN 2524-3438 (electronic)
Springer Proceedings in Earth and Environmental Sciences
ISBN 978-3-031-50261-3 ISBN 978-3-031-50262-0 (eBook)
https://doi.org/10.1007/978-3-031-50262-0

This Springer imprint is published by the registered company Springer Nature Switzerland AG
The registered company address is: Gewerbestrasse 11, 6330 Cham, Switzerland

Paper in this product is recyclable.

Preface

In the realm of Earth Sciences, where the profound understanding of our planet's resources and their sustainable utilization stands as a paramount concern, the exploration of mineral-based products and waste utilization has emerged as a pivotal area of research. Waste utilization has emerged as a pivotal facet in sustainable resource management, addressing environmental concerns and fostering economic viability. Recycling and repurposing methods are at the forefront, transforming discarded mineral materials into valuable resources, reducing landfill pressure, and minimizing environmental impact. Circular economy principles guide these initiatives, emphasizing the reintegration of mineral waste into production cycles. As we step into an era marked by rapid technological advancements and burgeoning industrialization, the significance of harnessing mineral resources efficiently and managing waste judiciously cannot be overstated. The mineral-based product industry is undergoing a paradigm shift fueled by technological innovation and sustainability imperatives. Current trends reflect a heightened emphasis on sustainable mining practices, integrating eco-friendly techniques, waste utilization, and renewable energy sources. Advanced materials development stands at the forefront, yielding innovative applications in electronics, renewable energy, and value-added products. Moreover, the surge in demand for battery minerals like lithium, cobalt, and nickel, driven by the electric vehicle revolution, has reshaped mining landscapes. Concurrently, a growing focus on recycling technologies underscores efforts to create a circular economy, mitigating resource depletion concerns.

In acknowledging this, it becomes imperative for us to find a sustainable path forward. The international seminar on "Prospects and Challenges of Mineral Based Products and Utilization of Wastes for the 'Make in India' Initiative," held from November 10–12, 2022, at Hotel Radisson Blu, was a pivotal moment. Attended by 120 delegates from all corners of India, this event served as a platform for deliberation and collaboration.

This book, a compilation of selected papers presented during the conference, delves into various facets of mining and its aftermath. The chapters are categorized into distinct sections: rare earth elements, ferrous minerals, non-ferrous minerals, industrial minerals, waste utilization and valorization, and other significant contributions. These papers, crafted by experts and curated by academicians and technocrats, represent a collective effort to explore sustainable mining practices.

The core principles highlighted in this book revolve around the optimal utilization of mined-out materials and the creation of value-added products from mining wastes. It is a testament to the collaborative spirit of researchers, professionals, and students who are dedicated to finding innovative solutions to the challenges posed by mining.

This volume is not just a compendium of scholarly work it is a beacon guiding us toward a future where the invaluable resources bestowed upon us by the earth are harnessed responsibly. We hope that the insights contained within these pages inspire further research, dialogue, and action, leading us toward a more sustainable and harmonious coexistence with our planet.

This book contains 25 chapters, the first chapter by Godbole et al. which offers a comprehensive overview of cutting-edge technologies utilized in the extraction of rare earth elements (REE). Deshpande et al. delved into the eco-environmental challenges associated with the utilization of REE. Jawadand et al. proposed a novel approach focused on reshaping the global supply chain of REE through urban mining. Singh et al. proposed novel ideas for the beneficiation and extraction of strategic scandium (Sc) from waste residues in the aluminum industry. Shrinivas Viladkar provides a comprehensive review including valuable insights and suggestions on the exploration of REE mineralization in Amba Dongar Carbonatite, Gujarat, India. Rajeev Bidwai sheds light on the Siwana Ring Complex in Western Rajasthan, highlighting its potential to become a REE hub.

Currently, ferrous elements have been subjected to various new techniques which represent significant advancements in the efficient utilization and conservation of ferrous minerals, ensuring both economic and environmental sustainability in the modern mining industry. Rath et al. conducted a detailed review of the Mineralogical Process Audit of slimes from the Kirandul Plant at Bailadila Iron Ore Mine in Chhattisgarh, India. The chapter by Subrahmanyam et al. focused on the flowability characteristics of iron ore. Chalavadi et al. discussed the topic of fine-size dry iron ore beneficiation using a Thin Bed Air Fluidized Dry Separator. Rao and Chaurasiya presented the role of NMDC in mineral conservation.

Non-ferrous minerals, ranging from low grade to high grade, are processed through beneficiation techniques to create a variety of mineral-based products. The chapter by Sinha et al. offers insights into the creation of high alumina synthetic aggregates using impure raw bauxite, outlining the development process. Bhukte et al. present an overview of the characterization and beneficiation of pyrophyllite, offering valuable insights into its properties and processing methods.

Industrial minerals are pioneers in the creation of mineral-based products and waste utilization, with current trends highlighting their forefront position in driving innovative solutions and sustainable practices. Jain and Sharma presented an interesting review of limestones with reference to the iron and steel industry. The chapter by Alok Sood adopts a pragmatic approach to Industrial Minerals Exploration. Basu et al. provide a comprehensive overview of the Atmanirbhar Bharat self-reliant initiative, focusing on the import substitution of raw materials in the refractory industry.

Sinha et al. present the concept of "R4 is the key" is presented, emphasizing the importance of reducing, reusing, recycling, and responsible waste management in sustainable practices. The chapter by Raghupatruni Rao discussed the idea of recovering placer industrial minerals from red sediments in badlands topography and adding value to them.

Waste utilization and valorization focus on creating value-added products from discarded materials, promoting sustainability and resource efficiency. Godbole et al. introduced an innovative approach, emphasizing the transformation of mining waste into valuable products through creative methods. Najar et al. explored the commercial and technical prospects associated with the untapped mineral values within the aluminum industry. The chapter by Rai et al. offers an Indian perspective on the potential applications of red mud in the cement industry, highlighting its innovative uses. Nandi and Shrirame outlined opportunities and methods for value-added bauxite products through effective beneficiation processes. K. S. Raju proposed sustainable techniques for dewatering and disposal of processing plant tailings, emphasizing environmentally conscious practices. Sit et al. introduced a novel approach for the utilization of internal waste (Fe <35%) from iron ore mines, showcasing innovative strategies. Deshpande et al. presented challenges and barriers in implementing value addition to mine waste, highlighting concerns regarding sustainability in the mining industry.

Current trends in mineral-based products and value addition focus on online analytical techniques, emphasizing real-time analysis and quality control for enhanced product quality and process efficiency. Mishra et al. present a novel approach to safe, accurate, and reliable online elemental analysis which is achieved through the unique Prompt Gamma Neutron Activation Analysis (PFTNA) technique, ensuring precision and efficiency in various industrial applications. We did our best to cover almost all known, potential, and possible avenues of current trends of mineral-based products and waste utilization. We believe this book will provide a useful reference to the students as well as researchers working in these areas.

Nagpur, India

Kirtikumar Randive
Ashok Kumar Nandi
Pradeep Kumar Jain
Sanjeevani Jawadand

Contents

Innovative Technologies for Recycling and Extraction of REE

Pratik Godbole, Kaustubh Deshpande, Sanjeevani Jawadand, M. L. Dora, Atul Selokar, Gopal Daware, Manoj Sahu, Ashok Kumar Nandi, and Kirtikumar Randive

Abstract Rare Earth Elements (REEs) have become indispensable components in the evolution of contemporary technologies, spanning applications in electronics and clean energy systems. However, the increasing demand for REEs has raised legitimate concerns regarding their sustainable sourcing and environmental repercussions. This abstract delves into the innovative technologies and methodologies employed in the recycling and extraction of REEs, offering a comprehensive overview of recent strides in this field. Conventional REE extraction typically entails energy-intensive processes and poses environmental challenges, including the generation of radioactive waste and chemical pollutants. In response, both researchers and industries are actively exploring alternative approaches that prioritize efficiency, sustainability, and environmental conscientiousness. This chapter provides profound insights into the evolving landscape of REE recycling and extraction, highlighting the potential for a future that is more sustainable and environmentally responsible in the utilization of these critical elements. The integration of these pioneering technologies holds promise for diminishing our reliance on primary mining, ameliorating environmental impacts, and ensuring a stable supply of REEs for emerging technologies.

Keywords Rare earth elements · Innovative technologies · Recycling methods · Extraction methods

---

P. Godbole · K. Deshpande · A. Selokar · G. Daware · K. Randive (✉)
Department of Geology, RTM Nagpur University, Nagpur, India
e-mail: randive101@yahoo.co.in

S. Jawadand
Shri Mathuradas Mohota College of Science, Nagpur, India

M. L. Dora
Geological Survey of India, Hyderabad 500068, India

M. Sahu
JSW Steel Limited, Bandra East, Mumbai, India

A. K. Nandi
Mineral Information & Development Centre (I) Pvt. Ltd., Nagpur, India

K. Randive et al. (eds.), *Current Trends in Mineral-Based Products and Utilization of Wastes: Recent Studies from India*, Springer Proceedings in Earth and Environmental Sciences, https://doi.org/10.1007/978-3-031-50262-0_1

1 Introduction

The global "Rare Earth" crisis in 2010, triggered by China's dominance with an over 80% share of the rare earth element (REE) market and its imposition of export restrictions, prompted a surge in exploration efforts for REEs worldwide, including India [1]. However, the Indian resource is characterized by significantly low ore grades and is often associated with radioactivity, making the extraction process lengthy, complex, and expensive. Furthermore, the Indian rare earth resource is primarily composed of Light Rare Earth Elements (LREE), with limited quantities of Heavy Rare Earth Elements (HREE) available for extraction. In the case of rare earths, a comprehensive value chain ecosystem is essential to transform reserves into finished products. This value chain comprises a multitude of processes and plants, including obtaining statutory clearances, mining operations, mineral beneficiation, rare earth extraction, separation, refining into oxide forms, metal extraction, and alloy production. Notably, in finished products, the usage of rare earths is typically in minuscule quantities. India has pre-existing facilities in place, spanning from mining operations to the separation and refining of rare earths into oxide forms. Moreover, India has developed the capability for metal extraction. However, the critical gap lies in the establishment of industrial-scale facilities for intermediate processes such as alloy production and magnet manufacturing. In these alloys, rare earths constitute a minor component, and a variety of other materials are also required for their production. Despite the rare earth sector being categorized as "free" from the stages of metal extraction onward, the absence of technology and infrastructure has hindered the development of industries in the intermediate segments [2].

The extraction of rare earth elements (REEs) from both primary and secondary ores is proving insufficient to meet the current demand. This inadequacy has led to a growing need for sustainable REE recovery from alternative sources to bridge the gap. The depletion of REE ores and the rising costs associated with traditional mining practices are compelling us to explore alternative avenues, especially the utilization of waste streams. These waste streams present a promising economic alternative for establishing additional sources of rare earth elements [3].

2 Innovative Technologies for Recycling and Extracting REE from Electronic Waste

Every year, a massive 50 million tonnes of electronic waste (e-waste) is generated worldwide, but regrettably, only a mere 20% of this e-waste is properly managed and processed. This leaves a significant portion of it largely untapped. This surplus of e-waste presents a substantial economic opportunity, with its material value alone estimated at a substantial $62.5 billion. As a result, e-waste emerges as a valuable resource ready for urban mining. The process of extracting resources from e-waste is notably less energy-intensive and environmentally friendly compared to traditional mining operations in the Earth's crust. Furthermore, it leads to a significant reduction in carbon dioxide emissions. Additionally, it offers unprecedented opportunities for global businesses and the global workforce. India plays a pivotal role in this context, ranking as the world's fourth-largest producer of e-waste. Each year, the country generates approximately 2 million tonnes of e-waste, but it only manages to recycle a mere 1.5% of the total output. This highlights the substantial potential for improving e-waste management and recycling efforts within India [4]. Following are some of the potential producers of E-waste.

1. Discarded Electronics: Old electronic devices such as computers, smartphones, and laptops contain trace amounts of rare earth elements in components like screens, magnets, and circuit boards [5].
2. Hard Drives: Hard drives in computers and other devices often contain neodymium magnets, which are a source of the rare earth element neodymium [6].
3. LCD Screens: Rare earth metals are used in the luminescent material in visual display unit either in display themselves (PDP AND OLED Technology) or the background illumination (LCD technology). The distinction can be made between the support matrix and the actual luminescent substances (activators) in luminescent material. The luminescent material contains Yttrium (Y_2O_3), Cerium ($CeMgAl_{11}O_{19}$) and lanthanum (La_2O_3) in largest proportion in 69.2, 8.5, 11.0% respectively followed by europium (4.9%), terbium (4.6%) and gadolinium (1.8%). Some luminescent substances may contain Eu, Tb, Gd, Sm, Er, Dy, Th, Lu quantitatively. Rare earth compounds used in plasma displays are (Y, Gd) BO_{33}: Eu, (Y_2O_3: Eu) [Red]; (Y, Gd) BO_3: Tb [Green]; $BaMgAl_{10}O_{17}$: Eu [Blue]. Cold cathode tubes are used for LCD background illumination in devices like notebooks, televisions, monitors etc. which contain Y, Eu, La, Ce, Tb, Gd, Pr etc. [7].
4. Lamps: Older fluorescent lamps contain europium and terbium, both of which are rare earth elements used in phosphors for lighting. Rare earth elements in LED light are generally used in luminescent substance which converts the high frequency blue or near UV light from the actual LED chip into continuous spectrum in the visible wavelength range. Cold white LEDs contain yellow luminescent substance which is Cerium doped Yttrium aluminium garnet or Gadolinium aluminium garnet while red luminescent substance in warm white LEDs contain activator ion Eu as a doping [6, 7].
5. Catalytic Converters: In some types of catalytic converters used in vehicles, cerium is used as a catalyst coating [7].
6. Batteries: Some types of batteries, particularly rechargeable ones, may contain small amounts of rare earth elements [8].
7. Magnets: Permanent magnets used in various electronics, such as speakers and motors, can contain rare earth elements like neodymium, praseodymium, and dysprosium [9].
8. Microelectronics: Some integrated circuits and microelectronics might have trace amounts of rare earth elements in the manufacturing process.

It is worth emphasizing that although these sources may contain small traces of rare earth elements, the overall concentration is typically low. Extracting these elements can be a complex process, often requiring specialized techniques for efficient and cost-effective isolation and recovery. Moreover, considering the environmental and ethical issues linked to rare earth element mining and processing, it is imperative to adopt extraction methods that prioritize sustainability. To mitigate the environmental impact and ensure the responsible extraction of these valuable materials, recycling and proper e-waste management play pivotal roles [6, 7]. Extracting rare earth elements (REEs) from electronic waste involves a series of processes aimed at separating and recovering these valuable elements. Here is a simplified overview of the typical steps involved (Fig. 1).

It is crucial to highlight that the extraction of rare earth elements (REEs) from electronic waste (e-waste) is a multifaceted and resource-demanding procedure. While it holds promise for reclaiming valuable materials and lessening environmental consequences, it necessitates meticulous handling of waste byproducts and chemicals to mitigate environmental risks. Furthermore, the advancement of recycling technologies and the implementation of governmental regulations are pivotal factors in fostering sustainable e-waste recycling methods [10].

2.1 Extraction Techniques

Following the initial steps of crushing and dismantling electronic waste, rare earth elements (REEs) are typically extracted using a combination of chemical and physical methods. While chemical extraction is commonly employed, there are also viable physical and electrical approaches for extraction. Let's delve into some of the primary chemical separation techniques used to extract and refine REEs from their source materials [11].

2.1.1 Hydrometallurgical Processes

These processes involve leaching the rare earth ore with suitable acids or alkalis to dissolve the rare earth elements into solution. Subsequent chemical separation techniques are then applied to separate and purify the elements from the leach solution. Several methods are involved in this including liquid metal extractions, molten slag extraction, liquid media extraction, and supercritical fluid extraction, but common techniques used after for selective leaching using mineral acid are ion exchange and solvent extraction [12, 27].

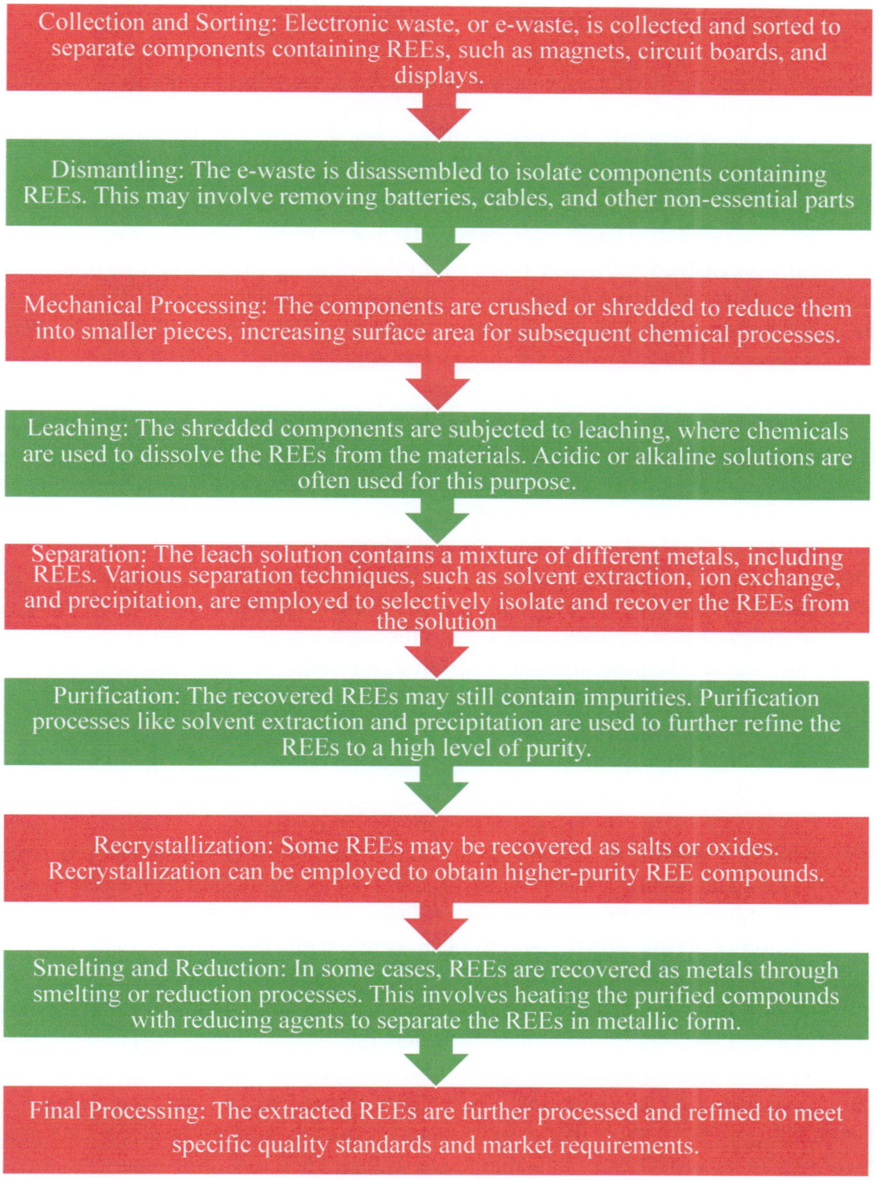

Fig. 1 Steps for recycling and extraction of REE from electronic waste [8]

2.1.2 Electrochemical Methods

Electrochemical methods are a valuable approach for selectively extracting rare earth elements (REEs) from electronic waste. These methods leverage variations in the reduction potentials of different REEs, allowing for the selective deposition or dissolution of specific elements at controlled electrode potentials. The process typically begins with the leaching or dissolution of crushed electronic waste using a suitable solvent, often involving large reactors with HCl or sulfuric acid. This step serves to extract the REEs from the waste material. Following this, an electrolysis process is employed, where a voltage is applied to electrodes. In this setup, the REE electrodes act as the anode, while a metal electrode functions as the cathode. During the electrolysis process, REEs are deposited onto the cathode, while impurities accumulate on the anode. To further purify the extracted REEs, they can be separated from the cathode through a combination of physical and chemical methods. These may include acid leaching or magnetic separation, followed by filtration of the leachate. Subsequent separation and purification steps, such as solvent extraction, ion exchange, or precipitation techniques, can be employed to isolate and refine the REEs [6]. The precipitation method entails introducing an appropriate reagent into the solution that contains rare earth elements, leading to the formation of insoluble compounds. The selection of the reagent depends on the rare earth elements that are the desired target. By manipulating factors such as pH and other conditions, it becomes possible to selectively induce the precipitation of specific rare earth elements from the solution while separating them from other elements [13].

2.1.3 Pyrometallurgical Processes

This extraction method is employed to recover rare earth elements (REEs) from high-grade waste sources. The process involves subjecting waste scraps containing REEs to high-temperature conditions in a furnace. This high temperature leads to the volatilization of REEs, either through heating processes like roasting and calcination or via chemical reactions. During this process, impurities in the form of slag are collected and separated from the valuable REEs. Pyrometallurgical extraction encompasses various techniques, including electrochemical processing, gas-phase extraction, gas–solid and solid–liquid processes, glass slag processes, roasting, molten slag extraction, liquid metal extraction, molten salt extraction, and smelting (typically used for rare earth oxides). These techniques are widely applied to recover a range of rare earth elements from electronic waste sources such as NdFeB magnets, waste printed circuit boards, and fluorescent lamp powder [11]. While pyrometallurgical extraction serves as one of the methods for rare earth element extraction, it does come with certain drawbacks. It generates significant quantities of flue gases, consumes substantial amounts of energy, and tends to be relatively costly. However, it has the advantage of being a less hazardous waste producer compared to hydrometallurgical processes. This is due to its high-temperature environment, which facilitates rapid reactions and makes it an efficient extraction process [6].

2.1.4 Biometallurgical Processes

Bioleaching is an extraction method for rare earth elements (REEs) that relies on the enzymatic activity of microorganisms or fungi, as well as the metabolic byproducts they produce [14]. In contrast to traditional hydrometallurgical and pyrometallurgical techniques, bioleaching stands out as an eco-friendly and cost-effective method that harnesses the power of microbes and their metabolic byproducts. The efficiency of this method is influenced by various factors including pH levels, solid concentration, energy source, agitation rate, irrigation rate, aeration rate, and inoculum concentration. Bioleaching can be conducted on a small scale in shake flasks, column reactors, or stirred reactors, and it can also be scaled up to commercial large-scale operations performed in heap systems, dump sites, agitated tanks, and vats, making it a versatile and adaptable approach for REE extraction [15].

There are two primary types of leaching in bioleaching: Direct and Indirect leaching. The mechanism of Direct bioleaching relies on two key factors.

(1) The growth and attachment of leaching microbes to the source surface are facilitated by the presence of extracellular polymeric substances. These substances consist of sugars, lipids, proteins, nucleic acids, or combinations thereof, forming a surrounding matrix around the microbial cells.
(2) Following attachment, the cells release various compounds, including organic acids, amino acids, peptides, lipids, enzyme complexes, cyanides, and more. These substances function as leaching agents or lixiviants. Notably, organic acids like citric, oxalic, tartaric, and gluconic acids are effective leaching agents in this technique. They enhance dissolution by facilitating proton donation in proton-promoted dissolution processes [16].

On the other hand, Indirect leaching is facilitated by planktonic microorganisms. In this process, dissolution is driven by the production of lixiviants, which can include Fe^{+3}, sulfuric acid, organic acids, and HCN. These biogenic leaching agents are responsible for dissolving metals and waste materials without the need for microbe attachment to the surface. Additionally, a combination of both approaches, known as cooperative leaching, can be employed as needed. In cooperative leaching, microbes attach to the surface of the metal-bearing material, while free cells in the solution simultaneously facilitate the dissolution process [15].

Bioleaching could be applied for:

Microbial Activity and Metal Extraction

Bioleaching hinges on the involvement of microorganisms like bacteria and fungi, which possess the capability to engage with metals present in waste materials. These microorganisms play a pivotal role in facilitating the dissolution of metals from the waste matrix using diverse mechanisms. These mechanisms encompass the production of organic acids, complexing agents, and enzymes that effectively break down mineral structures [17].

Bioleaching Process

In a controlled bioleaching process, the waste material is deliberately exposed to specific microorganisms. These microorganisms engage with the solid waste, initiating the release of organic acids and various metabolites that promote the solubilization of valuable metals, including REEs. As a result of this process, the leachate that emerges contains dissolved metals [14].

Purification and Recovery

The leachate, which holds dissolved REEs and other metals, can subsequently undergo conventional techniques like solvent extraction, precipitation, or ion exchange for the purpose of purification and metal recovery. These downstream processes play a crucial role in concentrating the metals into a suitable form that can be further refined and utilized [18].

2.2 Research on Efficient Extraction Techniques from Mining Tailings and Industrial Byproducts

There is ongoing research dedicated to improving the efficiency of several significant techniques used to recover rare earth elements (REEs) from various mine waste sources, including phosphogypsum, red mud, and cracking catalysts. These methods can be broadly categorized into four distinct groups based on their underlying techniques [19] (Fig. 2).

2.2.1 Extraction of REEs

Extraction or preconcentration methods are employed to selectively isolate and enrich rare earth elements (REEs) from complex matrices such as ores, mineral deposits, and industrial waste streams. These methods are necessitated by the typically low natural concentration of REEs, which makes direct extraction impractical. The utilization of extraction or preconcentration techniques greatly improves the efficiency and economic feasibility of subsequent processing stages. The processes outlined in Fig. 3 can be applied to extract REEs from waste materials [20].

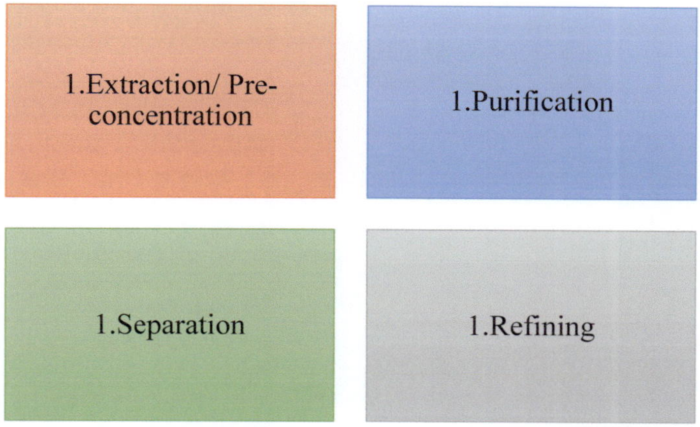

Fig. 2 Methods of REE extraction [19]

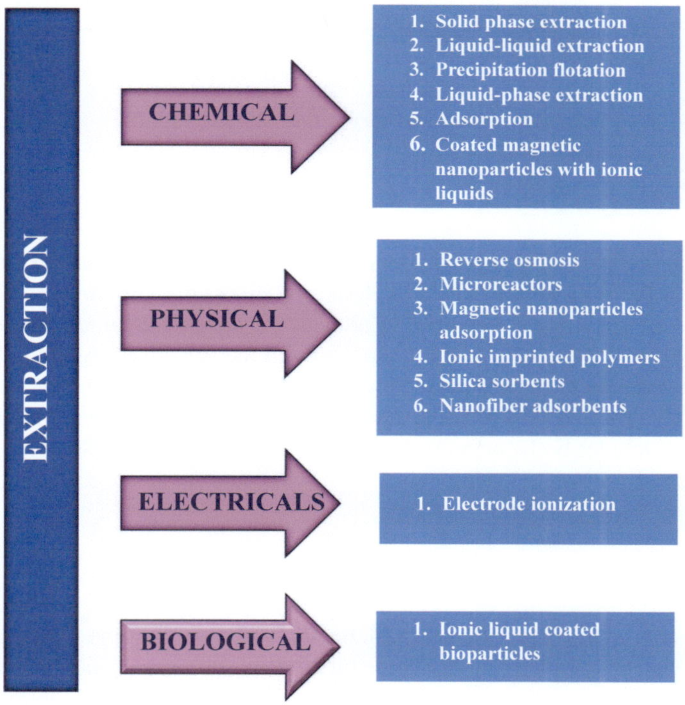

Fig. 3 Extraction techniques used for REEs [13]

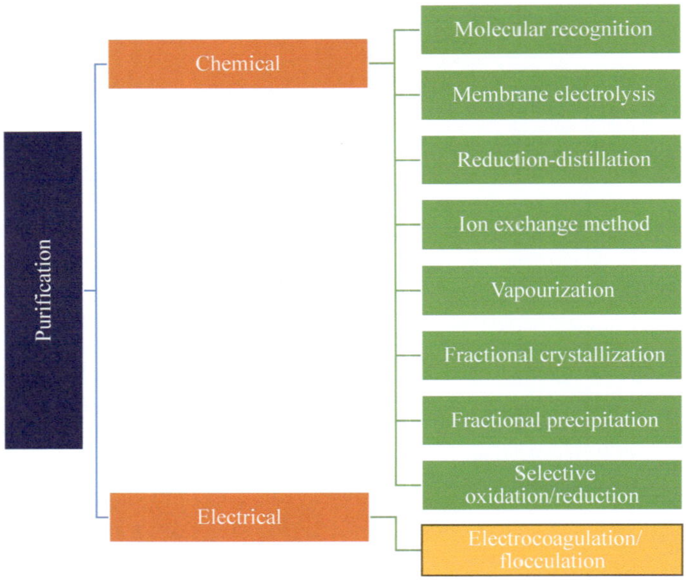

Fig. 4 Purification techniques in REEs recovery [13]

2.2.2 Purification Techniques in REE Recovery

Purification methods play a crucial role in the removal or reduction of impurities, allowing for the isolation of rare earth elements (REEs) in their purest forms. These techniques are specifically designed to address the varying chemical and electrical properties of both REEs and impurities. Through the selective targeting and elimination of unwanted elements, purification ensures that the final REE product meets the necessary criteria for its intended applications. The wide spectrum of purification approaches includes both conventional methods and innovative technologies, which are outlined as follows [21] (Fig. 4).

2.2.3 Separation Methods

The process of extracting Rare Earth Elements (REEs) from mine waste involves a wide range of separation methods. These methods are aimed at isolating and refining individual elements, effectively transforming waste materials into valuable resources. Some of the notable separation techniques include the following [20, 21] (Fig. 5).

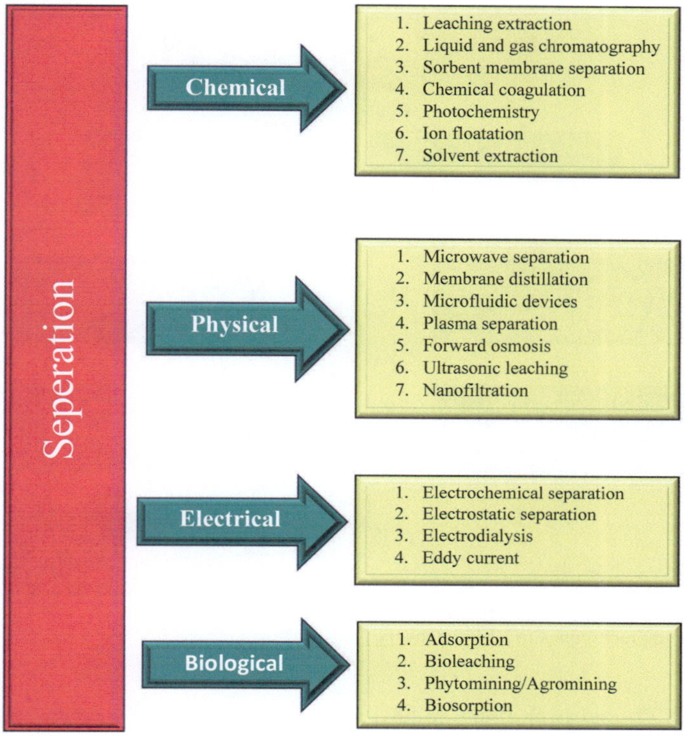

Fig. 5 Separation techniques in REEs recovery [13]

2.2.4 Refining Techniques

Refining techniques represent a pivotal phase in the process of recovering Rare Earth Elements (REEs) from intricate source materials like ores, industrial waste, and recycled products. This exploration of REE recovery refinement methods delves into various approaches, innovations, and advancements within the field, addressing the unique challenges associated with REE extraction. These challenges include the typically low concentrations of REEs, their proximity to other elements, and the consideration of environmental impacts [13] (Fig. 6).

These refining techniques are categorized based on their underlying mechanisms, which may involve physical, chemical, electrical, or biological processes.

Physical Methods

This category includes methods that leverage the physical differences among components, including factors such as size, density, and various other properties. Magnetic

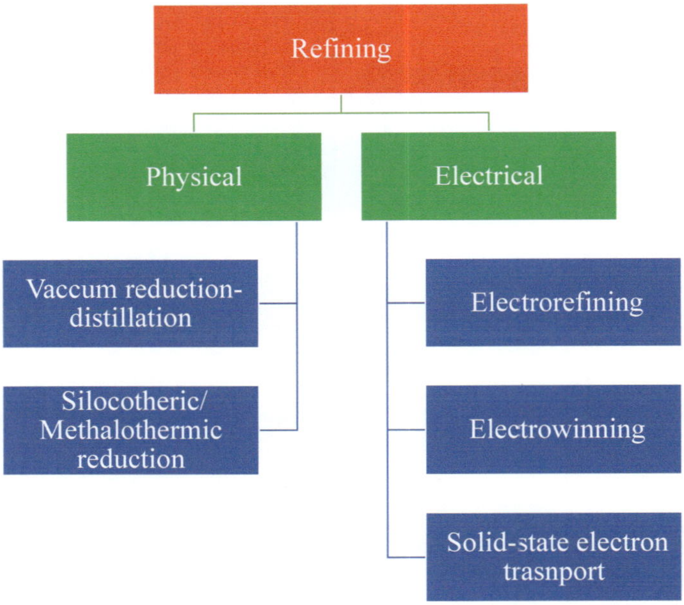

Fig. 6 Refining techniques in REEs recovery [13]

separation [22] and froth flotation [23] are frequently utilized to separate minerals containing REEs from mine waste by capitalizing on the differences in their physical characteristics. This enables efficient and effective separation of these minerals. This method is based on the principle that specific rare earth elements (REEs) oxides, including Nd, Dy, and Gd, exhibit magnetic properties. When a magnetic field is applied to a mixture containing both magnetic and non-magnetic materials, the magnetic components are attracted to the magnetic field and can be efficiently separated from the non-magnetic components [22].

In the study conducted by [23] in 2018 on **Froth floatation,** they examined the flotation behavior of rare earth oxides (REO) and the prevalent gangue minerals using sodium oleate and hydroxamic acid as collector agents. Additionally, they incorporated sodium silicate and starch as depressants in their experiments. Interestingly, when these depressants were present, sodium oleate exhibited superior selectivity for REO compared to hydroxamic acid. By employing a rougher scavenger-cleaner flotation setup with sodium oleate and the depressants, the researchers achieved a final concentrate that contained 63% REO, with a grade of 2.25% [23].

The **plasma separation** technique entails subjecting materials to elevated temperatures, which results in the breaking of chemical bonds, ultimately leading to the ionization of atoms, resulting in the formation of individual ions and electrons. Following this, particular elements can be chosen for selective extraction from the initial material and subsequently combined or deposited onto surfaces as necessary.

A notable advantage of this method is its chemical-free nature, rendering it both environmentally friendly and efficient as a one-step process [20].

The **membrane separation technique** (MST) stands out as a hybrid process that seamlessly combines extraction and stripping operations, eliminating the need for energy-intensive procedures like distillation, drying, or evaporation. Its remarkable energy efficiency, coupled with the ability to function without the use of substantial quantities of volatile organic compounds (VOCs), positions it as an environmentally-friendly and promising separation method. In contrast to conventional solvent extraction, liquid membranes (LMs) offer superior efficiency in swiftly separating and preconcentrating metal ions, thanks to their extensive mass transfer interfacial areas. An array of liquid membrane types has been developed and refined, including bulk liquid membrane (BLM), emulsification liquid membrane (ELM), supported liquid membrane (SLM), hollow fiber supported liquid membrane (HFSLM), electrostatic pseudo liquid membrane (ESPLM), among others [21].

Ultrasonic leaching is a method that utilizes an ultrasound processor to extract rare earth elements (REEs) from a solution. It accomplishes this by employing microscale cavitation to disrupt particles and chemical bonds, effectively releasing the REEs into the solution. This process eliminates the need for concentrated reagents, high temperatures, and prolonged extraction times. In a study, researchers investigated the impact of the power delivered to the extraction medium and the amplitude of ultrasound. They established optimal conditions, including a temperature of 70 °C and a carbonatite mass/volume ratio of 20 mg/mL. Remarkably high extraction efficiencies, reaching 82%, were achieved by using an ultrasound probe operating at 20 kHz for a duration of 15 min, an ultrasound amplitude of 40% (equivalent to 692 W dm^{-3}), and a diluted extraction solution composed of 3% v/v HNO_3 and 2% v/v HCl. Importantly, this approach yielded excellent extraction efficiency even when employing a diluted acid mixture and relatively low temperatures compared to conventional rare earth element extraction methods [24].

Nanofiltration is a method used to separate rare earth elements (REEs) through the use of membranes that selectively filter based on size exclusion or size discrimination properties. In contrast, the microwave separation technique distinguishes rare earth elements (REEs) by utilizing an aqueous solution containing REEs. This technique employs non-contact methods and on/off heating techniques [13].

Chemical Extraction Methods

Chemical methods involve utilizing solvents, acids, or other chemical agents to selectively dissolve and extract REEs from mine waste or ore. Hydrometallurgical processes, including leaching, solvent extraction, and ion exchange, are frequently employed to dissolve valuable REEs into a solution while excluding undesirable materials. These methods can be tailored to specific minerals, achieving efficient REE concentration [25].

Leaching method is the primary step for many methods of REEs recovery which involves extraction of mineral into solution using special reagents (acids, bases or salts) followed by separation, purification techniques. Leaching solvents commonly used are H_2SO_4-H_2O_2, aqua regia, thiourea, cyanide, HNO_3, NaOH, HCl [33]. But among all acid H_2SO_4 is preferred for leaching [26].

Solvent Extraction: Solvent extraction involves using a solvent that selectively binds to specific rare earth elements. This technique is often used in combination with ion exchange processes. The process involves leaching of REEs from scrapped waste with the help of acid or base solution called pregnant leach solution and from this leached solution containing REEs then using an organic solvent to selectively extract the rare earth ions. These ions are then stripped from the solvent using a suitable stripping agent such as Di-2-Ethylhexylphosphonic acid (P_2O_4) [27].

Ion exchange is a process where ions from a solution are exchanged with ions attached to a solid ion-exchange resin. This technique is useful for separating rare earth elements based on their ionic radii and charge. The ion-exchange resin can be designed to selectively bind certain rare earth ions, allowing for their isolation due to presence of specific functional group within it. In this method, solution of chemically treated waste bearing REEs reacts with specific resin which effectively adsorb REEs on its surface due to presence of specific functional group for REEs and remaining solution get separated. This adsorbate resin then washes, eluted with elution solution with higher concentration of replacing ion which have stronger affinity to functional group which result in detachment of REE from resin and introducing into new solution which will go for further studies [34].

Anionic exchange resin with secondary, tertiary and quaternary ammonium groups such as Dowex1X-8 (for Sc, Y, La, Ce, Pr, Nd, Eu, Sm, Lu, Yb, Er, Ho, Dy, Gd), Purolite RUA21207(target ion scandium) and cationic exchange resin with sulfonic acid group (eg. PuroliteC150: target ion Nd) Dowex 50WX8: target ion Y, La, Ce, Nd, Dy, Gd) and with carboxylic acid group (eg. CYBBER LX 280, Amberlite IRC 50) can be used for extracting rare earth elements [34, 35].

Photochemical separation relies on the light-absorbing properties of substances, utilizing visible light, ultraviolet, or infrared radiation. In this process, solvents and additives are employed to manipulate the charge transfer band, thereby altering the necessary light wavelengths for separation. This method offers distinct advantages, including cost-effectiveness, minimal emissions, and the ability to achieve high recovery efficiency and purity. However, it does come with some drawbacks, notably the requirement for large volumes of dilute solutions and the expense associated with using high-intensity light sources [13].

Liquid and gas chromatography operate on the fundamental principle that particles traverse a medium and are sorted based on their differential passage speeds. In their 2022 study, Dembowski and colleagues build upon prior research in this field by assessing the applicability of High-Speed Counter-Current Chromatography (HSCCC) for the separation of a specific subgroup of non-radioactive rare earth elements (Nd, Sm, Eu, Tb, and Y) at concentrations as low as 10^{-4} mol. They employ di-(2-ethylhexyl) phosphoric acid (HDEHP) dissolved in n-heptane as the stationary phase and hydrochloric acid as the mobile phase. A distinctive characteristic of HSCCC is its utilization of a liquid as the stationary phase, a design innovation that eliminates the need for a solid-state support by immobilizing one of the two immiscible phases within the rotating column [28].

Chemical coagulation is a process that entails aggregating non-settling particles, uniting them into larger and denser masses known as floc. This is achieved by introducing a chemical coagulant, such as metallic salts, into the water. Sometimes, a precipitant can be introduced before the coagulant to facilitate the bonding of the precipitates. The coagulant added possesses an opposite charge to the particles, resulting in the neutralization of charges. Sodium aluminate ($NaAlO_2$) was used as the coagulant to concentrate REEs from the acid mine drainage into a solid precipitate [29].

Biological/Bio-metallurgical Methods

Bio-metallurgical strategies encompass employing microorganisms or bio-based processes to recover REEs from mine waste. Specific bacteria and microorganisms possess the capability to selectively leach or accumulate REEs from ores or waste materials. This burgeoning domain, referred to as bioleaching or biohydrometallurgy, offers environmentally conscious and potentially cost-effective avenues for REE recovery [30].

Agromining represents a novel technique for extracting rare earth elements (REEs). It entails the cultivation of specific plants referred to as hyperaccumulators, which have a natural propensity to accumulate these precious elements within their tissues. Following cultivation, these plants are harvested and subjected to processing to extract the REEs they have absorbed. For example; Dicranopteris linearis (formerly referred to as D. dichotoma) stands out as a hyperaccumulator plant recognized for its ability to accumulate rare earth elements (REEs). It is particularly well-suited for REE phytoextraction due to its elevated levels of Nd and Pr, reaching as high as 900 and 300 $\mu g\ g^{-1}$, respectively. Additionally, it possesses significant concentrations of other light REEs such as La, Ce, and Sm, resulting in total REE concentrations of up to 3900 $\mu g\ g^{-1}$ [31].

Biosorption is a process that involves the capture or concentration of specific ions and molecules from aqueous solutions, leveraging various properties of biomass or biomolecules [14]. Dong et al. in 2021 have successfully demonstrated a biobased, all-aqueous method for extracting and separating rare earth elements (REEs) using the REE-selective lanmodulin protein. In this approach, lanmodulin is bound to porous support materials through thiol-maleimide chemistry, allowing for sequential REE purification and separation in a continuous flow system. Immobilized lanmodulin retains the favorable characteristics of its soluble counterpart, including exceptional REE selectivity, the capacity to bind REEs under low pH conditions, and robust stability across multiple adsorption/desorption cycles at low pH levels. Furthermore, this study showcases the potential of immobilized lanmodulin to achieve high-purity separation of the important clean-energy-critical REE pair Nd/Dy. It also demonstrates its capability to transform a low-grade leachate containing only 0.043 mol % REEs into distinct heavy and light REE fractions, achieving an impressive 88 mol % purity of total REEs in a single column operation while utilizing approximately 90% of the column's capacity. This groundbreaking ability to perform tandem extraction and grouped separation of REEs from highly complex aqueous feedstock solutions without the need for organic solvents represents a significant advancement for sustainable hydrometallurgy.

Electrochemical Methods

Electrochemical methods leverage electric currents to aid in the separation and retrieval of REEs. Electrometallurgy involves passing an electric current through a solution containing dissolved REEs, causing their deposition onto an electrode. Electrowinning and electrorefining exemplify electrochemical processes used for recovering metals, including REEs. The electrodialysis separation method separates rare earth elements (REEs) by directing the flow of ions through a semi-permeable membrane with the assistance of an electric potential. However, it is important to note that this process can incur higher operational costs when dealing with salinity levels exceeding 5000 ppm. Eddy current separators employ a magnetic field to separate metals based on their conductivity and density. While this method is effective in distinguishing between ferrous metals, non-ferrous metals, and non-metals, it presents challenges when attempting to separate rare earth elements [32].

It's vital to acknowledge that these categories are not mutually exclusive, and certain methods may amalgamate principles from multiple categories to enhance efficiency. As research advances, scientists and engineers are dedicated to devising more sustainable, efficient, and economically viable means of recovering REEs from mine waste, while accounting for both environmental consequences and financial feasibility.

3 Potential of Urban Mining and Recovering REEs from Discarded Products

Urban mining represents a promising avenue for recovering Rare Earth Elements (REEs) from discarded products. This concept recognizes that our cities are rich reservoirs of valuable materials contained within electronic waste (e-waste), end-of-life products, and discarded consumer goods (Fig. 7).

Light rare earth elements (REEs) play a crucial role in enhancing the mechanical properties of alloyed steel, aiding in desulfurization processes, binding trace elements in stainless steel, and contributing to the development of magnesium and aluminum alloys. REEs also hold significant importance in catalysis. For instance, Lanthanum

Fig. 7 Potential for urban mining and requirements [4]

(La) is utilized in fluid catalytic cracking (FCC) within oil refineries, a process that can boost oil refinery yields by as much as 7%. Cerium (Ce) finds its application in catalytic converters for automobiles [7].

When it comes to end-of-life fluorescent lamps, they serve as a valuable source of heavy rare earth elements (HREEs) such as Europium (Eu), Terbium (Tb), and Yttrium (Y). These HREEs are essential components in tri-phase phosphor lighting used in linear fluorescent lamp tubes, compact fluorescent lamps, and LCD backlights for flat panel displays [33]. Following their collection, these lamps undergo a sorting and recycling process, typically carried out by specialized recycling firms. These companies focus on extracting value from the predominantly glass and metal components of the lamps. Key activities within this recycling process include the removal of mercury (Hg) and the sale of reclaimed metals, polymers, and electronic boards. However, it's worth noting that certain materials do not find a place in this circular production chain and are instead destined for disposal at waste dumps. Subsequently, recycled materials are employed in the manufacturing of new lamps, contributing to the sustainability of the production chain. It is noteworthy that even within this highly efficient circular system, rare earth elements such as cerium (Ce), neodymium (Nd), europium (Eu), terbium (Tb), and yttrium (Y) can also be recovered and reused, further enhancing the resource efficiency of the entire process [8].

Each year, Ce and La are consumed in polishing applications, constituting approximately 10% of the total production of Rare Earth Oxides (REOs). During the polishing process and subsequent settling stages, the glass polishing powder accumulates impurities like SiO_2 and undergoes changes in its particle size distribution (PSD). This alteration in PSD, coupled with increased impurity content, diminishes its effectiveness in polishing, eventually rendering the product unusable and leading to its disposal in landfills, resulting in the loss of valuable natural resources. To address this issue and contribute to a circular economy while closing the rare earth metal loop, it is essential to recover Rare Earth Elements (REEs) from the polishing waste and reintegrate them into the polishing process or high-value applications. The contaminants SiO_2 and Al_2O_3 can be eliminated from the polishing waste through methods like NaOH roasting or NaOH leaching, thereby extending the lifespan of the polishing powder. However, it's worth noting that the alteration in PSD and the accumulation of other impurities impose limitations on the number of reuse cycles. Consequently, there is a pressing need for the selective recovery of REEs from the polishing powder to ensure its prolonged usability and sustainability [9, 33].

The research conducted by [9], an innovative approach was employed to achieve a near-total recovery (> 99%) of rare earth oxides, specifically La and Ce, from polishing waste. This was accomplished through a two-step process involving reductive acid leaching and alkali treatment. During the initial leaching phase, approximately 70% of the rare earths were successfully dissolved. The rare earth compounds that remained undissolved underwent a transformation into oxides or hydroxides through the application of an alkali treatment. Subsequently, in the second leaching step, these transformed compounds were dissolved in an acid solution, ensuring the complete retrieval of rare earths. The recovered rare earth oxides were subsequently utilized in the production of high-value Al-La-Ce alloys through fused salt electrolysis, resulting in in-situ alloy formation.

3.1 Bioremediation in Urban Mining

Bioremediation is an environmental process that harnesses the power of living organisms, including plants, bacteria, and fungi, to facilitate the restoration and cleanup of polluted environments. This approach is applicable to a range of sites, including mine lands, landfills, and dumping areas. When it comes to mine land reclamation and rare earth element (REE) extraction, bioremediation can assume a pivotal role in minimizing environmental damage while also enabling the recovery of valuable resources [36, 37] (Fig. 8).

Here is an elaboration of the bioremediation process for mine land reclamation with REE extraction:

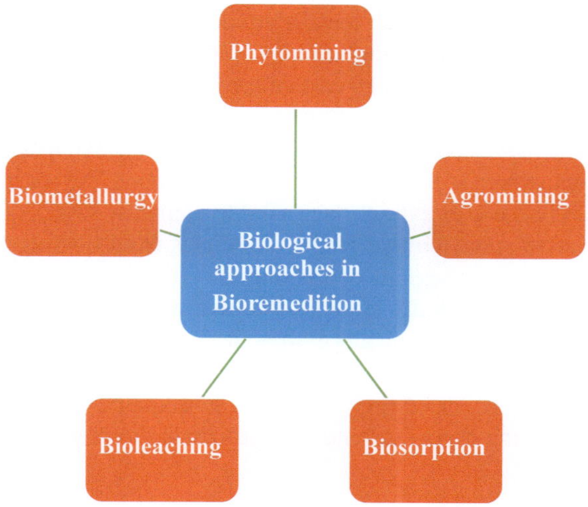

Fig. 8 Biological approaches in bioremedition [36, 37]

3.1.1 Site Assessment and Selection

The first step involves assessing the contaminated mine land to determine the extent of pollution and the types of contaminants present. This assessment helps in selecting suitable bioremediation strategies and identifying the appropriate organisms for the process.

3.1.2 Contaminant Identification

In the case of REE extraction, the primary contaminants of concern are rare earth elements themselves, along with potential co-contaminants from mining activities, such as heavy metals and other pollutants.

3.1.3 Phytoremediation

Phytoremediation involves phytomining, agromining and biometallurgy where plants used to extract or immobilize contaminants from the soi or act as detoxifier by removing it but for remediation purpose [36].

In the context of phytomining, specific plants, notably hyperaccumulators capable of extracting and accumulating significant quantities of rare earth elements (REEs) in their tissues, as well as indicator plants that accumulate metals in their above-ground parts and are used to identify mineral-rich soils, can contain substantial amounts of REEs or other metals [38]. In the research process, scientists must identify and cultivate these specialized plants directly on the mine site. These plants are intentionally grown in areas with metalliferous mine waste, even under potentially toxic conditions, without posing any physiological harm to the surrounding environment. As the plants absorb rare earth elements (REEs) through their root systems and transport them to the shoot system via vascular bundles, they are carefully harvested. The harvested plants are subsequently subjected to a combustion process to produce ash, which is then processed using methods like roasting, smelting, electrowinning, and others to extract the valuable elements [39]. Agromining represents an innovative approach that blends principles from both agriculture and mining to extract metals from regions containing low-grade ore deposits or mine waste. This method proves especially valuable in circumstances where conventional mining practices are either economically impractical or pose environmental challenges. Agromining closely resembles phytomining in that it involves cultivating hyperaccumulator plants and fungi, harvesting them, and subsequently subjecting them to a series of steps including drying, ashing, and processing to recover the desired (REEs) [40].

3.1.4 Bioleaching

Bioleaching is a microbial-driven process that relies on the activity of bacteria to extract metals from various sources, including ores. Acidophilic bacteria, which thrive in acidic environments, play a crucial role in dissolving rare earth elements (REEs) from mine waste. These bacteria produce organic acids that effectively leach the metals from the ore, making them more readily available for recovery [30]. Bioleaching can be applied to a range of industrial and urban waste materials, allowing microbes to facilitate the extraction of valuable elements.

Red Mud

Red mud, which is a byproduct of the Bayer process used in bauxite digestion, has the potential to serve as a secondary resource for rare earth elements (REEs), particularly scandium. It's estimated that approximately 2700 million tonnes of red mud residue have accumulated worldwide from the Bayer process, with an annual increase rate of 120 million tonnes. However, this residue can be hazardous due to its alkaline nature. To extract REEs from this potentially hazardous waste residue, a two-step process can be employed using the filamentous acid-producing fungus Penicillium tricolor RM-10, primarily for scandium extraction. This fungus is capable of producing citric and oxalic acids, which aid in the leaching process. Alternatively, other bacteria such as Acetobacter sp. (for lutetium and scandium) and Acetobacter tropicalis (for scandium) can be used in a one-step process. These microbes produce various organic acids like acetic, oxalic, and citric acids, which act synergistically to facilitate the extraction of REEs [5].

Phosphogypsum

Rare earth elements (REEs) are commonly found in association with phosphate deposits, particularly in phosphogypsum. Phosphogypsum is a waste material collected from the fertilizer industry, mainly generated through the wet phosphoric acid process. To extract REEs like Yttrium (Y), Cerium (Ce), Neodymium (Nd), Samarium (Sm), Europium (Eu), Ytterbium (Yb), and others from phosphogypsum, Gluconobacter oxydans is utilized as a biolixiviant-producing bacterium. This bacterium produces gluconic acid, which aids in the leaching process [5].

Cracking Catalysts

Fluid cracking catalysts, which originate from the oil refining and biofuel industries, are known to contain Cerium and lanthanum among other elements. To dissolve and recover rare earth elements (REEs) from this waste material, Gluconobacter oxydans, a bacterium capable of producing gluconic acid, is employed. It's noteworthy that this bacterium can also be cultured on agricultural waste as a more cost-effective alternative to commercial glucose. In addition to this, the fungus Aspergillus niger has been investigated for its effectiveness in leaching spent cracking catalysts, particularly for efficient recovery of lanthanum. A biochemical process utilizing Yarrowia lipolytica IM-UFRJ is used to recover REEs, making use of glycerin derived from biodiesel production. Furthermore, advanced bioleaching methods involving the presence of Fe(II) and the use of Acidothiobacillus ferroxidans have been applied to enhance lanthanum recovery [14, 17].

Fluorescent Lamps

Compact fluorescent lamps (CFLs) are end-of-life products, and their average composition includes glass (88 wt %), metals (5 wt %), plastic (4 wt %), lamp phosphor powder (3 wt %, which contains nearly 10% of rare earth phosphors bound in triband dyes), and mercury (0.005 wt %). The challenge with CFL disposal is that the removal of lamp phosphors containing rare earth elements (REEs) such as Yttrium (Y), Europium (Eu), Lanthanum (La), Cerium (Ce), Gadolinium (Gd), and Terbium (Tb) can result in the loss of valuable resources and environmental risks due to mercury content. To address this issue, a symbiotic mixed culture comprising Zygosaccharomyces lentus (yeast) and Komagataeibacter hansenii (acetic acid bacteria) sourced from tea Kombucha, along with K. xylinus, Lactobacillus casei, Yarrowia lipolytica, and Gluconobacter oxydans, can be utilized as a lixiviant. This microbial culture produces acetic acid and gluconic acid, which aid in the leaching of REEs from CFLs, mitigating both the resource loss and environmental risks [5].

Electronic Waste and Magnets

Magnets are commonly found in various post-consumer electronic devices and typically contain significant amounts of rare earth elements (REEs) such as Neodymium (Nd), Dysprosium (Dy), and Praseodymium (Pr). Acidithiobacillus ferrooxidans and Leptospirillum ferroxidans have demonstrated high leaching efficiencies when employed in the presence of sulfuric acid (H_2SO_4) [14].

3.1.5 Monitoring and Optimization

Continuous monitoring of the bioremediation process is of paramount importance. Key parameters including pH, temperature, nutrient concentrations, and microbial populations must be rigorously observed and adjusted as needed to maximize the process's efficiency.

3.1.6 REE Recovery

Once the REEs have been absorbed by the plants or leached by bacteria from the mine waste, the subsequent phase involves extracting these valuable elements from the plant biomass or the microbial solution. Various extraction methods, including chemical leaching or solvent extraction, can be utilized to recover the REEs.

Summary

In the quest for sustainable resource management, innovative technologies have revolutionized the recycling and extraction of Rare Earth Elements (REE) from electronic waste. These advancements focus on maximizing efficiency and minimizing environmental impact. The process begins with innovative recycling techniques, such as urban mining, emphasizing the importance of extracting REE from discarded electronic devices in urban areas. Hydrometallurgical processes play a crucial role, employing chemical solutions to selectively extract REE from electronic waste while minimizing ecological consequences. Bioleaching, a cutting-edge approach, utilizes microorganisms to dissolve metals, offering an eco-friendly and efficient extraction method. Alongside recycling methods, sophisticated extraction techniques like solvent extraction have emerged, employing specific organic solvents to selectively isolate REE, ensuring a sustainable supply while reducing the environmental footprint of extraction processes. These advancements underscore the pivotal role of innovative technologies in shaping the future of REE recycling and extraction, aligning with the broader goals of environmental conservation and responsible resource utilization.

References

1. Krishnamurthy, P. (2020). Rare Metal (RM) and Rare Earth Element (REE) Resources: World Scenario with Special Reference to India. Journal of the Geological Society of India, 95(5), 465–474, https://doi.org/10.1007/s12594-020-1463-7
2. Department of Atomic energy, 2023, Mining of rare earth elements, https://pib.gov.in/PressReleasePage.aspx?PRID=1914305
3. Balaram V. Potential Future Alternative Resources for Rare Earth Elements: Opportunities and Challenges. *Minerals*. 2023; 13(3):425, https://doi.org/10.3390/min13030425
4. Mudali U., Patil M., Saravanabhavan R., Saraswat V., Review on E-Waste Recycling: Part I—A Prospective Urban Mining Opportunity and Challenges, 2021, https://doi.org/10.1007/s41403-021-00216-z
5. Castro, L.; Blázquez, M.L.; González, F.; Muñoz, J.Á. Biohydrometallurgy for Rare Earth Elements Recovery from Industrial Wastes. Molecules 2021, 26, 6200, https://doi.org/10.3390/molecules26206200
6. Bhavan J., Joy J., Pazhami A., Identification and recovery of rare earth elements from electronic waste: Material characterization and recovery strategies, 2023, https://doi.org/10.1016/j.mtcomm.2023.106921
7. Buchert M. Manhart A., Bleher D., Pingel D., Recycling critical raw materials from waste electronic equipment, 2012, Page no. 5–6, 9, 19–20, 24, 37, 39, 41–42
8. Lukowiak A., Zur L., Tomala R., LamTran T., Bouajaj A., Strek W., Righini G., Wickleder M., Ferrari M., Rare earth elements and urban mines: Critical strategies for sustainable development, 2020, https://doi.org/10.1016/j.ceramint.2020.03.067
9. Borra, C.R., Vlugt, T.J.H., Yang, Y., Spooren J., Nielsen P., Amirthalingam M., Offerman S., Full length article: Recovery of rare earths from glass polishing waste for the production of aluminium-rare earth alloys, 2021, https://doi.org/10.1016/j.resconrec.2021.105766
10. Mudali, U.K., Patil, M., Saravanabhavan, R. Saraswat V. K Review on E-waste Recycling: Part II—Technologies for Recovery of Rare Earth Metals. Trans Indian Natl. Acad. Eng. 6, 613–631 (2021), https://doi.org/10.1007/s41403-021-00231-0
11. Ramprasad C., Gwenzi W., Chaukura N., Azelee N., Rajapaksha A., Naushad M., Rangabhashiyam S., Review :Strategies and options for the sustainable recovery of rare earth elements from electrical and electronic waste, 2022, https://doi.org/10.1016/j.cej.2022.135992
12. Hérès X, Blet V, Di Natale P, Ouaattou A, Mazouz H, Dhiba D, Cuer F. Selective Extraction of Rare Earth Elements from Phosphoric Acid by Ion Exchange Resins. *Metals*. 2018; 8(9):682, https://doi.org/10.3390/met8090682
13. Opare E. O., Struhs E., Mirkouri A., (2021), A comparative state of technology review, and future directions for rare earth element separation, Renewable and Sustainable energy reviews, 143, 110917, https://doi.org/10.1016/j.rser.2021.110917
14. Brown R., Mirkouei A., Reed D., Thompson V., Current nature-based biological practices for rare earth elements extraction and recovery: Bioleaching and biosorption, 2022, https://doi.org/10.1016/j.rser.2022.113099
15. Kara T., Kremser K., Wagland S., Coulon F., Bioleaching metal-bearing wastes and by-products for resource recovery: a review, 2023, https://doi.org/10.1007/s10311-023-01611-4
16. Park S., Liang Y., Bioleaching of trace elements and rare earth elements from coal fly ash, 2018, https://doi.org/10.1007/s40789-019-0238-5
17. Reed D., Fujita Y., Daubaras D., Jiao Y., Thompson V., Bioleaching of Rare Earth Elements from Waste Phosphors and Cracking Catalysts, 2016
18. Roberto F., and Schippers A., Progress in bioleaching: part B, applications of microbial processes by the minerals industries, 2022, https://doi.org/10.1007/s00253-022-12085-9
19. Pyrgaki, K.; Gemeni, V.; Karkalis, C.; Koukouzas, N.; Koutsovitis, P.; Petrounias, P. Geochemical Occurrence of Rare Earth Elements in Mining Waste and Mine Water: A Review. Minerals 2021, 11, 860, https://doi.org/10.3390/min11080
20. Gueroult R., Rax J. Fisch N., Opportunities for plasma separation techniques in rare earth elements recycling, 2018

21. L. Chen, Y. Wu, H. Dong, M. Meng, C. Li, Y. Yan, J. Chen, An overview on membrane strategies for rare earths extraction and separation, Separation and Purification Technology (2017), https://doi.org/10.1016/j.seppur.2017.12.053
22. Zinoveev, D.; Grudinsky, P.; Zhiltsova, E.; Grigoreva, D.; Volkov, A.; Dyubanov, V.; Petelin, A. Research on High-Pressure Hydrochloric Acid Leaching of Scandium, Aluminum and Other Valuable Components from the Non-Magnetic Tailings Obtained from Red Mud after Iron Removal. Metals 2021, 11, 469, https://doi.org/10.3390/met11030469
23. G.B. Abaka-Wood et al., The upgrading of rare earth oxides from iron-oxide silicate rich tailings: Flotation performance using sodium oleate and hydroxamic acid as collectors, Advanced Powder Technology (2018), https://doi.org/10.1016/j.apt.2018.08.01
24. Diehl L., Gatiboni T., Mello P., Muller E., Duarte F., Flores E., Ultrasound-assisted extraction of rare-earth elements from carbonatite rocks, 2017, https://doi.org/10.1016/j.ultsonch.2017.04.012
25. Peelman S., Kooijman D., Sietsma J., Yang Y., Hydrometallurgical Recovery of Rare Earth Elements from Mine Tailings and WEEE, 2018, https://doi.org/10.1007/s40831-018-0178-0
26. Jouni M., Lavallee A. Pabst T., Chung E., Kim R., Cheong Y., Neculita C., Sustainable Production of Rare Earth Elements from Mine Waste and Geoethics, 2022, https://doi.org/10.3390/min12070809
27. Pan, J.; Zhao, X.; Zhou, C.; Yang, F.; Ji, W. Study on Solvent Extraction of Rare Earth Elements from Leaching Solution of Coal Fly Ash by P204. Minerals 2022, 12, 1547, https://doi.org/10.3390/min1212154
28. Dembowski M., Rowley J., Boland K., Droessler J., Hathcoat D., Marchi A., Goff G., May L., Rare earth element separations by high-speed counter-current chromatography, 2022, https://doi.org/10.1016/j.chroma.2022.463528
29. Guanrong Song, Xingchao Wang, Carlos Romero, Huazhi Chen, Zheng Yao, Alfred Kaziunas, Rolf Schlake, Madhu Anand, Tom Lowe, Greg Driscoll, Boyd Kreglow, Harold Schobert, Jonas Baltrusaitis, Extraction of selected rare earth elements from anthracite acid mine drainage using supercritical CO2 via coagulation and complexation, Journal of Rare Earths, Volume 39, Issue 1, 2021, Pages 83–89, ISSN 10020721, https://doi.org/10.1016/j.jre.2020.02.007 1.
30. Klaus Bosecker, Bioleaching: metal solubilization by microorganisms, FEMS Microbiology Reviews, Volume 20, Issues 3–4,1997, Pages 591–604, ISSN 0168–6445, https://doi.org/10.1016/S0168-6445(97)00036-3.
31. Z, Laubie B, Morel JL, Tang YT, Qiu R, Simonnot M-O, Muhr L (2018) Recovery of rare earth elements from Dicranopteris dichotoma by an enhanced ion exchange leaching process. Chem Eng Process 130:208–213
32. Smith Y., Nagel J., Rajamani R., Eddy Current Separation for Recovery of Non-Ferrous Metallic Particles: A Comprehensive Review, https://www.sciencedirect.com/science/article/pii/S0892687518305 82XManuscript_a2672ffe5d7da2912700600f5d24242d6
33. Cardoso C.,Lopes C., Almeida J., Trindade T., Vale C., Pereira E., Recovery of Rare Earth Elements by Carbon-Based Nanomaterials—A Review, 2019, https://doi.org/10.3390/nano9060814
34. Ouardi Y. E., Virolainen S., Mouele E. S. M., Laatikainen M., Repo E., Laatikainen K., (2023), The recent progress of ion exchange for the separation of rare earths from secondary resources—A review, Hydrometallurgy, Volume 218, 106047, https://doi.org/10.1016/j.hydromet.2023.106047.
35. Smyshlyaev D., Kirillov E., Kirillov S., Bunkov G., Rychkov V., Botalov M., Taukin A., Yuldashbaeva A., Malyshev A., (2022), Recovery and Separation of Sc, Zr, and Ti from acidic sulphate solutions for high purity scandium oxide production: Laboratory and Pilot study, Hydrometallurgy, Volume 211, 105889, https://doi.org/10.1016/j.hydromet.2022.105889
36. Yan A, Wang Y, Tan SN, Mohd Yusof ML, Ghosh S and Chen Z (2020) Phytoremediation: A Promising Approach for Revegetation of Heavy Metal-Polluted Land. Front. Plant Sci. 11:359, https://doi.org/10.3389/fpls.2020.00359
37. Azubuike C., Chikere C., Okpokswasili G., Bioremediation techniques–classification based on site of application: principles, advantages, limitations and prospects, 2016, https://doi.org/10.1007/s11274-016-2137-x

38. Dinh T., Dobo Z., Kovacs H., Phytomining of noble metals – A review, 2021, https://doi.org/10.1016/j.chemosphere.2021.131805
39. Kumar. A., 2020, How Rare Earths dictate strategic interests of India in Defence, Nuclear, Space sectors, https://defence.capital/2020/09/12/how-rare-earths-dictate-strategic-interests-of-india-in-defence-nuclear-space-sectors/
40. Jally, Baptiste Laubie, Ye-Tao Tang, and Marie-Odile Simonnot, Processing of Plants to Products: Gold, REEs and Other ElementsBastien, Agromining: Farming for Metals Extracting Unconventional Resources Using Plants, Second Edition, https://doi.org/10.1007/978-3-030-58904-2

Eco-environmental Challenges Associated with Application of REE

Kaustubh Deshpande, Sanjeevani Jawadand, Pratik Godbole, Tushar Meshram, J. Vijaya Kumar, Kishor Deshmukh, Tejashree Raut, Pradeep Kumar Jain, and Kirtikumar Randive

Abstract The growing utilization of Rare Earth Elements (REE) in various industries presents both opportunities and challenges. This chapter explores the eco-environmental challenges associated with the application of REE. It delves into the current trends in the development of mineral-based products utilizing REE, emphasizing the significance of sustainable practices. The chapter investigates the application of rare earth minerals in the metallurgical industry, highlighting the complexities of extraction, processing, and disposal methods that impact the environment. Additionally, it discusses the economic concerns related to REE application and production, considering the balance between economic growth and environmental preservation. This chapter provides valuable insights into the intricate relationship between technological advancement, economic considerations, and environmental sustainability in the context of REE application, offering a comprehensive overview for researchers, policymakers, and industry professionals.

Keywords Rare earth elements · Applications · Challenges · Environmental challenges · Economic challenges

---

K. Deshpande · P. Godbole · T. Raut · K. Randive (✉)
Department of Geology, RTM Nagpur University, Nagpur, India
e-mail: randive101@yahoo.co.in

S. Jawadand
Shri Mathuradas Mohota College of Science, Nagpur, India

T. Meshram
Geological Survey of India, Nagpur, India

J. Vijaya Kumar
Ore Dressing Division, IMB, Bangalore, India

P. K. Jain
Indian Bureau of Mines, Nagpur, India

K. Deshmukh
Geological Survey of India, Hyderabad, India

K. Randive et al. (eds.), *Current Trends in Mineral-Based Products and Utilization of Wastes: Recent Studies from India*, Springer Proceedings in Earth and Environmental Sciences, https://doi.org/10.1007/978-3-031-50262-0_2

1 Introduction

Over the past thirty years, there has been a significant surge in the utilization of REEs and their alloys across a wide array of technological devices and applications. These include computer memory, DVDs, rechargeable batteries, catalytic converters in automobiles, exceptionally strong magnets, mobile phones, energy-efficient LED lighting, superconductors, additives for glass manufacturing, luminescent materials, compounds for phosphate binding, solar panels, and contrast agents for magnetic resonance imaging (MRI) [1]. REEs have become essential components in contemporary high-tech applications, including, tablets, laptops, hybrid/electric vehicles, wind turbines, and photovoltaic films, among others. They also find extensive use in well-established technologies like catalysts, glass manufacturing, ceramics, metallurgy, and alloy development. Furthermore, REEs play a pivotal role in various defence technologies, such as fighter jet propulsion systems, missile guidance systems, anti-missile defence mechanisms, space-based satellites, and advanced communication systems [2]. Additionally, Rare Earth Elements (REEs) are prominently employed in the development of high-strength magnets, the formulation of specialized metal alloys, as chemical oxidizing agents, and for reinforcing various other metals. They are also pivotal in automotive technology, notably in the production of catalytic converters. In the field of medical sciences, REEs are crucial components of portable X-ray machines, MRI contrast agents, nuclear medicine imaging technologies, and various applications in cancer treatment. Furthermore, REEs contribute significantly to the renewable energy sector, being essential in hybrid automobile systems, wind turbine manufacturing, low-energy lighting, fuel cells, magnetic refrigeration, and rechargeable batteries in hybrid electric vehicles (HEV), and as catalysts in the production of biofuels, among other uses [3, 4]. The surging worldwide demand for these elements has sparked considerable enthusiasm among exploration geochemists for technologies like hybrid vehicles, electric car batteries, and wind turbines. All of this unfolds against the urgent backdrop of addressing climate change and the challenges posed by global warming [1].

2 Current Trends in Development of Mineral-Based Products Using REE

Rare earth elements are considered highly essential in various industrial applications. Rare earths are used as modifiers regardless of quantity of REE found in heavy minerals such as xenotime and monazite [5]. In last three decades, REE have posed themselves as vital nutrient for technological advancements due to their unique properties [1]. Due to high specificity and irreplaceability, REEs are highly essential in commercial as well as industrial applications [6]. Discovery of new applications of REE in emerging fields, increase their demand [1, 6, 7]. 98% application of REE

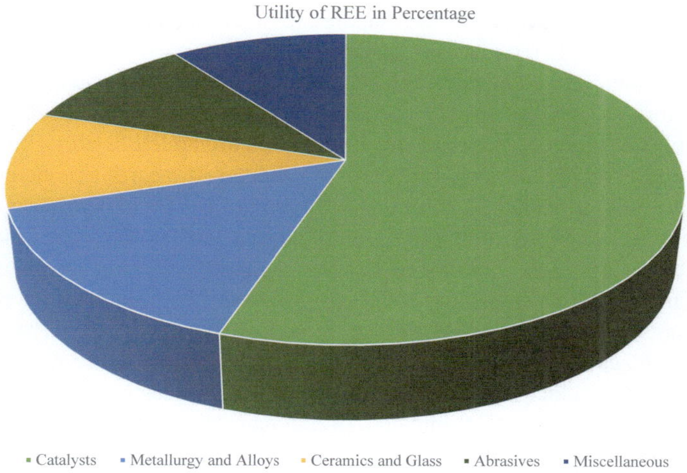

Fig. 1 Utility of REE [8]

is concentrated in mainly four sectors viz. Metallurgy, Chemical or Catalyst, Glass-based industry, and Abrasing media [6, 7]. Although, it is expected that trend of these applications will change in future. For example, application of REE in Metallurgy is increasing while, chemical or catalyst industry is on declining trend. It is also expected that, new industries can emerge [6] (Fig. 1).

3 Application of Rare Earth Minerals in Metallurgical Industry

Metallurgical industry has ranked second in consumption of REE. In metallurgy, ferrous metallurgy has a share of 86% while, remaining is of non-ferrous metallurgy [9].

Steel: Rare earths are utilized in steel industry to induce a refined microstructure and modify inclusions. Ability of REEs to scavenge oxide and sulphide impurities improves quality of steel and can impart properties such as toughness, fatigue resistance, corrosion resistance, and hot ductility. Addition of specific quantity of REE in steel is the key for the development of next generation steel manufacturing technologies [10]. Rare Earths possess unlimited solubility in molten iron, limited solubility in α-Iron (0.1–0.5%) and low solubility in γ-Iron (~ 0.04%). Increase in viscosity, precipitation of inclusion and reduction in surface tension of molten steel is the result of addition rare earth. Unlike alkaline earths, rare earths react not only in liquid phase but also, during crystallization. Rare earths like cerium and lanthanum have more affinity towards oxygen than, aluminium, zirconium, titanium, and silicon and greater affinity for sulphur than, zirconium and manganese. That is why, rare earth has a better hand as compared to other elements showing affinity for oxygen and sulphur [11]. It must be considered that; mixture of rare earth elements is added rather than individual rare earth. Addition of rare earth in steel depends upon various factors such as amount of RE, method and timing of introduction of RE, and also type of RE which is being added. Rare earths are added as either mischmetal or as rare earth silicide. Majorly rare earth is added as rare earth silicide but, use of mischmetal is trending [12]. Mischmetal is mixture of oxides of cerium, lanthanum, neodymium, and praseodymium [13]. Dissolved rare earth is distributed more along the grain boundaries rather than inner grain, eventually cleaning the grain boundaries and improving strength and toughness of steel [14]. Addition of rare earth increase transverse impact properties i.e.through-thickness ductility, bend performance of light-gauge hot-rolled material, overheating temperature of low-sulphur steel. When RE is added and RE/S ratio greater than 4, plastic property generated due to presence of MnS gets eliminated, giving hard and undeformable property to steel [12]. Addition of RE more than 21 ppm can deteriorate the preservative property of steel [10]. Addition of RE to the various types of steel have variety of effects. Addition of RE to pipeline steel will strengthen it for its utility in low temperature zone. Addition of RE to low alloy steel will improve its impact strength. Addition of RE to high alloy steel will not only increase its impact strength but also, heat resistance. Toughness and tensile strength of the structural steel can be improved by incorporation of RE. Wear resistance and hardness will be increased by addition of RE in rail steel [15].

Non-ferrous alloys: Addition of rare-earth to non-ferrous alloys implies their better performance. [16, 17]. Mischmetal addition is used for improvement of aluminium alloys for electrical application, to improve the wear resistance of highly leaded bearing bronzes, to develop cobalt superalloys having improved corrosion resistance and strength, to develop new generation magnesium alloys [17]. Rare earth elements are used for hardening in magnesium alloys [6].

Greenhouse gases (GHGs) are emerging as an environmental concern and their reduction in atmosphere is the need of the hour. Many of the industries have reduced their GHGs emission while, automobile and aeronautical industries are trying their best to reduce their burden. Increasing fuel efficiency and improved design and working is the key to reduce GHGs emission. An easily viable and cost-effective way of reducing GHGs emission is weight reduction of parts used in automobile and aeronautical industry [18]. Therefore, magnesium and its alloys are in the focus because, they are light weighted as compared to other alloys (Al-alloys: 2.7 g/cc and Steel: 7.86 g/cc) and, they have good strength to weight ratio, good castability, good damping capacity and are also non-toxic and recyclable. At the same time, Magnesium have limitations such as poor ductility, limited creep resistance, and most importantly it is highly corrosive [18, 19]. The rapid formation of oxide/hydroxide layer on magnesium is due to presence of huge amount of negative free surficial energy which is not able to prevent magnesium from corrosion. It is also easily soluble in water which weakens the metal. These problems can be solved by adding rare earth in Mg-alloys [20].

Magnesium has proven itself better over polymeric implants because, Mg is a constituent of physiological system and it is beneficial for bone strength and growth. But, on the other hand Mg in presence of chloride solutions and blood plasma has shown high degradation rate. Hydrogen bubbles are evolved during degradation which can coagulate near the implant, delaying healing process or separating of healed layers or blocking of blood vessels causing death. Therefore, alloying magnesium with REE such as neodymium and lanthanum has proven as an ideal option [21]. REE forms an oxide layer on the surface Mg-alloys, preventing release of Mg^{2+} inhibiting corrosion. REEs also form RE-hydrides with hydrogen present in Mg melt which prevents further mishap. Zinc along with REE is also used as biodegradable implants as zinc metal possesses low ultimate tensile strength (33.6 MPa), low yield strength (29.3 MPa) and less elongation (1.2%) [22].

Aluminium and REE alloy is another light metal-REE alloy which has wide applications [19]. To scavenge soluble hydrogen from Al melt, REEs are used which facilitates REE-hydrides promoting dehydrogenation. For removal of impurities such as Cu and Fe from alloy, REE is employed. These reduce deteriorating effect of Cu and Fe [19, 23]. Addition of REE in Al or Al alloy improves hardness of Al, corrosion resistance of Al, and it also alters mechanical as well as electrical properties of Al [23]. To attain high temperature alloy of titanium gadolinium, it is micro-alloyed with titanium. Gd or Dy are added to Titanium alloys for strengthening the titanium alloys [24].

A rechargeable Ni-MH (Nickel-Metal hydride) battery is used in automotive industry which is used to power electric motor while discharging and is charged when specified amount of alloy hydride' s hydrogen is oxidized. Role of rare earth is to provide adequate amount of space for storage, rapid desorption, and rapid-reabsorption of this hydrogen. 10% of the world's REEs are used in this application. Ni-MH batteries are available in various shapes and sizes. Ni-MH batteries has sale of around one billion per year [25]. Ni-MH batteries are widely used in Hybrid Electric Vehicles (HEV) which are cheaper, safer and simpler than Lithium batteries. Ni-MH batteries can be used to store electricity produced by renewable resources. Most of the Ni-MH batteries use $LaNi_5$-type negative electrode [26]. Various little substitutions like La and Ni in $LaNi_5$ batteries increase efficiency and life cycle of Ni-MH batteries. Small amount of Ni is replaced with Co which not only reduces hydrogen storage capacity but also increases life cycle of the battery. At the same time, replacement of La with Ce-Nd mischmetal along with Co improves performance of Ni-MH batteries [27]. Small amount of Mg can also replace La in $LaNi_5$ batteries which increases electrode discharge capacity. Although Ni-MH batteries have strong competition with Li-ion batteries, they are lead performers in the field for large-scale energy storage. More research is needed to improve long term cycling of the Ni-MH for withstanding in the competition of huge demand of batteries [26].

Economic importance is increasing in technology for the transition from fossil fuel energy to green energy. It is led by development of new technologies where permanent magnets are mainly RE-based magnets. These magnets are mostly used in wind turbines and electric vehicles [28]. The most important and unique property of permanent magnets is that, they possess an ability to retain a huge magnetic flux in absence of applied magnetic field [29]. $RE_2Fe_{14}B$ magnets opened new era of research and development in the field of permanent magnets, where Nd is main constituent used in $RE_2Fe_{14}B$ [30]. Along with $Nd_2Fe_{14}B$, $SmCo_5$ and $SmCo_{17}$ were developed in the last century [29]. $Nd_2Fe_{14}B$ magnets are widely used as they are powerful magnets available at room temperature but, they demagnetize at high temperature which can be improved with the addition of Dy and Tb [31]. Dy and/or Tb are partially substituted beyond 5 wt%. These Dy and/or Tb substituted NdFeB magnets are used in HEVs for working of their motors and generators [32]. But, consumption of Nd in $Nd_2Fe_{14}B$ magnets led to inflation in the prices of Nd, Pr, and Dy which led to the development of substitution with highly abundant REEs such as La and Ce [33]. Substitution of La and Ce individually against Nd in $Nd_2Fe_{14}B$ magnets deteriorate the quality of magnets as properties of La and Ce are not compatible. But, use of La-Ce mischmetal (MM) along with Nd and Pr (RE) with limiting MM/RE about 21.5% has showed promising effect. Use of mischmetal reduces cost of manufacturing of Nd-Fe-B magnets as well as it creates secular use of REE [33, 34]. Other than NdFeB magnets SmCo and GdCo magnets are also widely used in field where high performance and high resistance to demagnetization is required at high temperatures with higher stability [32]. Demand of REE in magnets will not shrink in future but, will be seen increasing as world is focusing on more and more green technologies and hence, cost efficiency will be the most important factor in case of REE and other complementary material used [28].

4 Applications of Rare Earth Minerals in Chemical/Catalytic Industry

Chemical industry plays an important role in the progress of nation in the fields of industrial development, designing and development of energy sources. Electrodes play vital role in the development of energy sources. These electrodes are required to withstand high temperatures in the range of 1500–2300 °C, abrasion, thermal shock, alkali attack and also electrical and thermal losses which are required to be minimum. These requirements are fulfilled by rare earth-based chemical materials [6]. Rare-earth based catalysts are mainly used in petrochemical and oil refining industries which constitutes 95–100% of total world market of catalyst. World market of catalyst is around 17.5 billion dollar per year and is increasing with the rate 15–20% per year [35]. Application of some important catalysts are given in Table 1.

Zeolite based catalyst brought revolution in the petroleum refining industry which was further modified with the introduction of rare earth such as La, Ce, Nd, and Pr and ammonium through exchange followed by calcination. It has been found that, smaller the cation higher will be the dealumination. Dealumination is the phenomenon in which aluminium is removed from the framework of the zeolites when exposed to high temperatures. Adding rare earth to the zeolites reduces dealumination in high temperature hydrothermal conditions. Addition of RE to the zeolite increases

Table 1 Application of some widely used REE-based catalysts [36]

Catalyst	Application
Lanthanum chloride and lanthanum nitrate	In petroleum refining which speeds up the production of petrol and diesel from large molecule distillation residues
Cerium oxide with platinum, palladium and rhodium	In automobile industry where it converts $CO_{(g)}$, Volatile Organic Compounds (gaseous hydrocarbons), nitrogen oxides in automotive engines to innoxious $CO_{2(g)}$, $H_2O_{(g)}$, and $N_{2(g)}$ gases
Ceria and Lanthana	Enhancing hydrogen production, synthetic gas production, and ethylbenzene dehydrogenation
Neodymium versatates and neodymium organophosphates	Enhancing speed of organic polymerization
Samarium iodide	Enhancing speed of plastic decomposition, dichlorination of polychlorinated biphenyls, and dehydrogenation and dehydration of ethanol

hydrogen transfer and hence drop in olefin concentration, causing minimization of petrol Research Octane Number. RE in zeolites also acts as deactivator of vanadium compounds. Although RE containing zeolite-based catalysts are beneficial, further research is still needed [37]. RE-based Single Atom Catalyst (SAC) has emerged as new innovation in photo/electrocatalysts. SAC are the catalysts whose catalytical sites are single atoms and hence, have applications in the fields of energy conversion and storage. Ce-based SAC was first which developed in 2018 and worked efficiently after which different SACs such as La-, Er-, Dy-, Pr-SAC were developed. RE-SACs have upper hand against transition and precious metal SACs due to their high coordination number, strong atom-support interactions, and unique spin–orbit coupling which favours better activity and selectivity, high catalyst durability, and various potential applications such as photocatalytic nitrogen reduction reaction, photo/electrocatalytic carbon dioxide reduction reaction, and electrocatalytic oxygen reduction reaction [38]. Rare earth tailings can also be used as catalyst after alkali and acid treatment which enhances the rate of denitrification. Such utility of tailings will help to solve the problem of increasing burden of tailings [39]. While focusing towards green energy, hydrogen attracts researchers as burning of hydrogen produces only water and hence, hydrogen can be used as green fuel. Hydrogen is produced from methane, methanol and ethanol reforming reactions in which C–C bond is broken. Co-based catalysts such as Al_2O_3, ZnO, MgO, CeO_2, ZrO_2 etc. have shown diversity in activity, product distributions, and resistance to coking and sintering. $Co/MgAl_2O_4$ catalysts modified by the addition of Ce has shown highest hydrogen production [40]. REE-based catalysts have ability to provide a breakthrough in the field of creating low-temperature energy saving and environment-friendly catalytic process which suggests that, use of REE is unavoidable and especially of Cerium. Other than catalysts REEs are also used in chemical industry for the manufacturing of pigments, lacquers and paints [35].

Alloys such as Mg and Al alloys are replacing steel in automotive and aeronautical components as they save much of the energy but, their corrosing property is still a concern. Most common methods such as anodization, metallic paintings, vapor deposition methods, chemical conversion coatings, and organic coatings are used to prevent corrosion, out of which chemical conversion coating is most common. Earlier Cr-based conversion coatings were used but, Cr is genotoxic carcinogen which led to investigation for non-toxic, environment friendly conversion coatings [41, 42]. Similarly, in the pigment industry use of Cd-, Pb-based pigments is being avoided due to environmental policies. This created space for environment friendly pigments [43, 44] and coatings which has been fulfilled by the rare earths [41].

La-, Ce-, Sm-, Nd-, Gd-, Pr-conversion coatings are utilized to prevent corrosion but, mostly Ce-based conversion coatings are preferred as they are available at affordable prices, are environmentally accepted, offer good corrosion resistance, compatible with organic and inorganic additives, high solubility in water, and most importantly REEs are not biotoxic in nature [45]. The formation of rare earth (RE)-based conversion coatings is influenced by several key factors, including: (1) the parameters of the bath and deposition process (such as the concentration of RE-ions, pH level, temperature, agitation, and duration of coating application), (2) the condition of the substrate (including polishing techniques, pre-treatment methods, alloy composition, and microstructure), and (3) the characteristics of co-additives (such as coating accelerants, pH adjusters, and co-conversion agents). These co-conversion agents, which can be either organic or inorganic in nature, are applied simultaneously to enhance the effectiveness of the conversion coatings [46]. Research is still ongoing on RE-based corrosion coatings for preventing corrosion. Cerium is in the lime light and significant amount of research is performed on lanthanum systems. Most of the work is focused on Al [42] and Mg-alloys [45] and work on other metals is in progress [42].

Pigments are the fine powders which disperse in the media forming heterogeneous mixture which is a part of various decorative and preventive coatings which are applied on plastics, ceramics, glasses etc. Pigments are classified into natural and synthetic. Pigments which are used to colour ceramics are most important as they possess thermal as well as chemical stability and hence, most of the pigments used in ceramic industry are inorganic. These inorganic pigments are produced at industrial level with the help of chromophore elements. But, environmental restrictions necessitated research in the field of pigments [44, 47]. Leather industry is also witnessing transition as various metal ions and compounds utilized in manufacturing are banned in number of countries due to environmental concern [44]. This need gave rise to the innovation of RE-based pigments. Unique spectroscopic and magnetic properties of REEs attracted researchers to develop RE-based pigments. In RE-based pigments mixed oxide forms a major class where, REEs (Ce, Pr, and Nd) are introduced into stable host oxides to develop colour [44]. CeO_2-based pigments are classified as inorganic colorants recognized for their exceptional thermal stability. When doped with praseodymium, they exhibit a distinctive orange-pink hue. At an elevated temperature of 1300 °C, lanthanum oxide (La_2O_3) and praseodymium dioxide (PrO_2) dissolve into cerium oxide, resulting in a cohesive blend of the three oxide compounds. Remarkably, the pigments derived from the CeO_2-PrO_2-La_2O_3 mixture produce captivating red tones when employed in ceramic glazes. Possessing a fluorite structure, these pigments are both heat and chemical resistant, and they hold promise as environmental-friendly inorganic colorants. But, separation prices for REEs are high and these pigments do not tend to be economically viable. Therefore, RE-oxides are preferred in this context [43]. Rare earth phosphate is the main material which is easily available can be used in manufacturing of pigments used in plastics and paints. $CePO_4$ with monazite structure when doped with Ca^{2+} gives green to yellow colour [48]. $PrPO_4$ doped with calcium exhibits remarkable brightness, accompanied by discernible green and yellow constituents. These materials hold significant potential as pigment candidates for coloring paints, plastics, and high-temperature ceramics. Their application could lead to the substitution of existing hazardous metal-based pigments, such as chromium oxide pigment, with safer alternatives [47]. The mixed rare earth oxides, a by-product generated following the extraction of radioactive thorium from monazite sands, have become a viable option for an environmentally friendly colorant. This colorant aligns with present pigment demand while offering the advantage of utilizing materials that would otherwise be considered waste [43].

5 Application of Rare Earth Minerals in Glass and Ceramics Industry

Glasses, glass–ceramics, and ceramics are the three important engineering materials which are useful in number of multifunctional and industrial application [49]. Glasses manifest as transparent, amorphous states resulting from the supercooling of liquids. These inorganic formations emerge from the fusion of materials and subsequently solidify into a rigid state devoid of their crystalline structure. Over the course of numerous years, glasses incorporating diverse ions have garnered significant interest due to their promising utility in electrochemical, electronic, and electro-optic devices [50]. Variety of glasses such as silicate, germanate, chalcogenide, and fluorite glasses and etc. are utilized in formulation of electro-optic devices [51]. Rare earths possess distinctive characteristics that set them apart from other optically active ions. They exhibit emission and absorption spectra with narrow wavelength bands, and these spectral transitions are relatively unaffected by the surrounding host material. Despite their weak transition intensities, rare earth ions demonstrate prolonged lifetimes of metastable states and generally achieve high quantum efficiencies, with the exception of aqueous solutions. These inherent properties arise from the specific states involved in these optical processes, collectively contributing to the exceptional performance of rare earth ions across a multitude of optical applications [52]. The advancement of optical materials through the incorporation of rare earth elements holds paramount significance within the realms of science and technology such as sensor devices, solid state lasers, medicine and environmental monitoring, active optical fibers, and

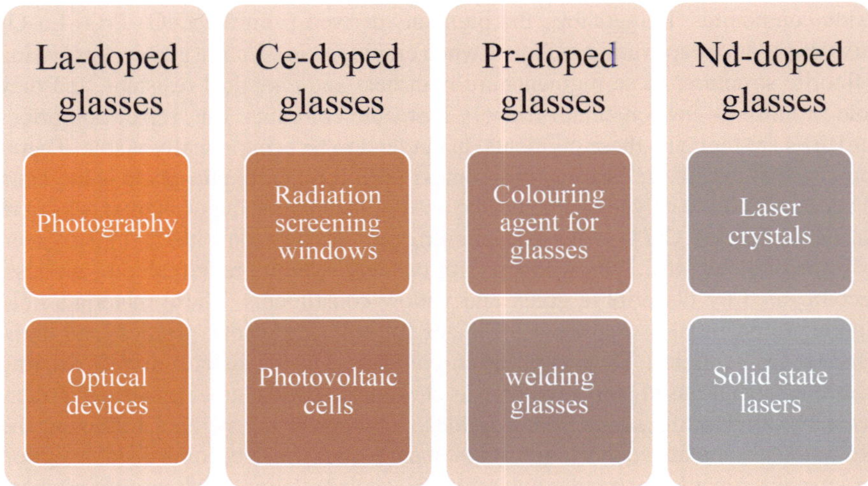

Fig. 2 Application of RE-doped glasses in various fields [6]

phosphors for cathode ray tube, displays, plasma monitors, and lightning applications [50] (Fig. 2).

Bulk metallic glasses (BMGs) represent a nascent class of advanced engineering materials, possessing inherent and captivating attributes. These encompass exceptional hardness, substantial resistance to corrosion, and an impressive capacity for enduring large elastic strains [53]. These glasses have dimension more than 1 mm which is obtained at much lower cooling rates of 100–1 K/s [54]. These glasses are used due to their propertiesand have wide applicability such as electronics, aviation, aerospace, machinery, microelectronics, and are also used in medical sciences [55]. Using conventional casting method glasses incorporated with Pd, Zr, Ti, Ni, Cu, Co, Fe, and Mg were manufactured with critical cooling rates less than 1000 K/s and thickness more than 1 mm [56]. Generally, BMGs are manufactured by selecting a single metal as a base [57]. But, it has been found that, more variation in atomic sizes enhances the glass forming ability. Therefore, use of alloys was suggested for the formation of BGMs [58]. General procedure of BMG manufacturing is to select metals with different sizes leading to high viscosity and slow crystallization of supercooled liquid [59]. The rare earth elements have chemical comparability, approximately same atomic sizes, electronic structures and elasticity moduli which can be proved as the good system [54]. The rare earth elements are either used as minor additions to alloy or are used as main constituent during manufacturing of BMGs. Due to similarity in properties of REs, the ratio of REs in the BMGs can arbitrarily be changed which is not the case with transition elements. This property is used in formulating the properties gradually by altering the ratios of REs. Also, REBMGs are preferred because they are easily deformable into any desired shape [60]. Research for LREE-based BMGs is less than that of HREE-based BMGs [56]. This is because, HREEs have high elasticity modulus than LREEs and therefore,

HREE-based BMGs could exhibit higher thermal stability [58]. REBMGs are used as microstructures, hard magnets and magnetic refrigerants etc. Ce-BMG which possess thermoplastic property are used manufacturing of high-performance microstructures and micro-machines. La-BMGs are used in the field of superconductivity. Nd-BMGS are employed as hard magnets while, Gd-, Ho-, Dy-, Tb-, and Er-BMGs are used as magnetic refrigenrants and demand for REBMGs will be increasing in near future [60].

Apart from glasses, Ceramics are the multiphase material having amorphous and crystalline phases with varying ratios between 20 and 95% with superior properties such as resistance to thermal shock, thermal expansion resistance, mechanical strength, optical properties etc. over glass. Ceramics are manufactured either by controlling crystallisation mechanisms of specially designed bulk glasses using single or two step heat treatment or by combined sintering and crystallisation processing route for fine powdered glass [61, 62]. Ceramics are doped either with transition metals or rare earths which must be added less than 0.1% so that, properties of glass ceramics are not deteriorated [63]. These ceramics are applied in six major fields viz., cooking appliances, dinner sets, electronics, medicines and dentistry, tough glass ceramics and optical materials [64]. Addition of rare earths as stabilizer increases toughness and strength and also reduces production cost by reducing sintering temperature and hence, can be used as semiconductor sensors, microwave dielectric, piezoelectric ceramics [65]. and also in manufacturing of opto-electronic devices [64]. Use of Rare earths such as La and Ce in alumina ceramics is for increasing toughness and reducing sintering temperature respectively. Similarly addition of Ce and Nd in zirconate ceramics can increase bending strength and toughness. Mostly Ce, La, and Nd are added to the non-oxide ceramics which plays role as sintering aids, to reduce sintering temperature and promote calcination to improve performance of the commodities [65].

Synthetic garnets are important constituents for the development of lasers which has similar structure to the natural garnet. The general chemical formula for these synthetic garnets is $R_3A_2D_3O_{12}$ where, A and D sites are occupied by Al or Ga or In while, R is occupied by RE [66]. Rare earth aluminium garnets ($RE_3Al_5O_{12}$) are the most suitable for laser crystalline material and also in high temperature ceramic composites especially yttrium aluminium garnets(YAGs). ReAG are also used in adiabatic demagnetization refrigirators [67]. ReAG doped with lanthanides are utilized as scintillators and phosphors. YAG are considered as most studied in the garnet class which is not used in pure form but, is doped with rare earth ions for its utility. When RE is introduced in YAG it replaces the Y in the YAG [68]. Numerous YAG compositions are developed with either addition of new lanthanides or addition Ga or In in place of Al [69]. YAG doped with Ce is used as phosphors in manufacturing of LED lights. YAG doped with Nd and Yb are used in solid state lasers. La doped YAGs are used in optical glasses, rechargable batteries etc. [68]. Er doped YAG lasers are used in the field of dermatology [70], Dentistry [71], Vetarnary sciences [72] etc. Dy doped YAG is used as thermographic phosphor and Tm doped YAG can alternatively be used as thermographic phosphor [73]. In recent years, metal halide perovskites with formula ABX_3, where A is monovalent cations such as

CH_3NH_3+, $HNCHNH_2$+, Cs+; B represents divalent cations such as Pb^{2+} and Sn^{2+} and X represents halogen anions which is used in commodities such as solar cell, LEDs, laser, photodetectors, etc. These materials are also doped with lanthanides for optical and electrical properties and additional research should be supplemented in development of RE doped perovskite [74].

6 Application of Rare Earth Minerals in Abrasive Industry

An abrasive grain is hard and tough substance which contains numerous pointed protrusions which are used for cutting purposes [75]. Worldwide usage of abrasives is approximately more than 16 kt per year. 13% of world's REE is utilized as abrasive material which may fluctuate according to the inflation or deflation in cerium prices. Ceria or cerium oxide is used as abrasive for polishing glass replacing silica, alumina, and zirconia [76]. Such polished glasses are used in variety of commodities such as decorative glass, ophthalmological glasses, LCDs, decorative crystals, electronic devices etc. [76, 77]. Ceria has fastest polishing rate among other abrasive materials [78]. Cerium is preferred over abrasives such as silica, alumina and zirconia because of the chemo-mechanical mechanism of Cerium Oxide polishing. Cerium oxide (CeO_2) is commonly subjected to heat treatment to achieve a softer consistency than certain other types of glass. This property enables it to deliver superior material removal rates and impeccable surface finish during the polishing process when compared to harder materials like $A12O_3$ and ZrO_2, which primarily rely on mechanical polishing mechanisms [79]. Ultra-smooth and defect free glasses are achieved by Chemical Mechanical Polishing (CMP) process [80]. CMP is widely used in manufacturing of electronic devices. Electronic devices may contain materials which are softer than the ceria generating permanent scratch defects which results in delamination and formation of puddles. To avoid such defects, composite particles containing nanoparticles of ceria dispersed within cross-linked, polymeric microspheres formed by copolymerization is used for polishing which will be able to deliver scratch less surfaces [81]. Literature on the CMP is variable and most of the times confusing in nature due to confidentiality by industries as what other additives they use in this process along with ceria therefore, understanding complete mechanism is a challenge [76].

In this section we have briefly discussed application of REE in four major fields. Application of REE in industrial as well as commercial appliances are very wide and is beyond this chapter.

7 Environmental and Economic Challenges in Production of REE

Emerging era of green technologies have enhanced importance of rare earth elements and eventually production of rare earth elements [82]. When observed carefully these rare earths are not actually rare earth, their abundance is earth's crust is more than that of precious metals. Cerium is most abundant element among the rare earths which is abundant than copper or lead. Problem is their extraction from their host minerals due to similarity in their chemical properties [83]. These chemical properties along with their physical, structural properties have attracted them to be applicable in various industrial as well as commercial applications and more importantly in clean energy technologies due to policies of reducing carbon emissions [84]. This increased mining and production of REEs has impacted on environment, [82] on economies due to monopoly of China over REEs, which constitutes around 90% of world's REE. 2010 crisis has alarmed the world about supply chain of REEs [83].

8 Rare Earths: Are They Really Green?

Rare earths which have numerous applications in clean energy technologies as well as hi-tech technologies, which has resulted into increased extraction of REEs [82]. Although REE are abundant and around 200 REE containing minerals are explored, only three minerals have proved their potentiality for REE extraction [83]. Research suggests that, each REE and mineral deposit is exceptional in its own properties which requires development of extraction route for those specific deposits [82]. This creates a jumble of reports which are on research methodologies, assumption, and investment models which prevents common investment value and similarity in REE prices, structure, and expenditure [85]. Due to huge application potential, they also attract illegal miners without obtaining proper permits which also impacts the environment [8].

Generally, metals have four stages of supply chain which are shown in Fig. 3.

While discussing these stages of supply chain for REE, third and fourth stages are more complicated [8] This is because, uniqueness of every rare earth and mineral deposit and this variance creates problem such as consumption of time, consumption

Fig. 3 Four stages of supply chain [8]

of huge energy and use of chemicals possessing significant environmental concern [86]. While considering REEs as pollutants, gadolinium, lanthanum, cerium and samarium are the worst. While comparing impact of REEs, base metals and precious metals, precious metals have highest environmental impact while, base metal is least and REE belong in between them [87]. But, there are lack of researches regarding the deteriorating effects of REE on environment as well as human health [88].

While discussing impact REE, main contributors are mining and beneficiation process [86]. Mining of REE is majorly of three types viz, open pit, underground, and in-situ leaching. For open pit mining, common method of removing overburden, mining, milling, crushing, grinding, separation or concentration is involved. If rock is hard underground mining is adopted whereas wet-dredging or dry mining is adopted for placer deposits [89]. Whichever may be the mining method, it generates solid waste which is called as tailings and liquid waste. Ores of REE, commonly contain radioactive elements which are threat to the environment and human health [88, 90]. The mine waste generated harbors substantial amounts of radionuclides, posing a substantial threat to the well-being of human populations, wildlife, and plant ecosystems when exposed to these highly hazardous materials [91]. Mining methods also consumes huge amount of water which creates problem of wastewater flowing in streams [89]. From host rock, REE mobilize through acid mine drainage which have pH less than 5 and they are highly toxic and corrosive and may contain radioactive material. Along with this, REEs also mobilize through acid mine drainage [88]. REEs are transported through adsorption on colloidal particles and suspended particles are responsible for the distribution of REEs in water bodies [92]. Adsorption of REE on Fe and Al hydroxides is main phenomenon for transportation of REEs [88]. Biotite which adsorbs REE, can be used as a potential geo-indicator for studying concentration REE in water [93]. Release of particulates or dust from tailing dump, conveyor systems, transportation, or by other means can be carried away by winds and can be accumulated in the leeward areas, including surface soils and water bodies. Aerosols which are mixture of dust and chemical containing vapour which is released through actions such as cutting, drilling, and blasting of parent rocks. Such released matter may cause irritation, toxication, or they may be carcinogenic depending on the physiochemistry of the released particles [90].

Not only from mining operations but also, manufacturing of commodities and disposal of industrial and commercial products leads to release of REEs in the environment. REEs have reported from the solid waste and wastewaters, whose concentration decreases away from source area [94]. The cumulative concentration of Rare Earth Elements (REEs) in soil typically ranges from 100 to 200 mg/kg, with discernible increase attributed to anthropogenic activities. The inherent low mobility of REEs stems from their strong affinity for binding to soils and sediments. Additionally, an intriguing source of REE introduction into soils is the utilization of lanthanides as dietary supplements for animals to enhance their growth. It is worth noting that certain REEs exhibit solubility in water, allowing them to migrate within the soil matrix, potentially contaminating groundwater and other aquatic ecosystems. For instance, the widespread application of Gadolinium (Gd) in medical devices has led to heightened Gd concentrations in groundwater and adjacent water bodies [95]

Table 2 Human health hazards of some selected REEs [99]

Rare earth	Health disorder
Lanthanum	Neural tube defects, orofacial clefts in infants
Cerium	Endomyocardial fibrosis, lung and bladder cancer, hypertension
Neodymium	Orofacial clefts in infants
Samarium	Interstitial lung fibrosis
Europium	Affects embryonal kidney cells
Gadolinium	Astrocytoma
Terbium	Interstitial lung fibrosis
Erbium	Cytotoxity
Ytterbium	Cytotoxity

Presence of REEs in water and sediments may cause deteriorating effects on aqueous biological systems which enter bodies of aqueous animals through body surface and food ingestion for example, La attacks particularly on nervous system while, Ce and Eu are accumulated in liver tissues [92]. Uptake of REEs especially La by plants replacing Ca in plant tissues shows adverse effects on cell membrane reducing its penetrability and influencing release of proton from cells. Presence of La affect's electron transfer from NADH to oxygen in plant membranes, reducing lipid peroxidation [96].

Uptake of REEs through contaminated foodstuff, medical treatments and occupational exposure of humans may show adverse effects. Concentration of REEs in human body can be investigated through nails, hairs, milk, serum, and sperms [94]. Exceedingly high REE concentration is in the vicinity of REE mining and processing areas and also near solid and electronic and electric waste dumping sites [97]. REE concentration in hair of males and females is studied in people, until the age of 60 years [98]. Effects of some selected REEs on human health is given in Table 2.

From above it can be concluded that, The use of REEs will not shrink in future but will be increasing exponentially. Excessive use of REEs in technologies will disrupt the biogeochemical cycles of environment turning green elements into red elements [95].

9 Economic Concerns of REE Application and Production

Although deposits of REE exist in all continents but, 50% of deposits are concentrated in China and extract its 90% of the REE deposits and hence, has monopoly over rare earth. Monopoly of China is not due to presence of huge number of deposits but, due to its research and development in REE extraction [85]. Although, there are numerous REE containing minerals only three are viable for extraction and also extraction techniques vary according to the deposit [83]. Complexity of mineral from which an specific element to be extracted and availability of variety of minerals at the same site makes it difficult to extract desired element and utilize huge amount of energy and water resource [8]. Also, uniqueness of every deposit creates need of researching new methodologies for extraction [82]. Furthermore, LREE are more abundant than HREE and also REEs with odd atomic numbers are more abundant than, even atomic number which is called as Oddo-Harkins rule. [8, 100]. Therefore, setting of project in which mining, processing, and extracting is expensive which increases cost of REE. Expenditure on project is divided in two parts viz., initial investment which is required to start the project and operational investment which is required to run the project. Operational cost is inversely proportional to REE content in mineral i.e., more the content lesser the operational cost, which eventually reduces total cost of REE [85]. Uncertainty in the concentration of REE also creates a new expenditure of stockpiling of REE which is available in huge concentration [101].

Demand of REE is high and supply of that REE from natural resources itself will create the problem of demand and supply and thereby, will increase the prices of specific REE. This is known as Balance problem [102]. Both high demand and low supply of specific REE such as Nd and Dy; low demand and high supply of specific REE such as Ce, Pr, Gd, Tm, Yb, and Lu leads to inflation in prices of collective REEs [100]. Economic concerns also arise due to restrictions imposed by particular nation in the view of geopolitical or environmental issues. Such monopolistic practices not only control the prices of REEs but also, creates uncertainties in prediction of future prices [103]. Imposing restrictions will surely create the price difference in REE containing commodities which are produced outside the monopolistic nations such as China. Increasing dependence of world for REE containing commodities over China thereby increasing economic concerns of other countries whose problems regarding REEs are at very primary level i.e., regulatory issues, research and development etc. [104].

Summary

This chapter explores the environmental challenges tied to Rare Earth Element (REE) application, focusing on three key areas. Firstly, it analyses the current trends in mineral-based product development utilizing REE, emphasizing sustainable practices. Secondly, it investigates the intricate use of rare earth minerals in the metallurgical industry, shedding light on extraction complexities and environmental impacts. Lastly, the summary delves into economic concerns, balancing the benefits of REE application with environmental preservation. This chapter provides critical insights into the eco-environmental challenges of REE utilization, serving as a valuable resource for researchers, policymakers, and industry experts.

References

1. Balaram, V. (2019). Rare earth elements: A review of applications, occurrence, exploration, analysis, recycling, and environmental impact. Geoscience Frontiers, 10(4), 1285–1303, https://doi.org/10.1016/j.gsf.2018.12.005
2. Deshmane, V., Islam, S., & Bhave, R. R. (2019). Selective Recovery of Rare Earth Elements from a Wide Range of E-Waste and Process Scalability of Membrane Solvent Extraction. Environmental Science & Technology, https://doi.org/10.1021/acs.est.9b05695
3. Jeyagoapal, 2019, Geological Perspectives of Rare Earth Elements in India with a Special Reference to Tamilnadu: A Review, Journal of Chennai Academy of Sciences 4, 57–65 (2021)
4. Atwood, 2012, The Rare Earth Elements. Edited by David A. Atwood. © 2012 John Wiley & Sons, Ltd. ISBN 978-1-119-95097-4
5. Subasinghe CS, Ratnayake AS, Roser B, Sudesh M, Wijewardhana DU, Attanayake N, Pitawala J. (2022) Global distribution, genesis, exploitation, applications, production, and demand of industrial heavy minerals. Arab J Geosci.;15(20):1616, https://doi.org/10.1007/s12517-022-10874-0. Epub 2022 Oct 15. PMCID: PMC9568896.
6. Jha A. R. (2014), Rare earth materials properties and application, Taylor & Francis group, LLC, 331pp.
7. Preinfalk, C., Morteani, G. (1989). The Industrial Applications of Rare Earth Elements. In: Möller, P., Černý, P., Saupé, F. (eds) Lanthanides, Tantalum and Niobium. Special Publication No. 7 of the Society for Geology Applied to Mineral Deposits, vol 7. Springer, Berlin, Heidelberg, https://doi.org/10.1007/978-3-642-87262-4_19
8. Ganguli R., and Cook D. R. (2018), Rare earths: A review of the landscape, MRS Energy & Sustainability: A review Journal, 1–16pp., https://doi.org/10.1557/mre.2018.7
9. Volkov, A., Stulov, P., Leont'ev, L.I., & Uglov, V.A. (2020). Analysis of the Use of Rare-Earth Metals in Ferrous Metallurgy in Russia and the World. *Steel in Translation, 50*, 363–374.
10. Pan, F.; Zhang, J.; Chen, H.-L.; Su, Y.-H.; Kuo, C.-L.; Su, Y.-H.; Chen, S.-H.; Lin, K.-J.; Hsieh, P.-H.; Hwang, W.-S. (2016) Effects of Rare Earth Metals on Steel Microstructures. *Materials* , *9*, 417, https://doi.org/10.3390/ma9060417
11. Smirnov, L.A., Rovnushkin, V.A., Oryshchenko, A.S. *et al.* Modification of Steel and Alloys with Rare-Earth Elements. Part 1. *Metallurgist* **59**, 1053–1061 (2016), https://doi.org/10.1007/s11015-016-0214-x
12. Waudby P. E. (1978), Rare earth additions to steel, International metal reviews, 74–98pp, https://doi.org/10.1179/imtr.1978.23.1.74
13. USGS, 2014, The Rare earth elements-vital to modern technologies and lifestyles, https://doi.org/10.3133/fs20143078
14. Wang L., Lin Q., Ji J., Lan D., (2006), New study concerning development of application of rare earth metals in steel, Journal of alloys and compounds, 408–412, 384–386pp.
15. Wang, C. (2023). Role and Application of Rare Earth Elements in Steel. In: Theory and Application of Rare Earth Materials. Springer, Singapore, https://doi.org/10.1007/978-981-19-4178-8_8
16. Wang, C. (2023). Role of Rare Earths in Non-ferrous Metals and Alloys. In: Theory and Application of Rare Earth Materials. Springer, Singapore, https://doi.org/10.1007/978-981-19-4178-8_10
17. Hirschhorn, I.S. Recent Applications of the Rare Earth Metals in Nonferrous Metallurgy. *JOM* **22**, 40–43 (1970), https://doi.org/10.1007/BF03355666
18. Calado LM, Carmezim MJ and Montemor MF (2022) Rare Earth Based Magnesium Alloys—A Review on WE Series. *Front. Mater.* 8:804906, https://doi.org/10.3389/fmats.2021.804906
19. Rhamdani A. R., Rhamdani M. A., Brooks G., Pownceby M.., Thaha Y. N., Abbott T., Grandfield J., & Hartley C., (2023), The production of rare earth based magnesium and aluminium alloys—A Review, Mineral processing and extractive metallurgy review, https://doi.org/10.1080/08827508.2023.2243009

20. Chai SS., Zhang W., Yang J., Zhang L., Theint M., Zhang X., Guo S., Zhou X., Ma X., (2023), Sustainabilty applications of rare earths from metallurgy, ,magnetism, catalysis, luminescence to future electrochemical pseudocapacitance, energy storage, RSC sustainability, https://doi.org/10.1039/D2SU00054G
21. Trivedi P., Nune K., Misra D. K., (2016), Degradation behaviour of magnesium-rare earth biomedical alloys, Advance performance materials, volume 31, issue 12, https://doi.org/10.1080/10667857.2016.1213550
22. Li H. , Wang P. , Lin G. , Huang J. , (2021), The Role of Rare Earth Elements in Biodegradable Metals: A Review, Acta Biomaterialia , https://doi.org/10.1016/j.actbio.2021.05.014
23. Nie Z., Jin T., Fu J., Xu G., Yang J., Zhou J. and Zuo T., (2002), Research on rare earth in aluminium, Material science forum, vols 396–402, 1731–1736pp., http://www.scientific.net/MSF.396-402.1731
24. Khorev T., (2011), Alloying Titanium alloys with rare earth metals, Russian engineering research, vol. 31, no. 11, 1087–1094pp., https://doi.org/10.3103/S1068798X11110104
25. Lucas, J., Lucas, P., Mercier, T.L., Rollat, A., & Davenport, W.G. (2015). Rare Earths in Rechargeable Batteries, https://doi.org/10.1016/B978-0-444-62735-3.00010-3
26. Yartys V., Noreus D., Latroche M., (2016), Metal hydrides as negative electrode materials for Ni-MH batteries, Applied Physics A., https://doi.org/10.1007/s00339-015-9538-9
27. Sakai T., Hazama T., Miyamura H., Kuriyama N., Kato A., and Ishikawa H., (1991), Rare-earth based alloy electrodes for a nickel-metal hydride battery, Journal of the less common metals, 172–174, 1175-1184pp.
28. Trench A., and Skyes J., (2020), Rare earth permanent magnets and their place in the future economy, Engineering, 6, 115–118pp., https://doi.org/10.1016/j.eng.2019.12.007
29. Mohapatra J., and Liu J., (2018), Rare-earth permanent magnets: The past and future, Handbook of magnetic materials, volume 27, 1–57pp., https://doi.org/10.1016/bs.hmm.2018.08.001
30. Herbst J., (1991), $R_2Fe_{14}B$ materials: Intrinsic properties and technological aspects, Rev. Mod. Phys., 63, 819, https://doi.org/10.1103/RevModPhys.63.819
31. Smith B. and Eggert R., (2018), Costs, Substitution, and Material use: The case of Rare Earth Magnets, Environ. Sci. Technol.52, 6, 3803–3811pp., https://doi.org/10.1021/acs.est.7b05495
32. Dent, P.C. (2012). Rare earth elements and permanent magnets (invited). *Journal of Applied Physics, 111*, https://doi.org/10.1063/1.3676616
33. R. Lai, R. Chen, W. Yin, X. Tang, Z. Wang, C. Jin, D. Lee, A. Yan, (2017) High performance (La, Ce, Pr, Nd)-Fe-B die-upset magnets based on misch-metal, Journal of Alloys and Compounds, https://doi.org/10.1016/j.jallcom.2017.06.290
34. Lu Q., Niu J., Liu W., Yue M., and Altounian Z., (2017), Enhanced coercivity of spark plasma sintered (La,Ce)FeB magnets, IEEE magnetics conference (INTERMAG), https://doi.org/10.1109/INTMAG.2017.8008060
35. Brykin, A.V., Artemov, A.V. & Kolegov, K.A. Analysis of the market of rare-earth elements (REEs) and REE catalysts. *Catal. Ind.* **6**, 1–7 (2014), https://doi.org/10.1134/S2070050414010048
36. Lucas, J., Lucas, P., Mercier, T.L., Rollat, A., & Davenport, W.G. (2015). Rare Earth Catalysts, https://doi.org/10.1016/B978-0-444-62735-3.00009-7
37. Sousa-Aguiar E., Trigueiro F., and Zotin F., (2013), The role of rare earth elements in zeolites and cracking catalysts, Catalysis Today, https://doi.org/10.1016/j.cattod.2013.06.021
38. Wang X., Zhu Y., Li H., Lee J., Tang Y., and Fu G., (2022), Rare-earth Single Atom Catalysts: A new frontier in Photo/Electrocatalysis, Small Methods, Volume 6, Issue 8, 1–23pp., https://doi.org/10.1002/smtd.202200413
39. Wang Z., Huang Y., Luo H., Gong Z., Zhang K., Li N., and Wu W., (2019), Denitrification performance of rare earth tailings-based catalysts, Green process synth, 8, 865–872pp., https://doi.org/10.1515/gps-2019-0057
40. Barroso M., Gomez M., Aruua L., Abello M., (2014), Co catalysts modified by rare earths (La, Ce or Pr) for hydrogen production from ethanol, International Journal of Hydrogen Energy, 39, 8712–8719pp., https://doi.org/10.1016/j.ijhydene.2013.12.043

41. Castano C., Fahrenholtz W., O'Keefe M., (2020), Ceria-based coatings and pigments, In Metal oxides, Cerium Oxide (CeO_2): Synthesis, Properties and Applications, 211–257pp., https://doi.org/10.1016/B978-0-12-815661-2.00006-2
42. Hughes A. E., Harvey T. G., Birbilis N., Kumar A., Buchheit R. G., (2014), Coatings for corrosion prevention based on rare earths, In Woodhead Publishing Series in Metals and Surface Engineering, Rare Earth-Based Corrosion Inhibitors, Woodhead Publishing, 186–232pp., https://doi.org/10.1533/9780857093585.186
43. Sreeram K., Kumeresan S., Radhika S., Sundar V., Muralidharan C., Nair B., Ramasami T., (2008), Use of mixed rare earth oxides as environmentally benign pigments, Dyes and Pigments, 92, 243–248pp., https://doi.org/10.1016/j.dyepig.2006.08.036
44. Sreeram K., Aby C., Nair B., Ramasami T., (2008), Colored cool colorants based on rare earth metal ions, Solar energy materials and Solar cells, 92, 1462–1467pp., https://doi.org/10.1016/j.solmat.2008.06.008
45. Saji V., (2019), Review of Rare-earth-based conversion coatings for magnesium and its alloys, Journal of Materials Research and Technologies, 8(5), 5012–5035pp., https://doi.org/10.1016/j.jmrt.2019.08.013
46. Saji, V.S. (2022). Rare-Earth Conversion Coatings. *Conversion Coatings for Magnesium and its Alloys*, https://doi.org/10.1007/978-3-030-89976-9_8
47. Sivakumar V. and Varadaraju U., (2005), Environmentally benign novel green pigments: $Pr_{1-x}Ca_xPO_4$ (x= 0–0.4), Bull. Mater. Sci. Vol. 28, No. 3, 299–301pp., https://doi.org/10.1007/BF02711264
48. Divya V. and Divya J., (2021), Environmentally benign rare earth pigments: Effect of Calcium dopant and tuning of bandgaps for different color hues, Pigment & Resin technology, https://doi.org/10.21203/rs.3.rs-749523/v1
49. Pisarski WA. (2022) Rare Earth Doped Glasses/Ceramics: Synthesis, Structure, Properties and Their Optical Applications. Materials (Basel). Nov 16;15(22):8099, https://doi.org/10.3390/ma15228099 PMID: 36431585; PMCID: PMC9698667.
50. Yasaka, P., & Kaewkhao, J. (2015). Luminescence from lanthanides-doped glasses and applications: A review. *2015 4th International Conference on Instrumentation, Communications, Information Technology, and Biomedical Engineering (ICICI-BME)*, 4–15, https://doi.org/10.1109/ICICI-BME.2015.7401304
51. Falconi, Mario Christian & Loconsole, Antonella Maria & Prudenzano, F.. (2022). Modeling of rare-earth-doped glass devices and optical parameter indirect evaluation. Optical Materials. 132. 112835, https://doi.org/10.1016/j.optmat.2022.112835
52. Miniscalco W., (2001), Optical and Electronic properties of Rare earth ions in glasses, Taylor and Francis, https://doi.org/10.1201/9780203904657.CH2
53. Williams, E., & Lavery, N.P. (2017). Laser processing of bulk metallic glass: A review. *Journal of Materials Processing Technology, 247*, 73–91, https://doi.org/10.1016/J.JMATPROTEC.2017.03.034
54. Li S., Wang R. J., Wang W. H., (2006), Bulk Metallic Glasses based on rare-earth elements in lanthanum series, Journal of Non-crystalline solids, 352, 3942–3946pp., https://doi.org/10.1016/j.jnoncrysol.2006.05.039
55. Gao K., Zhu X. G., Li W. H., Xu X., Pan B. T., Li W. R., Zhou W. H., Li L., Huang W., Li Y., (2022), Recent development in the application of bulk metallic glasses, Journal of Material Science & Technology, 131, 115–121pp., https://doi.org/10.1016/j.jmst.2022.05.028
56. Li S., Zhao D. Q., Pan M. X., Wang W. H., (2005), A bulk metallic glass based on heavy rare earth gadolinium, Journal of Non-crystalline solids, 351, 2568–2571pp., https://doi.org/10.1016/j.jnoncrysol.2005.07.005
57. Liu X. F., Wang R. J., Zhao D. Q., Pan M. X., Wang W. H., (2007), Bulk metallic glasses based on binary cerium and lanthanum elements, Applied Physics letters, 91, 041901, https://doi.org/10.1063/1.2766656
58. Li S., Xi X. K., Wei Y. X., Luo Q., Wang Y. T., Tang M. B., Zhang B., Zhao Z. F., Wang R. J., Pan M. X., Zhao D. Q., Wang W. H., (2005), Formation and properties of new heavy rare-earth based bulk metallic glasses, Science and Technology of advanced materials, 6, 823–827pp., https://doi.org/10.1016/j.stam.2005.06.019

59. Wei Y. X., Zhang B., Wang R. J., Pan M. X., Zhao D. Q., Wang W. H., (2006), Erbium- and Cerium-based bulk metallic glasses, Scripta Materialia, 54, 599–602pp., https://doi.org/10.1016/j.scriptamat.2005.10.044
60. Luo, Q. and Wang W. H. (2009). Rare earth based bulk metallic glasses. Journal of Non-Crystalline Solids. 355. 759–775, https://doi.org/10.1016/j.jnoncrysol.2009.02.006
61. Alzahrani A.S. (2022), A review of glass and crystallizations of glass-ceramics, Advances in materials physics and chemistry, 12, 261–288pp., https://doi.org/10.4236/ampc.2022.121018
62. Singh, S. P., and Sontakke A. D., 2021. "Transparent Glass Ceramics" *Crystals* 11, no. 2: 156, https://doi.org/10.3390/cryst11020156
63. Bodišová, Katarína & Galusek, Dušan & Švančárek, Peter & Pouchly, Vaclav & Maca, Karel. (2015). Grain growth suppression in alumina via doping and two-step sintering. Ceramics International. 41, https://doi.org/10.1016/j.ceramint.2015.05.162.
64. Goncalves M. C., Santos L. F., Almeida R. M., (2002), Rare-earth doped transparent glass ceramics, Comptes Rendus Chimie, Volume 5, Issue 12, 845–854pp., https://doi.org/10.1016/S1631-0748(02)01457-1
65. Guanming Q., Xikun L., Tai Q., Haitao Z., Honghao Y., Ruiting M., (2007), Application of Rare earths in advanced ceramic materials, Journal of Rare Earths, Vol. 25, Suppl., 281–286pp.
66. Frukacz Z. and Pawlak D. A., (2001), Garnets, Growth of, Encyclopedia of materials: Science and Technology (Second Edition), 3455–3463pp., https://doi.org/10.1016/B0-08-043152-6/00616-1
67. Shevchenko E. V., Charnaya E. V., Khazanov E. N., Taranov A. V., Bugaev A. S., (2017), Journal of alloys and Compounds, 717, 183–189pp., https://doi.org/10.1016/j.jallcom.2017.05.106
68. Zhanturina, N., Sergeyev D., Aimaganbetova Z., Zhubaev A., and Bizhanova K., 2022. "Structural Properties of Yttrium Aluminum Garnet, Doped with Lanthanum" *Crystals* 12, no. 8: 1132, https://doi.org/10.3390/cryst12081132
69. Kim JM, Goel M, Pathak A. Epithelial ingrowth - Nd:YAG laser approach. Clin Exp Ophthalmol. 2014 May-Jun;42(4):389–90, https://doi.org/10.1111/ceo.12240. Epub 2013 Oct 23. PMID: 24119137.
70. Modena DAO, Miranda ACG, Grecco C, Liebano RE, Cordeiro RCT, Guidi RM. Efficacy, safety, and guidelines of application of the fractional ablative laser erbium YAG 2940 nm and non-ablative laser erbium glass in rejuvenation, skin spots, and acne in different skin phototypes: a systematic review. Lasers Med Sci. 2020 Dec;35(9):1877–1888, https://doi.org/10.1007/s10103-020-03046-7. Epub 2020 May 29. PMID: 32472427.
71. Watanabe H., Kataoka K., Iwami H., Shinoki T., Okagami Y., and Ishikawa I., (2003), In vitro and in vivo studies on application of erbium: YAG laser for dentine hypersensitivity, International Congress Series, 1248, 455–457pp., https://doi.org/10.1016/S0531-5131(02)01347-X
72. Hympanova L, Rynkevic R, Mori Da Cunha MGMC, Diedrich CM, Blacher S, De Landsheere L, Mackova K, Krofta L, Roovers JP, Deprest J. The ewe as an animal model of vaginal atrophy and vaginal Er:YAG laser application. Menopause. 2020 Nov 23;28(2):198–206, https://doi.org/10.1097/GME.0000000000001679. PMID: 33235032.
73. Jovicic G, Zigan L, Will S, Leipertz A. Luminescence Properties of the Thermographic Phosphors Dy^{3+}:YAG and Tm^{3+}:YAG for the Application in High Temperature Systems. *Zeitschrift für Physikalische Chemie*. 2015;229(6): 977–997, https://doi.org/10.1515/zpch-2014-0603
74. Li, B.; Tian, F.; Cui, X.; Xiang, B.; Zhao, H.; Zhang, H.; Wang, D.; Li, J.; Wang, X.; Fang, X.; et al. Review for Rare-Earth-Modified Perovskite Materials and Optoelectronic Applications. Nanomaterials 2022, 12, 1773, https://doi.org/10.3390/nano12101773
75. Salonitis, K. Fysikopoulos, A. & Chryssolouris, G. (2014). Abrasive Material, In book: CIRP Encyclopedia of production engineering, https://doi.org/10.1007/978-3-642-20617-7_6416
76. Lucas, J., Lucas, P., Mercier, T.L., Rollat, A., & Davenport, W.G. (2015). Polishing with Rare Earth Oxides Mainly Cerium Oxide CeO_2, https://doi.org/10.1016/B978-0-444-62735-3.00012-7

77. Hendrick J. B. and Sinha S. P., (1994), Cerium-based polishing compounds: discovery to manufacture, Journal of alloys and compounds, 207/208, 377–382pp
78. Hoshino T., Kurata Y., Terasaki Y., Susa K., (2001), Mechanism of polishing of SiO_2 films by CeO_2 particles, Journal of Non-crystalline solids, 283, 129–136pp.
79. Sabia R. and Stevens J., (2000): Performance Characterization of Cerium Oxide Abrasives for Chemical-Mechanical Polishing of Glass, Machining Science and Technology: An International Journal, 4:2, 235–251pp., https://doi.org/10.1080/10940340008945708
80. Chen Y., Zuo C., Li Z., Chen A., (2018), Design of Ceria grafted mesoporous silica composite particles for high efficiency and damage-free oxide chemical mechanical polishing, Journal of alloys and compounds, Volume 736, 276–288pp., https://doi.org/10.1016/j.jallcom.2017.11.112
81. Countinho C. A., Mudhivarthi S. R., Kumar A., Gupta V. K., (2008), Novel Ceria-polymer microcomposites for chemical mechanical polishing, Applied Surface Science, 255, 3090–3096pp., https://doi.org/10.1016/j.apsusc.2008.08.093
82. Zapp P., Schreiber A., Josefine M., Kuckshinrichs W., (2022), Environmental impacts of rare earth production, MRS Bulletin, Volume 47, 267–275pp., https://doi.org/10.1557/s43577-022-00286-6
83. Charalampides, G., Vatalis, K., Karayannis, V., & Baklavaridis, A. (2016). Environmental Defects and Economic Impact on Global Market of Rare Earth Metals. IOP Conference Series: Materials Science and Engineering, 161, 012069, https://doi.org/10.1088/1757-899x/161/1/012069
84. Dang D. H., Thompson K. A., Ma L., Nguyen H. Q., Luu S. T., Duong M. T. N., Kernaghan A., (2021), Towards the circular economy of rare earth elements: A review of abundance, extraction, applications, and environmental impacts, Archives of Environmental Contamination and Toxicology, https://doi.org/10.1007/s00244-021-00867-7
85. Jaroni M. S., Friedrich B., Letmathe P., (2019), Economic feasibility of rare earth mining outside China, Minerals, 9, 576, 1–18pp., https://doi.org/10.3390/min9100576
86. Weng Z. H., Jowitt S. M., Mudd G. M., Haque N., (2013), Assessing rare earth element deposit types and links to environmental impacts, Applied Earth Science (Trans. Inst. Min. Metall. 8), Vol. 122, No. 2, 83–96pp., https://doi.org/10.1179/1743275813Y.0000000036
87. Ikhlayel M., (2017): Evaluation of the environmental impacts of rare earth elements production, International Journal of Environmental Studies, https://doi.org/10.1080/00207233.2017.1341737
88. Migaszewski Z. M., and Gałuszka A., (2015) The Characteristics, Occurrence, and Geochemical Behavior of Rare Earth Elements in the Environment: A Review, Critical Reviews in Environmental Science and Technology, 45:5, 429–471, https://doi.org/10.1080/10643389.2013.866622
89. Haque N., Hughes A., Lim S., and Vernon C., (2014), Rare earth elements: Overview of mining, mineralogy, uses, sustainability, and environmental impact, Resources, 3, 614–635pp., https://doi.org/10.3390/resources3040614
90. Filho W. L., (2016), An analysis of the environmental impacts of the exploitation of rare earth metals, Rare earth industry, Technological, Economic, and Enviornmental implications, 269–277pp., https://doi.org/10.1016/B978-0-12-802328-0.00017-6
91. Randive K., Godbole P., Jawadand S., Chopra V., Dora M. L., Dhoble S., (2023) Radioactive waste management in India: Present status and future perspective, Editor(s): Nishikant A. Raut, Dadasaheb M. Kokare, Bharat A. Bhanvase, Kirtikumar R. Randive, Sanjay J. Dhoble, 360-Degree Waste Management, Volume 2, Elsevier, 2023, Pages 273–298, https://doi.org/10.1016/B978-0-323-90909-9.00005-8
92. Malhotra N., Hsu H., Liang S., Roldan M., Lee J., Ger T., Hsiao C., (2020), An updated review of toxicity effect of the rare earth elements (REEs) on aquatic organisms, Animals, 10, 1663, 1–27pp., https://doi.org/10.3390/ani10091663
93. Singh P., Yadav J. K., Jigyasu D. K., Singh S., Kumar N., Singh I., Singh M., (2021), Biotite as a geoindicator of rare earth element contamination in Gomati river basin, Ganga alluvial plain, northern India, Environ Monit Assess, 193:361, 1–24pp., https://doi.org/10.1007/s10661-021-09105-y

94. Gwenzi W., Mangori L., Danha C., Chaukura N., Dunjana N., Sanganyado E., (2018), Science of the total environment, 636, 299–313pp., https://doi.org/10.1016/j.scitotenv.2018.04.235
95. Gonzalez V., Vignati D. A. L., Leyval C., Giamberini L., (2014), Environment International, 71, 148–157pp., https://doi.org/10.1016/j.envint.2014.06.019
96. Hu Z., Richter H., Sparovek G. & Schnug E., (2004): Physiological and Biochemical Effects of Rare Earth Elements on Plants and Their Agricultural Significance: A Review, Journal of Plant Nutrition, 27:1, 183-220, https://doi.org/10.1081/PLN-120027555
97. Pagano G., Thomas P. J., Nunzio A. D., Trifuoggi M., (2019), Human exposures to rare earth elements: present knowledge and research prospects, Environmental Research, 171, 493–500pp., https://doi.org/10.1016/j.envres.2019.02.004
98. Wei B., Li Y., Li H., Yu J., Ye B., Liang T., (2013), Rare earth elements in human hair from a mining area of China, Ecotoxicology and Environmental safety, 96, 118–123pp., https://doi.org/10.1016/j.ecoenv.2013.05.031
99. Voncken J. H. L. (2016), Economic aspects of the rare earth, In book, The Rare earth elements, Springer briefs in earth sciences, 107–114pp., https://doi.org/10.1007/978-3-319-26809-5_6
100. Bruoziotis A. A., Giarra A., Libralato G., Pagano G., Guida M., Trifuoggi M., (2022), Toxicity of rare earth elements: An overview on human health impact, Front. Environ. Sci., Volume 10, https://doi.org/10.3389/fenvs.2022.948041
101. Binnemans K. and Jones P. T., (2015), Rare earths and the balance problem, J. Sustain. Metall., 1:29–38pp., https://doi.org/10.1007/s40831-014-0005-1
102. Binnemans K., Jones P. T., Muller T., Yurramendi L., (2018), Rare earths and the balance problem: How to deal with changing markets ?, Journal of Sustainable Metallurgy, https://doi.org/10.1007/s40831-018-0162-8
103. Massari S. and Ruberti M., (2013), Rare earth elements are critical raw materials: Focus on international markets and future strategies, Resources policy, 38, 36–43pp., https://doi.org/10.1016/j.resourpol.2012.07.001
104. Mancheri N. A., (2015), World trade in rare earths, Chinese export restrictions, and implications, Resources policy, 46, 262–271pp., https://doi.org/10.1016/j.resourpol.2015.10.009

Rare Earth Elements: Reshaping the Global Supply Chain with Urban Mining

Sanjeevani Jawadand, Pratik Godbole, Kaustubh Deshpande, Rajkumar Meshram, Sneha Dandekar, M. L. Dora, Shubhangi Lanjewar, Krutika Jangale, and Kirtikumar Randive

Abstract The growing demand for Rare Earth Elements (REEs) in various high-tech industries has led to a re-evaluation of global supply chains. This chapter explores the innovative concept of urban mining, a practice that seeks to recycle REEs from electronic waste and obsolete devices. By doing so, urban mining not only mitigates the challenges posed by traditional mining methods but also fosters a sustainable approach to REE sourcing. This paradigm shift in the supply chain dynamics is critical, given the indispensable role of REEs in modern technologies. Through a comprehensive overview, this chapter delves into the significance of urban mining, highlighting its potential to reshape the global REE supply chain and pave the way for a more sustainable and resource-efficient future.

Keywords Rare earth elements · Supply chain · Urban mining · Sustainability

1 Introduction

In the ever-evolving landscape of global technology, Rare Earth Elements (REEs) stand as the bedrock of innovation, permeating every aspect of modern life. These 17 elements, crucial for high-tech industries, have sparked a paradigm shift in the global

---

S. Jawadand · K. Jangale
Shri Mathuradas Mohota College of Science, Nagpur, India

P. Godbole · K. Deshpande · S. Dandekar · S. Lanjewar · K. Randive (✉)
Department of Geology, RTM Nagpur University, Nagpur, India
e-mail: randive101@yahoo.co.in

R. Meshram
Geological Survey of India, Central Region, Nagpur, India

M. L. Dora
Geological Survey of India, Hyderabad, India

K. Randive et al. (eds.), *Current Trends in Mineral-Based Products and Utilization of Wastes: Recent Studies from India*, Springer Proceedings in Earth and Environmental Sciences, https://doi.org/10.1007/978-3-031-50262-0_3

supply chain dynamics, especially with the emergence of urban mining. The escalating demand for REEs, driven by their vital role in renewable energy, electronics, and defense sectors, has compelled the exploration of alternative supply sources. Urban mining, a novel approach that focuses on recycling REEs from electronic waste and discarded products, is reshaping the way these elements are sourced and utilized. This transformative method not only addresses the scarcity concerns but also promotes sustainability, making it a cornerstone in the future of REE supply chains.

2 Rare Earth Element Resources in India and Global Scenario

2.1 Global Scenario and Chinese Monopoly

As of the latest available data, the global rare earth element (REE) reserves, concentrated primarily in key nations including China, Brazil, Vietnam, Russia, and India, are estimated to collectively total around 130 million tons (US Geological survey, 2023). These reserves are predominantly sourced from four geological environments: carbonatites, alkaline igneous systems, ion-adsorption clay deposits, and placer deposits enriched with monazite and xenotime. China, possessing approximately one-third of the world's REE reserves, maintains its preeminent position as the global leader in both the exploration and production of rare earth elements [1] (Fig. 1).

In 2010, a conflict between Japan and China led to restrictions on shipments to Japan, causing a worldwide spike in prices (Fig. 2 and 3) and prompting a global race to secure rare earth element supplies. The temporary surge in prices revived dormant mines, highlighting the Western nation's inability to promptly bridge the supply gap, and raising doubts about China's reliability as a long-term REE supplier. Post that, China has continued to strengthen its global dominance and is now using it as a significant geopolitical tool. An illustrative example of this occurred in May 2020 when President Xi Jinping publicly visited a magnet facility. Shortly afterward, an editorial from Xinhua, the government's official state-run press agency, cautioned that the United States was jeopardizing the supply of crucial materials essential for maintaining its technological prowess by engaging in a trade war against China. A month later, Beijing issued a threat of sanctions against U.S. defence contractors in response to the approval of a $620 million deal to provide missile components to Taiwan. These actions underscore China's willingness to leverage its control over critical resources to exert influence on the global stage [3]. Clearly, the current situation highlights the urgent need to explore alternative resources to reduce dependence on China (Fig. 4).

China's utilization of quotas, export duties, and other strategies to manage the production and distribution of rare earth elements (REEs) has resulted in inflated

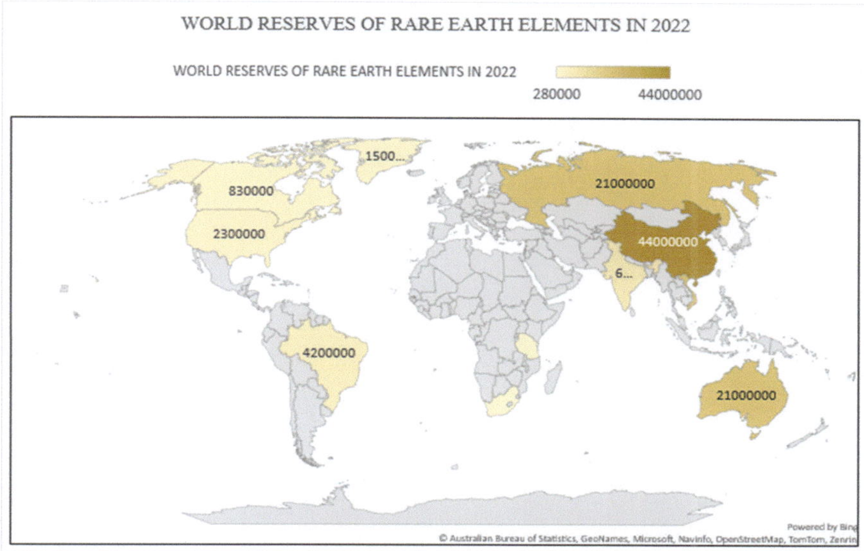

Fig. 1 World Reserves of rare earth elements in 2022 [2]

prices. This deliberate price manipulation is intended to exert economic pressure on the United States and other nations, compelling them to relocate their operations, employment opportunities, and technology advancements to China. China and India have engaged in at least 16 rounds of border discussions without reaching a resolution to the border issue [7]. As depicted in Fig. 2 and 3, there was a noticeable surge in rare earth element (REE) prices in 2011, attributed to geopolitical tensions. Given the current state of the India-China relationship, which is not characterized by trust and goodwill, relying on imports of REEs from China becomes unreliable. This underscores the urgency to fortify our supply chains and establish secondary reserves of REEs.

2.2 *Overview of India's Rare Earth Mineral Deposits and Reserves*

The rare earth resources in India are reportedly the fifth largest in the world [8]. India's rare earth acquisition amounts to approximately 6.9 million metric tons, constituting approximately 2.8% of the global total rare earth element reserves [9]. In India, rare earth metals are found in primary deposits associated with igneous and hydrothermal processes, as well as secondary deposits that have been concentrated through sedimentary processes and weathering [1]. Accessory minerals like zircon, monazite, xenotime, allanite, sphene, and apatite have a tendency to accumulate rare

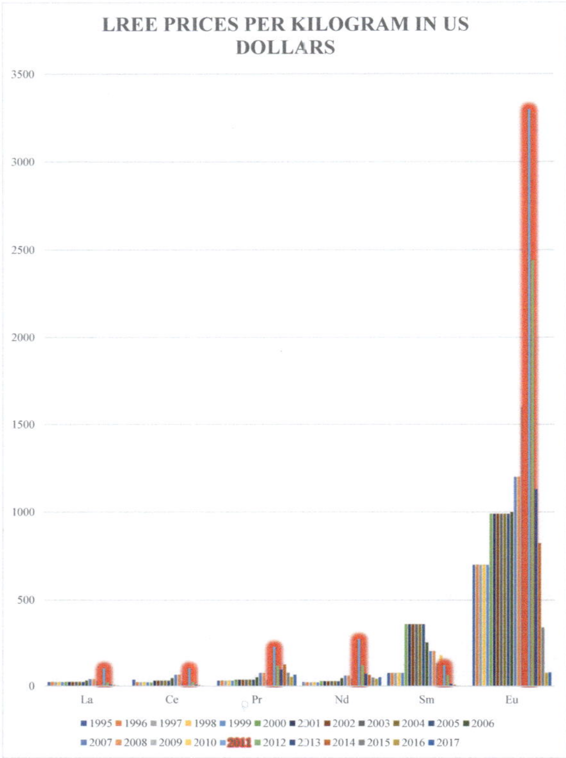

Fig. 2 Showing the prices of lower rare earth elements with price rise induced due to Chinese restrictions on export of REEs [4]

earth elements (REE) to a much greater extent than the major rock-forming minerals such as feldspars, micas, pyroxenes, and amphiboles [10].

Decades of exploration have uncovered substantial reserves of Light Rare Earth Elements (LREE) within monazite and associated minerals found in the coastal sands of Kerala, Tamil Nadu, Odisha, and Andhra Pradesh. IREL, a public sector enterprise under the Department of Atomic Energy (DAE), has been extracting REE from certain deposits such as Chavara in Kerala, Manavalakurichi in Tamil Nadu, and Chhatrapur in Odisha for several decades. An estimated 12 million tonnes of monazite resources are believed to be present along the Indian coastline [11]. The resource assessments for monazite in both beach and inland placer deposits saw an increase from 12.47 million tonnes in 2016 to 12.73 million tonnes in 2020 (Indian Mineral Yearbook, 2022). However, additional resources have also been identified, including ocean bottom sediments, waste rocks within closed mines, river sediments, industrial waste materials like red mud, coal, and fly ash, all of which have been found to contain substantial quantities of these valuable elements [1].

Exploration of fresh avenues for locating Rare Earth Element (REE) resources in India has shown promising early signs. These emerging domains encompass ion adsorption clays, quartz pebble conglomerates (QPC), iron oxide breccia complexes,

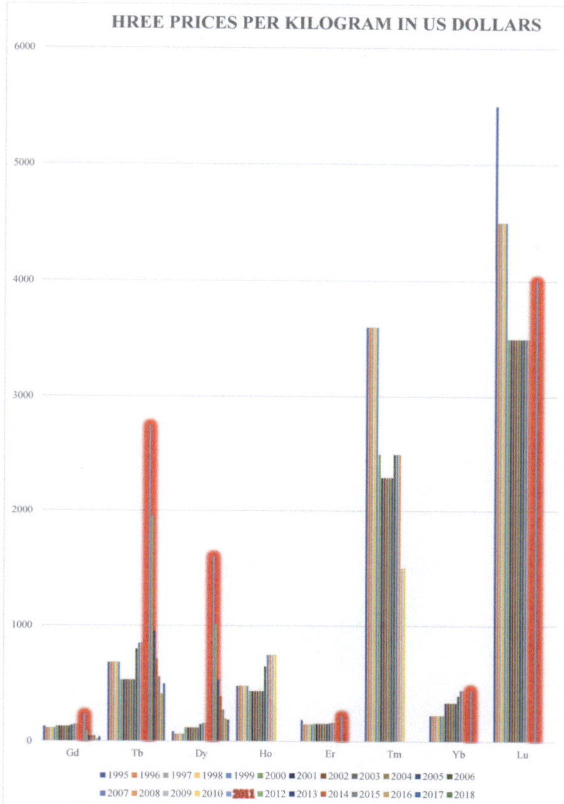

Fig. 3 Showing the prices of higher rare earth elements with price rise induced due to Chinese restrictions on export of REEs [5]

phosphorite deposits, coal ash, laterite formations, red mud, tin slag residues, phosphate ore bodies, tailings from Pb–Zn–Cu ore processing, blast furnace slag derived from steel plants, and electronic waste. These diverse resource pools hold potential for REE recovery. The initial findings indicate positive prospects, and a sustained commitment to assessing the viability of extracting REEs from these sources could yield substantial benefits in the long term. By diversifying REE supply sources, India can enhance its standing in the global REE market while also advancing sustainability and resource efficiency [11] (Figs. 5 and 6).

2.3 Current Development of REEs in India and Supply Trends

At present, India's rare earth element (REE) sector is predominantly overseen by a single government entity, Indian Rare Earths Limited (IREL). IREL is primarily

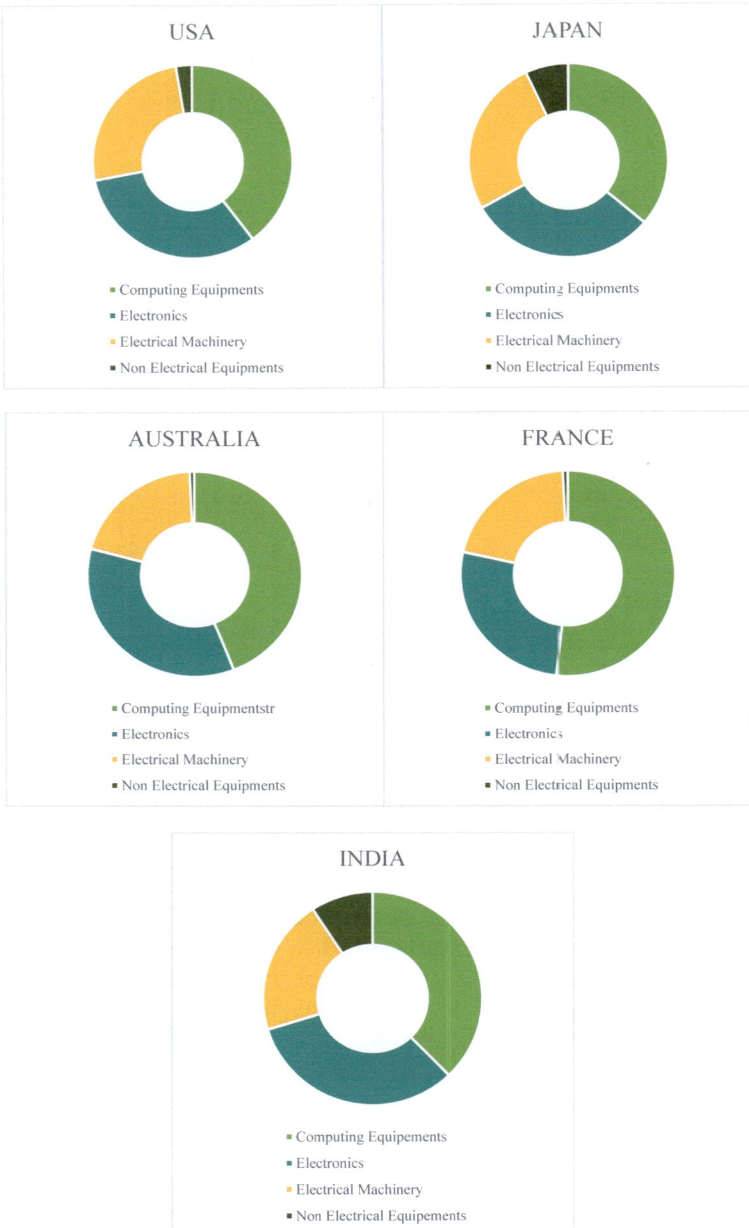

Fig. 4 Dependency of major players in REE market on China for selected high-tech imports (%) [6]

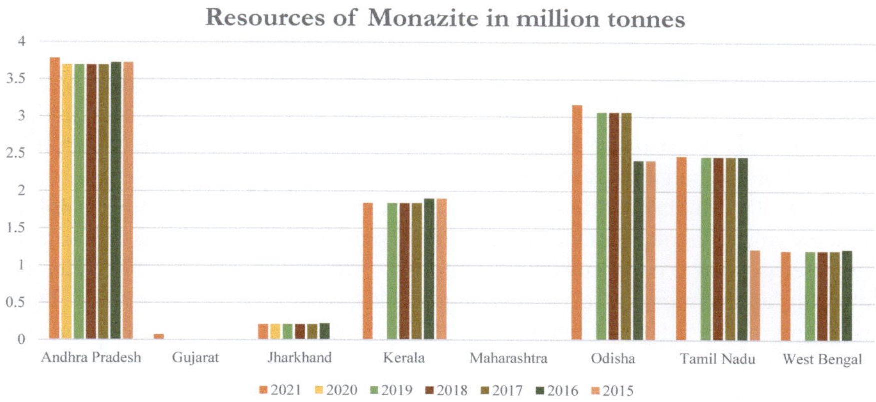

Fig. 5 Resources of monazite in million tonnes [4]

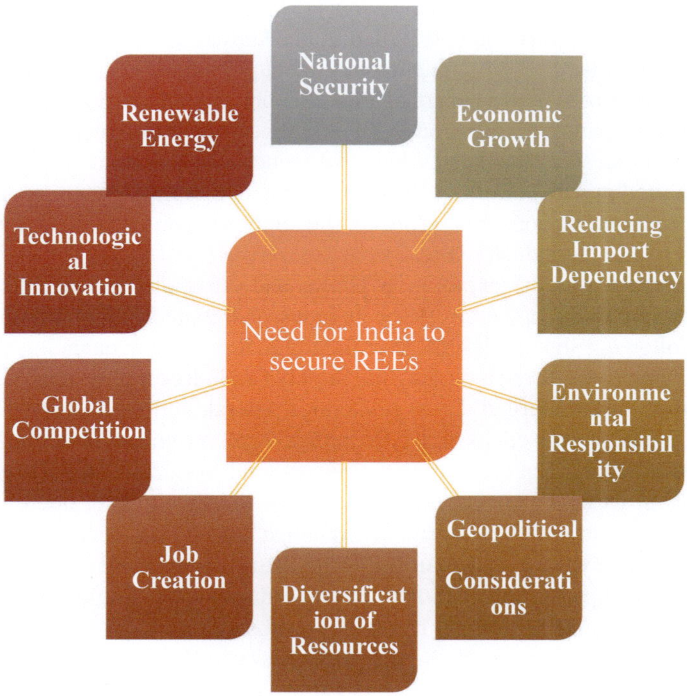

Fig. 6 Primary factors for which India needs to secure its own REE resources

engaged in the production of rare earth oxides, which are subsequently exported to international firms for further processing and the manufacturing of final products. This operation constitutes the upstream segment of the industry, characterized by relatively lower production costs but yielding modest returns. On the other hand, India's midstream and downstream REE industries are underdeveloped, which contributes to the nation's reliance on imported rare earth magnets [12].

In year 2021, India imported 11.5 million dollars and exported 56.9 million dollars in rare earth compounds. During the period spanning from 2020 to 2021, India witnessed robust export growth in Rare-Earth Metal Compounds, with Vietnam emerging as the most rapidly expanding market, valued at $14.9 million. Japan and Germany also demonstrated substantial growth, with export figures reaching $6.74 million and $6.66 million, respectively, during this timeframe. In addition to this, India experienced notable growth in its import markets for Rare-Earth Metal Compounds, with China leading the way at $2.05 million. Germany and the United States also demonstrated significant growth, with imports reaching $737,000 and $362,000, respectively, during this period [13]. A newly formed partnership, the Minerals Security Partnership (MSP), led by the United States and consisting of 11 nations, has the objective of strengthening the supply chains for critical minerals. India, although not initially a part of this coalition, is actively engaged in diplomatic efforts to secure its inclusion in the initiative [14]. India's capacity to expand downstream processing has been constrained since 2019 when the Indian government prohibited private companies from engaging in mineral exploration. This decision was prompted by concerns regarding illegal mining, unauthorized exports, and corruption within the industry (Fig. 7).

India, currently in the exploration phase, and Australia, the world's second-largest producer in the rare earth elements industry, initiated cooperation on critical minerals following the signing of their memorandum of understanding (MoU) in 2020. Collaborative efforts, including the Supply Chain Resilience Initiative (SCRI) and the Quad, have emerged across bilateral, trilateral, and multilateral platforms. In 2022, Australia revised its India Economic Plan, extending it until 2035 and elevating the Mining Equipment, Technologies, and Services (METS) sector as a top priority in its engagement with India. Following this update, one of the keys focuses of the Australia-India Business Exchange (AIBX) has been the mining and resources sector. By 2022, approximately 40 Australian companies were reported to be actively participating in the Indian mining industry [15]. Situated in Visakhapatnam, Toyotsu Rare Earths India Limited is a subsidiary of Japan's Toyoto Tsusho Corporation. It specializes in the refinement of rare earths, procuring its rare earth concentrate from the Indian Rare Earths Limited (IREL). This partnership exemplifies the international collaborations aimed at processing and harnessing India's rare earth resources [8]. In a recent development, three Indian state-owned entities, namely the National Aluminium Company Limited (NALCO), Hindustan Copper Limited (HCL), and Minerals Exploration Corporation Limited (MECL), have come together to form a joint venture called Khanij Bidesh India Limited. This joint venture is dedicated to the exploration, acquisition, and refining of strategic rare earth metals, with a primary focus on Latin American countries [16]. India could receive financial

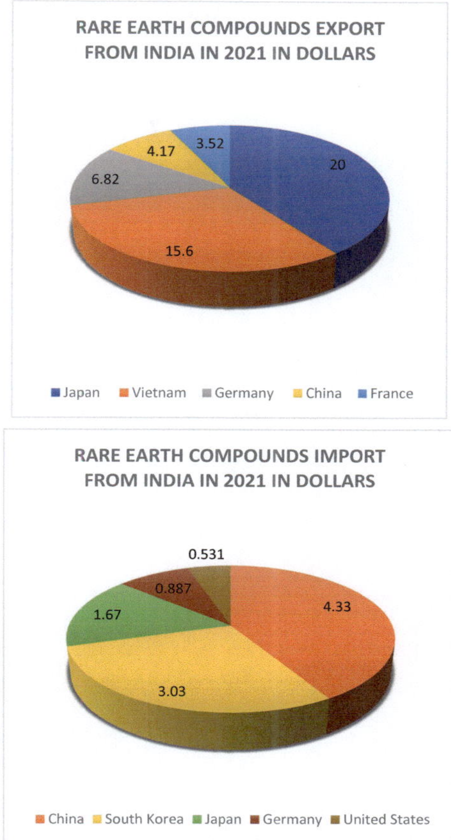

Fig. 7 Rare earth compound import–export in India in 2021 [13]

support from the United States and France for its rare earth element (REE) production project. However, before seeking international cooperation, India must address certain internal issues. To bolster its upstream REE industry, two crucial steps need to be taken. First, India should promptly formulate a comprehensive policy for the REE sector with practical goals. Second, it should consider amending the Atomic Mineral Concession Act of 2016, which currently designates all beach sand mines containing over 0.75 percent Monazite for government-owned companies [6].

The Confederation of Indian Industry (CII) has proposed the establishment of an 'India Rare Earths Mission' staffed by industry professionals, akin to the India Semiconductor Mission. This mission would prioritize rare earth mineral exploration, particularly as a crucial element of the government's Deep Ocean Mission plan. This proactive approach aims to encourage private sector participation in mining these valuable minerals and boost India's rare earth resources [17] (Fig. 8).

Fig. 8 Supply trends in India for rare earth elements

3 Integrating Domestic REE Resources into India's Manufacturing Ecosystem

China is the main player which controls supply of REE to the world which are used numerous industrial as well as commercial application [18]. The relationship between China and India does not conform to traditional models of political integration, military protection, or domination. Consequently, it is improbable that the existing economic synergies between the two nations will result in deeper technological and industrial collaboration. Instead, these synergies are more likely to foster conventional advancements in trade and investment exchanges [19]. It is need of hour for India to pave its own way for REEs amid the Chinese actions. India ranks 4th according to the DAE in case of rare earth oxide content in India and India is mining and extracting REEs from more than 50 years. Monazite is the major mineral from which REE is extracted in India. India extracts rare earth oxides which includes oxides of yttrium, cerium, neodymium, praseodymium, and lanthanum [4]. Monazite in India lacks in sufficient quantities of Heavy rare earths [20].

Currently, most of the REE based products are imported into India readymade form. India and China are countries who started REE mining approximately at same time but, with time China successfully strengthened their REE ecosystem and India remained as a REE raw material supplier. Change in direction and strategy is requirement of India which significantly alter India from infant to mature phase. Neodymium, europium, dysprosium, terbium and yttrium are considered as drivers of green energy technology. Among these, HREEs such as terbium and dysprosium are not easily available in India and if available their extraction will consume huge capital [21]. But, these metals can be easily replaced with La and Ce in Nd-Fe-B magnets [22, 23] Pr can replace Nd in Nd-Fe-B magnets [20], La-SAC can be used in place of Dy-SAC [24] Dy-BMG, Er-BMG, and Tb-BMG can be replaced by Gd-BMG, and Ce is most researched element amongst REE [25] Current approach of India must focus on exploration programs along with the development of strategic policies for rare earth to take advantage of tarnished global image of China. Primarily focus of India must be on the minerals which have high applicability and thereby strategic value. Secondly, India should observe global demand and should produce REEs accordingly. Thirdly, India must be self-reliant in case REE extraction and refinement [21].

While focusing on the exploration and conventional mining of REE, India must see growing heaps of waste in its backyard as opportunity. If this waste is utilized to create new source of REEs, it will not only reduce strain on the mining sector and thereby mother earth but also, reduce toxic effects of REE released into the environment. These sources are called as secondary sources [1] (Fig. 9).

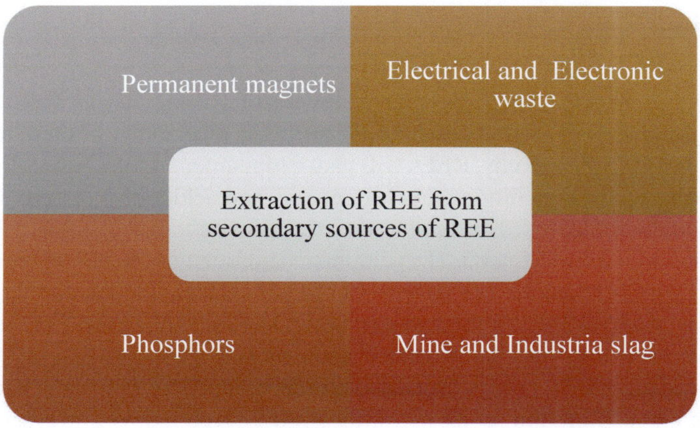

Fig. 9 Potential secondary sources of REE [20]

According to Basel convention, countries should develop their own setup form recycling of waste and must stop exporting their waste to the developing nations. India should also take steps for such waste and should focus on research and development of cost-effective method for recycling. Substitution can also be a good option and India must search of substitutes of REE [20].

Extraction methods such as precipitation, filtration, liquid–liquid extraction, solid–liquid extraction, ion-exchange, super-critical extraction, electrowinning, electrorefining, electroslag refining etc. are some of the conventional methods which are not currently viable. On the other hand, extraction of REE with help of biological factors such as algae, fungi, bacteria, resin, activated carbon, plants can pose as a cost-effective and better tool for extraction of REE [26]. As grade of REE is low in waste, generation of revenue will be high as compared to using conventional methods such as pyrometallurgy and chemical leaching techniques. Utilization of food waste, sewage sludge, and agricultural waste for enhancing microbial growth instead of use of sucrose, tryptone, and glucose will further reduce the cost of extraction REE [27]. Similarly, phytomining of the waste particularly mine waste is also an economically and technically viable process which requires further research and development to make it employable in common practice [28].

If India can formulate a comprehensive national strategy for rare earth elements and foster effective collaboration among research institutions, mission-oriented organizations, and industrial producers of rare earth products, there remains a promising opportunity for India to establish a significant presence in the global rare earth industry. However, this endeavor will necessitate a substantial commitment of financial resources, time, and concerted effort, and its realization will inevitably be a gradual process.

4 Government Initiatives, Research Efforts and Urban Mining

The scarcity or limited geographic sources for essential minerals can create vulnerabilities in the supply chain and potentially disrupt the availability of these resources. In the coming global economy, REEs will play a fundamental role, particularly given India's dedication to transitioning to clean energy and achieving net-zero emissions by 2070. This has heightened the importance of critical minerals. The Rajya Sabha has approved the Mines and Minerals (Development and Regulation) Amendment Bill of 2023, which seeks to amend the Mines and Minerals (Development and Regulation) Act of 1957. Bill focuses on the implementation of an exploration license specifically tailored for deep-seated and critical minerals is in the works. The introduction of exploration licenses for deep-seated and critical minerals is on the horizon. Despite the mining and exploration sector permitting 100% foreign direct investment (FDI) through an automatic route, there has not been substantial FDI inflow into these sectors. Globally, junior mining companies with specialized expertise are actively involved in mineral exploration, particularly for deep-seated and critical minerals like gold, platinum group minerals, rare earth elements, and more. Consequently, there is an immediate requirement to incentivize and attract foreign direct investment in these sectors [29]. In accordance with Gazette Notification No. 26/2015–2020 issued on 21.08.2018, the export of rare-earth compounds that fall under the classification of Beach Sand Minerals (BSM) is subject to control. Specifically, these compounds include Ilmenite, Rutile, Leucoxene (a mineral containing Titanium), Zircon, Garnet, Sillimanite, and Monazite (which contain Uranium and Thorium). The regulation of these exports is outlined under Serial Number 98A of Chapter 26 within Schedule 2 of the ITC (HS) Classification [4].

Creation of a dedicated Department of Rare Earths (DRE) under the purview of the Ministry of Petroleum and Natural Gas is the need of the hour. This department would be responsible for formulating and overseeing India's rare earth policy while also playing a pivotal role in fostering investments and promoting research and development within the sector. As an alternative approach, we could consider reprioritizing the mission of the Department of Atomic Energy (DAE) to align with similar objectives. The DAE could leverage its accumulated expertise, developed over the years through entities like IREL, to drive advancements in the rare earth sector. The approaches to enhance India's rare earth sector, and the ultimate strategy could involve a blend of prioritizing minerals, focusing on demand, and emphasizing geostrategic considerations [21].

Academic institutions and central research and development organizations have a crucial role in creating a curriculum to train experts and qualified engineers in the techniques of extracting and processing rare earth elements (REEs). A leading institution such as the Indian School of Mines in Dhanbad or other distinguished academic centres could be designated to establish a Centre of Excellence for Rare Earth Elements (REEs). This centre would collaborate with both public and private sector entities to pioneer the development of environmentally-friendly technologies for the future. The cornerstone of these efforts should be public sector investment, given the inherent complexities associated with foreign investment, which warrant careful consideration. Furthermore, the government may contemplate the establishment of a new entity, potentially named the Department of Rare Earths, staffed with experts in the field. This department would be responsible for shaping policies and executing the necessary actions to strategically guide and oversee the development of the REE sector [30].

Urban mining has emerged as the forefront of environmental protection efforts, and in transitioning economies such as India, it holds a pivotal position in the pursuit of sustainability [31]. It offers a methodical approach to overseeing the inventories of human-made resource reserves and the management of waste, encompassing products and structures. This approach is geared towards achieving long-term environmental preservation, resource sustainability, and economic advantages [32]. Having already secured its place as the globe's third-largest consumer of resources, India is poised for a future where its material needs continue to surge. Projections indicate that by 2050, these requirements could escalate to a staggering 25 billion tons. This growth is expected to be most pronounced in the categories of fossil fuels, metals, and minerals [33].

In view of recycling program, the government is required to play an essential role of creating, promoting, sustaining, and regulating commercially viable technologies and in combining governmental and non-governmental firms [34]. Government of India has urged manufacturer, producer, refurbisher, or recycler to enroll themselves on the Central Pollution Control Borad. Duties of manufacturer and refurbisher is to collect e-waste during manufacturing and refurbishing respectively. Advertising the e-waste collection drive is the duty of producer. In accordance with prevailing regulations, all manufacturers, producers, refurbishers, and recyclers are permitted to retain electronic waste (e-waste) for a maximum duration of 180 days. Furthermore, they are obligated to meticulously document all transactions involving the sale, transfer, and storage of e-waste. These records must be readily accessible for inspection. Additionally, e-waste storage practices must align with the currently enforced rules and guidelines [35] (Figs. 10 and 11).

Over the past decade, rare earth elements (REEs) have assumed a strategic significance for the majority of G20 nations, with their consistent availability now a vital economic and national security concern. Despite a multitude of national, regional, and multilateral efforts, the challenge of excessive geographical concentration and the associated geopolitical complexities in supply chains remain unresolved. Throughout this period, numerous bilateral and plurilateral initiatives have emerged with the primary objective of diversifying and revitalizing global supply chains for critical minerals. Nevertheless, these existing initiatives must evolve towards inclusivity in their pursuit of an alternative vertically integrated rare-earth mining to manufacturing structure. Any endeavor to reshape global supply chains also necessitates meticulous planning, seamless coordination, and addressing cost and efficiency issues, especially among the world's foremost economies represented by the G20 states [36] (Fig. 12).

Summary

The chapter explores the transformative impact of urban mining on the global supply chain of Rare Earth Elements (REEs). With REEs being essential in modern technologies, the article emphasizes the challenges posed by their traditional mining methods and the increasing demand. Urban mining, a novel practice focusing on recycling REEs from electronic waste and obsolete devices, emerges as a sustainable solution. By addressing scarcity issues and promoting environmentally friendly practices, urban mining is reshaping the REE supply chain. The chapter underscores the significance of this approach in fostering a more sustainable and resource-efficient future for high-tech industries worldwide.

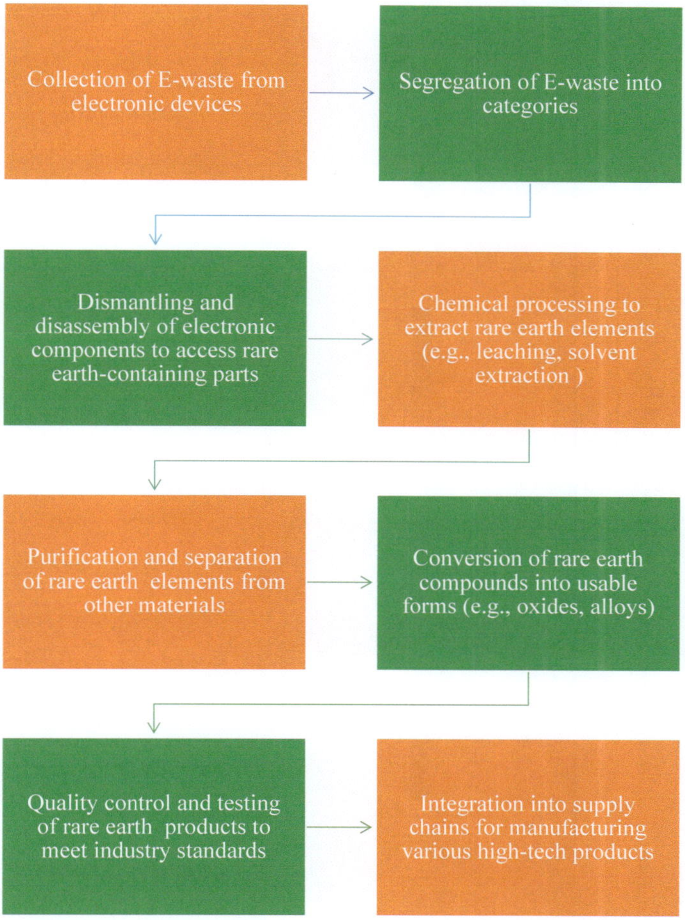

Fig. 10 Steps to procure REE by Urban mining in India

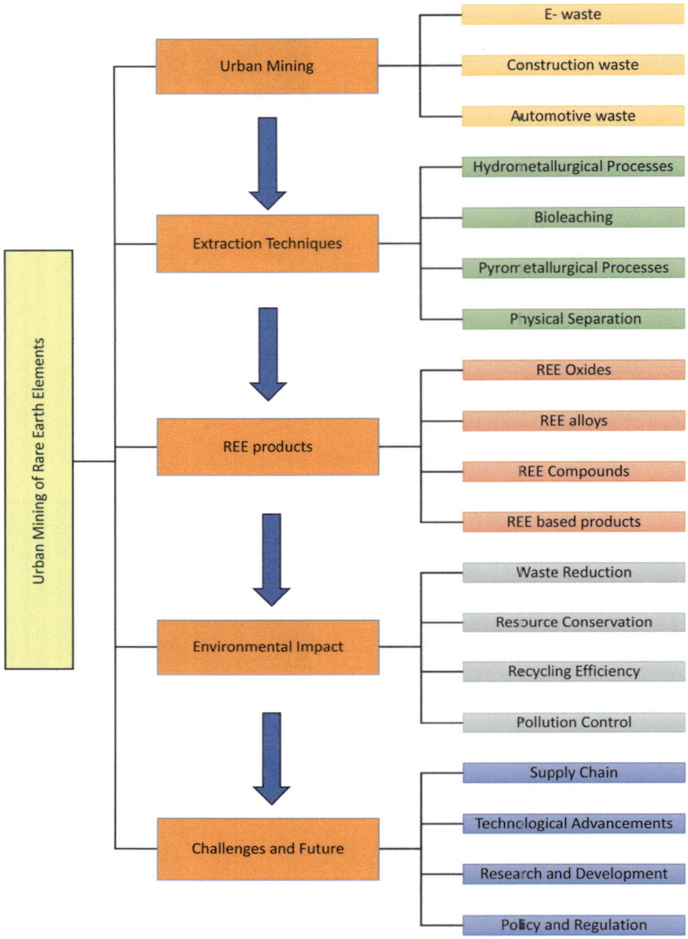

Fig. 11 Flow chart showing steps of Urban mining

Fig. 12 Recommendation to the G20 Nations for productive future of REEs [36]

References

1. Balaram, V. (2019). Rare earth elements: A review of applications, occurrence, exploration, analysis, recycling, and environmental impact. Geoscience Frontiers, 10(4), 1285–1303. doi: https://doi.org/10.1016/j.gsf.2018.12.005
2. US Geological survey, 2023 https://pubs.usgs.gov/periodicals/mcs2023/mcs2023.pdf
3. Gustavo Ferreira & Jamie Critelli, "China's Global Monopoly on Rare-Earth Elements," Parameters 52, no. 1 (2022): 57-72, doi:https://doi.org/10.55540/0031-1723.3129.
4. Indian Minerals Yearbook 2021 Edition RARE EARTH edition 54 to 60. (2015 to 2021) https://ibm.gov.in/writereaddata/files/16821576166443b030a640bRare_Earths_2021.pdf
5. USGS (2017), Minerals Yearbook [Advance release], 60.1–60.17pp., (1995–2017) https://www.usgs.gov/centers/national-minerals-information-center/rare-earths-statistics-and-information
6. RIS policy brief, 2021, Accessing Rare Earth Elements and Prospects for Manufacturing; https://www.ris.org.in/sites/default/files/Publication/Policy-brief-108.pdf
7. Jha A. R. (2014), Rare earth materials properties and application, Taylor & Francis group, LLC, 331 pp
8. Department of Atomic energy, 2023, Mining of rare earth elements, https://pib.gov.in/PressReleasePage.aspx?PRID=1914305
9. Randive K., and Jawadand S., (2020), Mineral economics an Indian perspective, Nova Science Publishers Inc., 43–65 pp
10. Randive, Kirtikumar, Jami Vijay Kumar, Abhijeet Bhondwe, and Shubhangi Lanjewar. 2014. "Understanding the Behavior of Rare Earth Elements in Minerals and Rocks." Gond. Geol. Mag 29 (January), 29-37

11. Singh, Y. (2019). Rare-earth Resources of India. Journal of the Geological Society of India, 94(1), 109–109. doi:https://doi.org/10.1007/s12594-019-1276-8
12. Gupta. M, 2022; Prospects for Rare Earth Elements Trade Between India and Australia, India Briefing, https://www.india-briefing.com/news/prospects-for-rare-earth-elements-trade-between-india-and-australia-24491.html/
13. Rare earth- metal compounds in India, Observatory for economic complexity https://oec.world/en/profile/bilateral-product/rare-earth-metal-compounds/reporter/ind
14. Mishra. M, 2022 Explained: What are rare earth elements, and why is India keen to join a global alliance to ensure their supply?, Indian express,https://indianexpress.com/article/explained/explained-sci-tech/explained-rare-earth-elements-india-global-alliance-supply-8070563/
15. Mishra. N., 2023, India-Australia rare earth supply chain collaboration, https://www.orfonline.org/expert-speak/india-australia-rare-earth-supply-chain-collaboration/
16. Kumar. A., 2020, How Rare Earths dictate strategic interests of India in Defence, Nuclear, Space sectors, https://defence.capital/2020/09/12/how-rare-earths-dictate-strategic-interests-of-india-in-defence-nuclear-space-sectors/
17. The Hindu, 2022, Industry urges govt. to establish 'India Rare Earths Mission' to reduce reliance on China, https://www.thehindu.com/business/industry-urges-govt-to-establish-india-rare-earths-mission-to-reduce-reliance-on-china/article66187382.ece
18. Voncken J. H. L. (2016), Economic aspects of the rare earth, In book, The Rare earth elements, Springer briefs in earth sciences, 107–114 pp. https://doi.org/10.1007/978-3-319-26809-5_6
19. Huchet J., (2008), Between geostrategic rivalry and economic competition, China perspectives, 50–67pp. https://doi.org/10.4000/chinaperspectives.4073 ISSN: 1996-4617
20. Ministry of mines (2012), Rare earths and energy critical elements: A roadmap and strategy for India, 1–56 pp
21. Kanisetti A., Pareek A., and Ramchandran N., (2020), A rare earths strategy for India, Takshashila discussion document 2020–16, V 1.0, 24, 1–32 pp. https://takshashila.org.in/research/4xk40ror7j1i18cnx7nl64cn2qnevw
22. R. Lai, R. Chen, W. Yin, X. Tang, Z. Wang, C. Jin, D. Lee, A. Yan, (2017) High performance (La, Ce, Pr, Nd)-Fe-B die-upset magnets based on misch-metal, Journal of Alloys and Compounds. https://doi.org/10.1016/j.jallcom.2017.06.290
23. Lu Q., Niu J., Liu W., Yue M., and Altounian Z., (2017), Enhanced coercivity of spark plasma sintered (La, Ce)FeB magnets, IEEE magnetics conference (INTERMAG). https://doi.org/10.1109/INTMAG.2017.8008060
24. Wang X., Zhu Y., Li H., Lee J., Tang Y., and Fu G., (2022), Rare-earth Single Atom Catalysts: A new frontier in Photo/Electrocatalysis, Small Methods, Volume 6, Issue 8, 1–23 pp. https://doi.org/10.1002/smtd.202200413
25. Haque N., Hughes A., Lim S., and Vernon C., (2014), Rare earth elements: Overview of mining, mineralogy, uses, sustainability, and environmental impact, Resources, 3, 614–635 pp. https://doi.org/10.3390/resources3040614
26. Das N. and Das D., (2013), Recovery of rare earth metals through biosorption: An overview, Journal of Rare earths, Vol. 31, No. 10, 933–943 pp. https://doi.org/10.1016/S1002-0721(12)60382-2
27. Dev S., Sanchan A., Dehghani F., Ghosh T., Briggs B., Aggarwal S., (2020), Mechanisms of biological recovery of rare-earth elements from industrial and electronic wastes: A review, Chemical Engineering Journal, 397, 124596, 1–17 pp. https://doi.org/10.1016/j.cej.2020.124596
28. Dinh T., Dobo Z., Kovacs H., (2022), Phytomining of rare earth elements—A review, Chemosphere, volume 297, 134259, https://doi.org/10.1016/j.chemosphere.2022.134259
29. Parliament Passes Mines and Minerals (Development & Regulation) Amendment Bill, 2023, ministry of mines, 2023, https://mines.gov.in/admin/storage/app/uploads/64d0d561570791691407713.pdf
30. A., 2023, The quagmire of Rare Earth Elements: The strategic overtones and their critical importance

31. Manu Sharma; Sudhanshu Joshi; Kannan Govindan (2021). Issues and solutions of electronic waste urban mining for circular economy transition: An Indian context. Journal of Environmental Management. https://doi.org/10.1016/j.jenvman.2021.112373
32. Tunsu, Cristian; Petranikova, Martina; Gergorić, Marino; Ekberg, Christian; Retegan, Teodora (2015). Reclaiming rare earth elements from end-of-life products: A review of the perspectives for urban mining using hydrometallurgical unit operations. Hydrometallurgy, S0304386X15300311–. https://doi.org/10.1016/j.hydromet.2015.06.007
33. Arora, Rachna; Paterok, Katharina; Banerjee, Abhijit; Saluja, Manjeet Singh (2017). Potential and relevance of urban mining in the context of sustainable cities. IIMB Management Review, S0970389617303294. https://doi.org/10.1016/j.iimb.2017.06.001
34. Mudali U., Patil M., Saravanabhavan R., Saraswat V., Review on E-Waste Recycling: Part I—A Prospective Urban Mining Opportunity and Challenges, 2021. https://doi.org/10.1007/s41403-021-00216-z
35. E-waste management rules (2022), Gazette of India, Minstry of Environment, Forest and Climate Change, https://cpcb.nic.in/uploads/Projects/E-Waste/e-waste_rules_2022.pdf
36. Resilient Rare Earths: Pathways for the G20, 2023 https://www.orfonline.org/wp-content/uploads/2023/06/TF1_132_RareEarths.pdf
37. Wang Z., Huang Y., Luo H., Gong Z., Zhang K., Li N., and Wu W., (2019), Denitrification performance of rare earth tailings-based catalysts, Green process synth, 8, 865–872 pp. https://doi.org/10.1515/gps-2019-0057

Beneficiation and Extraction of Strategic Scandium (Sc) from Aluminium Industry Waste Residue

Upendra Singh, Jyoti Pendam, Sonali Thawrani, Mayur Tirpude, and Anupam Agnihotri

Abstract Bauxite residue is generated during Bayer's process, which produces alumina from bauxite. In accordance with available land space, the residue is typically disposed of at an industrial site in man-made mud ponds. REEs are present in significant amounts in bauxite residue, and India generates a sizable amount of it (>10 MT). As a result, they contain enormous amounts of REEs that can serve as a valuable secondary source for these elements. REEs are rarely found in concentrated deposits, so their primary production is constrained. Scandium is the most strategically important and national priority among the various REEs. There are no known scandium deposits with concentrations higher than 100 mg/kg. As a result, materials like Bauxite residue with a scandium content of 50 mg/kg or higher can be regarded as ore. A list of 12 essential minerals that could be crucial for the fulfillment of the "Make in India" program and the sustainable development of the Indian economy was released by the Council on Energy, Environment, and Water. With more than 90% of the global market's supply of rare earth oxide (REO), China is the world's top supplier. Less than 2% of the world's REO is produced in India. The development of different processing techniques for the extraction of REEs from this residue has been identified as key to this issue. The developed process will aid in the handling of such a significant amount of waste, thereby minimizing statutory and regulatory scrutiny over waste treatment. Physical beneficiation and process know-how for the successful separation and recovery of REEs from bauxite residue are presented in this study.

Keywords Beneficiation · Pre-concentration · Rare earth oxides · Bauxite residue

---

U. Singh (✉) · J. Pendam · S. Thawrani · M. Tirpude · A. Agnihotri
Jawaharlal Nehru Aluminium Research Development and Design Centre (JNARDDC), Nagpur-23, India
e-mail: singhu@jnarddc.gov.in

K. Randive et al. (eds.), *Current Trends in Mineral-Based Products and Utilization of Wastes: Recent Studies from India*, Springer Proceedings in Earth and Environmental Sciences, https://doi.org/10.1007/978-3-031-50262-0_4

1 Introduction

Bauxite residue also known as Red mud is the by-product of the Bayer's Process, which produces alumina from bauxite. In general, around 1–2 metric tonnes of red mud is produced during alumina production. To produce alumina, bauxite is digested in a concentrated caustic solution at high temperatures and pressure [1]. The volume of red mud produced by alumina refineries has been increasing globally in an awful manner. It has become a severe problem to discard such a large amount of waste. Each year, over a million tonnes of red mud is generated globally. In India, more than 8 million tonnes of red mud are produced annually. Due to its toxic and caustic properties, red mud poses a very serious environmental issue. It has a pH that ranges from 11 to 12 [2]. Red mud is the combination of six major oxides Fe_2O_3, Al_2O_3, SiO_2, TiO_2, CaO and Na_2O, and trace elements such as V, Ga, P, K, Zn, Mg, and REEs [3].

The International Union of Pure and Applied Chemistry (IUPAC) defines the REE as a group of 17 elements consisting of scandium (Sc), yttrium (Y), and the 15 lanthanides (elements 57–71) is known as the REE [4]. Due to their distinctive physicochemical characteristics, rare earth elements (REEs) are crucial components of numerous cutting-edge, environmentally friendly innovations. The REEs are employed in the creation of catalysts, rechargeable NiMH batteries, light phosphors, permanent magnets, and other products. Since other materials cannot match the performance of REE in high-tech items, their demand is skyrocketing, and some are even categorized as vital elements. Although REEs are not uncommon, the supply is unstable due to geology paucity, extraction challenges, and reliance on suppliers in nations with unstable political systems. It would be preferable to transform red mud into a valuable secondary raw material because it contains a significant proportion of rare earth [5].

The leach recoveries from the direct leaching of red mud into an inorganic acid solution are not adequate. A number of beneficiation procedures are used to enrich the REE content in order to make a rare earth project commercially feasible. The generated pre-concentrate can subsequently be put to use in leaching and solvent extraction processes. The objective of this work was to create an essential and useful technique for recovering scandium from red mud. The current research describes a method for recovering REEs from red mud. Figure 1 displays a flow chart with a succession of steps depicted in block form.

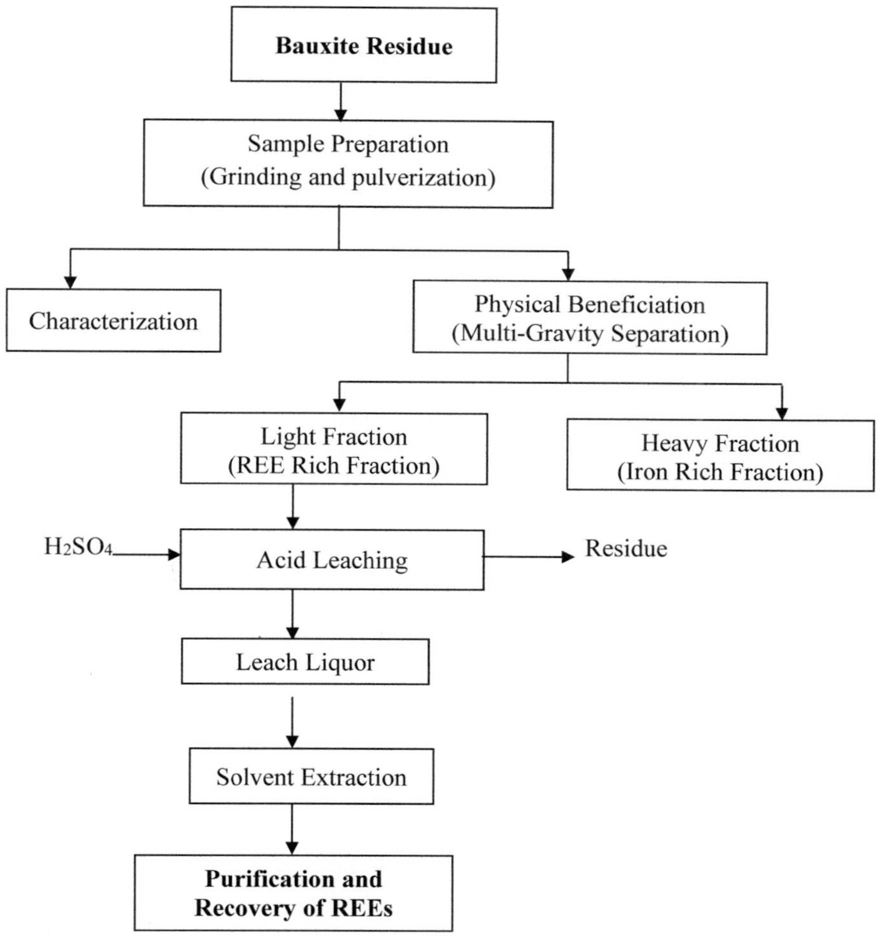

Fig. 1 Process flow chart for recovery of REEs from Bauxite residue

2 Materials and Methods

2.1 Reagents and Material

Sulphuric acid (H_2SO_4) Hydrochloric acid (HCl), and Nitric acid (HNO_3), are all of excellent reagent grade (J.T. Baker, USA). Milli-Q water (MERCK, USA) was used throughout the preparation of solutions. All the other reagents used were of analytical reagent grade. All the laboratory glass wares were of Class A grade.

The bauxite residue from the Indian alumina refinery was used for this study. The chemical analysis of the major component present in red mud is mentioned in

Table 1 Characterization of Indian red mud before preconcentration

Fe_2O_3 (%)	Al_2O_3 (%)	TiO_2 (%)	Na_2O (%)	SiO_2 (%)	LOI (%)
37.5	20.9	14.0	6.5	8.5	10.4

Table 2 Rare earth element concentration of Indian red mud

Concentration (mg/kg)															
Sc	La	Ce	Pr	Nd	Sm	Eu	Gd	Tb	Dy	Ho	Er	Tm	Yb	Lu	Y
48.0	58.5	98.1	8.1	31.9	5.2	1.62	5.2	0.59	3.0	0.55	1.50	0.21	1.35	0.18	9.8

Table 1. Red mud contains a high amount of iron and aluminum. Chemical analysis of rare earth elements is given in Table 2. The REEs with the highest quantities in the bauxite residue are scandium, lanthanum, and cerium. The material was sent for physical beneficiation prior to the leaching test. Hydrocylone (Tega Industries Ltd., India) was used for the beneficiation process. All the leaching experiments were performed in a flask fitted with a reflux condenser.

2.2 *Instrumentation*

The concentration of scandium and matrix elements was analyzed and calculated using Optical emission spectrometry (ICP-OES iCAP 7400 Duo Thermo Scientific) and subsequently confirmed by mass spectroscopy analyses (Perkin Elmer ICP-MS NexIon-2000). Standard solutions (1000 mg/L) used in ICP analyses were supplied by Reagecon Solutions, Thermo Fisher Scientific, USA. The raw material was characterized also by X-ray fluorescence spectroscopy (XRF) and wet chemical techniques.

2.3 *Experimental*

The hydro cyclone test was carried out under optimized conditions: Pressure = 20 psi, Vortex diameter = 11 mm, Apex diameter = 6.4 mm, Slurry Conc. = 20%. Pressure, Apex diameter, and vortex diameter were chosen as the major parameters for optimization.

Multi-gravity beneficiation test was carried out under suitable conditions: Shake Speed = 4.9 cps, Shake Amplitude = 15 mm, Drum speed = 120 rpm, Tilt Angle = 4° from the horizontal axis, Slurry Conc. = 33.3%. Drum speed, shake amplitude, and tilt angle were chosen as the major parameters.

Three different acids namely HNO_3, HCl, and H_2SO_4 were used for the leaching experiments. Different experiments were conducted to optimize the effect of different

Table 3 Experimental matrix for leaching of REEs from bauxite residue

Parameter	Conditions
Acid	HNO_3, HCl, and H_2SO_4
Acid concentration	1 M, 2 M, 3 M
S/L Ratio	1/10, 1/20
Time	2–5 h

acid concentrations, solid-to-liquid ratio (S/L), and time on REE extraction. The detailed experimental conditions for the leaching experiments are shown in Table 3.

According to Eq. (1), the percentage of dissolution of the rare earth elements (REE) and main elements was estimated.

$$\%\ \text{Recovery of REE} = (C_M/C_{M,in}) * (\ m_S/\ m_L) \quad (1)$$

when C_M represents element concentration in the leachate determined with ICP-OES, $C_{M,in}$ represents the concentration of the element in the original red mud, m_S represents the initial mass of the red mud and m_L represents the mass of water utilized for the leaching experiments.

The extraction tests were performed in a separation funnel and at room temperature (Fig. 2). After disengagement of phases, the raffinate phase was separated out and after filtering (using a syringe filter 0.42 micron) sent for ICP-OES/MS analysis. The concentration of Sc, Fe, Al, and others in the organic phase was evaluated using the mass balance of metals in the aqueous solution before and after extraction. The detailed experimental matrix for solvent extraction studies is given in Table 4.

Scrubbing of the organic phase after liquid–liquid extraction was carried out using 2 M HCl and 3 M H_2SO_4/3% H_2O_2 before stripping. The objective was to remove

Fig. 2 Lab-scale solvent extraction experiments

Table 4 Experimental matrix for solvent extraction

Parameter	Conditions
pH	0.5–4.5
D2EHPA concentration	0.1–0.2 M
Contact time	2–10 min
O: A Ratio	1:1–1:10

the trace impurities (Fe, Ti, Al) from the organic phase without losses of rare earth elements. 2 M NaOH solution was used to precipitate the scandium. The precipitate was filtered out of the aqueous phase, and placed in an oven at 60–70 °C for 24–48 h, and then the resultant product was calcined at 700–800 °C for 1 h to produce the desired oxide as Sc_2O_3.

3 Results and Discussion

3.1 *Beneficiation*

Selective enrichment of the REE was seen in finer fraction of Hydrocylone and MGS. Table 5 summarizes the optimized condition beneficiation experiment results. Hydrocycloning produced an overflow preconcentrate with 49% REE enrichment, 22% reduction in iron content, and 435% recovery. With MGS, the REE grade of the preconcentrate was increased by 65% with 70% recovery and the iron concentration of this fraction was lowered by 27%.

3.2 *Leaching with Mineral Acid*

Various leaching experimental test parameters i.e. acid concentration, S/L ratio, and leaching time were studied. The experimental condition with a leaching time of 2 h,

Table 5 Characterization of red mud before and after physical beneficiation

Test performed	Sc (mg/kg)	La (mg/kg)	Ce (mg/kg)	Fe_2O_3 (%)	Weight recovery (%)
Red mud without beneficiation	48.0	58.5	98.1	37.5	–
Red mud after hydrocyclone (O/F)	51.2	69.5	121.0	29.2	43
Red mud after MGS (Light fraction)	55.0	70.5	127.0	27.5	70

S/L ratio of 1/10, temperature of 95–100 °C, and H_2SO_4 concentration 2 M showed the best leaching efficiency for rare earths with relatively low dissolution of iron. The concentration of acid has a major influence on the leaching efficiency. Sc, La, and Fe leaching increased with increasing acid concentration. However, the solubility of cerium decreases continuously with an increase in sulfuric acid concentration. Decreasing the solubility of cerium sulfate can be elucidated by the "common ion effect" which states that an increase in SO_4^{2-} ion concentration will result in a reduction in cerium sulfate solubility. The effect of concentration on the solubility of Sc is most pronounced in H_2SO_4 up to 2 M, and it is almost negligible above 2 M H_2SO_4. The leaching efficiencies for Sc, La, and Ce were 60%, 68%, and 76%, respectively, under optimal conditions.

3.3 Solvent Extraction and Purification

In the solvent extraction study, scandium was extracted from the leached solution using D2EHPA solution [bis(2-ethylhexyl) phosphoric acid] in white kerosene (commercial). The extraction of Scandium included the simultaneous extraction of contaminants like Fe, Ti, and Al. The extraction orders of metals: Sc (III) > Ti (IV) > Fe (III) > Al (III). Due to the affinity of scandium towards the D2EHPA solution, around 95% scandium was extracted in the organic phase from aqueous leach liquor. Iron and titanium were scrubbed from the organic phase with 2 M of HCl and 3 M of H_2SO_4 + 3–5% of H_2O_2 respectively. The Sc_2O_3 product with > 80% purity and > 95% recovery was obtained by precipitation of scandium with sodium hydroxide followed by calcination of scandium hydroxide precipitate.

4 Conclusion

- An efficient and cost-effective process to extract rare earth elements (REEs, mainly Sc, La, and Ce) from Indian red mud has been investigated.
- The present work will help to find a long-term solution for the effective utilization of red mud.
- This process will transform red mud as a secondary source for REE.
- A conceptual/experimental flowsheet to recover scandium from red mud has been proposed for future process development and production.

References

1. Chenna Rao Borra, Yiannis Pontikes, Koen Binnemans, and Tom Van Gerven, (2015) Leaching of rare earths from bauxite residue (red mud), Minerals Engineering, 1–8.
2. Sucharita Patel, Pal BK (2015) Current Status of an Industrial Waste: Red Mud an Overview, IJLTEMAS, IV(VIII), August 2015.
3. Abhilash, Shivendra Sinha, Pratima Meshram, Pandey BD, Behera, PK, Satpathy BK Red Mud: A Secondary Resource for Rare Earth Elements, IBAAS, III, 148–162.
4. The EURARE Project: Development of a Sustainable Exploitation Scheme for Europe's Rare Earth Ore Deposits.
5. Borra CR, Blanpain B, Pontikes Y, Binnemans K, Van Gerven T (2016) Recovery of Rare Earths and Other Valuable Metals from Bauxite Residue (Red Mud): A Review. J. Sustain. Metall. **2**, 365–386.

Exploration for Rare Earth Elements Mineralization in Amba Dongar Carbonatite, Gujarat, India: An Overview and Some Suggestions

Shrinivas G. Viladkar

Abstract Total REE reserves in India are about 6% of the Global reserves, however, this figure will definitely go up after obtaining the results of explorations in call carbonatite complexes of India. The Amba Dongar carbonatite was discovered in 1963 by Sukheswala and Udas. They were the first to note the importance of REE in this complex and reported an average of 1.04% RE_2O_3 in carbonatites. Over the past seven years, large-scale exploration has been launched by the two prime organizations; Atomic Minerals Divisions, DAE, India, and the Geological Survey of India for Rare Earth Elements. Since then, many publications have discussed REE deposits related to the Amba Dongar Carbonatite and have given the estimate of total REE deposits in Amba Dongar. Recent publications report REE ores of around 105 Mt tons at 3% RE_2O_3. So far, the exploration has been restricted to only the Southwestern part of Amba Dongar, and that too only in easily accessible areas. The writer of this paper is therefore, of the opinion that at this early stage of exploration, the estimated averages are far from the total budget of REE in the Amba Dongar complex. REE minerals occur both as magmatic and hydrothermal stages and are located all over the outcrops of carbonatite breccia, sövite, and ankeritic carbonatite albeit in certain areas of rich mineralization. Amba Dongar diatreme has very complex geology and structure and therefore it calls for a more detailed study including a geophysical survey before the total assessment of the REE budget is presented. Some suggestions have been made in this regard for precise exploration targeting.

Keywords Amba Dongar · Sövite · REE · Ankerite · Exploration

S. G. Viladkar (✉)
Carbonatite Research Centre, Amba Dongar, Kadipani, Gujarat 317 009, India
e-mail: sviladkar@gmail.com; srinivas@iiserb.ac.in

Earth and Environmental Science Department, IISER, Bhopal, Madhya Pradesh 462066, India

K. Randive et al. (eds.), *Current Trends in Mineral-Based Products and Utilization of Wastes: Recent Studies from India*, Springer Proceedings in Earth and Environmental Sciences, https://doi.org/10.1007/978-3-031-50262-0_5

1 Historical Background

Interest in the carbonatite-hosted REE deposits peaked worldwide in recent years. Exploration of such deposits in the Amba Dongar complex intensified recently when the Atomic Minerals Directorate for Exploration & Research (AMD) started drilling in this area in 2015. Explorations for REE and critical metals in Indian carbonatite complexes have been the subject of some papers in Indian journals recently (e.g., Nagabhushanam[1], Krishnamurthy [2], Tirupathi [3]). These and some other publications raised many questions and provided an opportunity for the writer to address the important issues related to the estimation of reserves that are based on very limited areas.

The Amba Dongar carbonatite diatreme is situated in the westernmost part of the Deccan Basalt Province and the recent U–Pb dating on apatites from carbonatites gave an age of 65.4 ± 2.5 Ma placing it coeval with the magmatic activity with Deccan volcanism [4]. A large reserve, 11.6 million tonnes of fluorspar ore is associated with the carbonatite and was mined by the Gujarat Mineral Development Corporation (GMDC) from 1964 till 2015. In addition, it also contains potentially viable REE and Nb deposits.

A preliminary geological investigation by Sukheswala and Udas [5] generated immense interest in the mineralization of U, REE, and fluorite. They reported an average of 1.04% of R_2O_3 in Amba Dongar carbonatite. Realizing the economic importance of ankeritic carbonatite Late D.S. Sharma, a Geologist from AMD carried out sampling of the entire ankeritic carbonatite plug in the southwestern part of Amba Dongar village using the quarter-coning method. The present author had an opportunity to help Late Sharma in his endeavor.

Viladkar [6] and Viladkar and Wimmenauer [7] established the presence of magmatic monazite, bastnäsite, cerite, barite, and thorite on the basis of thin section examination of sövite as well as ankeritic carbonatites. Noteworthy concentrations of LREE were reported earlier by the writer from various samples of sövite and ankeritic carbonatite (these analyses were reported in Table 3 of Viladkar [6]). It is significant to note that ankeritic carbonatite samples show much greater enrichment (37,500 ppm) over sövite (13,922 ppm).

Further detailed work on the REE contents of the different carbonatites, fenites, and nephelinites was carried out by Viladkar and Dulski [8] who also reported a similar trend of REE enrichment from sövite to ankeritic carbonatite. The chondrite normalized pattern of these rock types is shown in Fig. 1. Unfortunately, these values published in 1981 and 1986 went unnoticed by some earlier reviewers. In 2009, Doroshkevich, et al. [9] analysed all REE-minerals such as florenceite, parisite, cerite, synchecite, monazite, and bastnäsite from sövites close to the ankeritic carbonatite exposures in the south-western part of Amba Dongar.

Exploration for REE by the AMD and Geological Survey of India (GSI), the prime organizations of the Government of India started around 2015. A large number of boreholes were drilled mainly in sövite and ankeritic carbonatite but restricted only to the south-western part of the Amba Dongar complex and subsequently their estimated

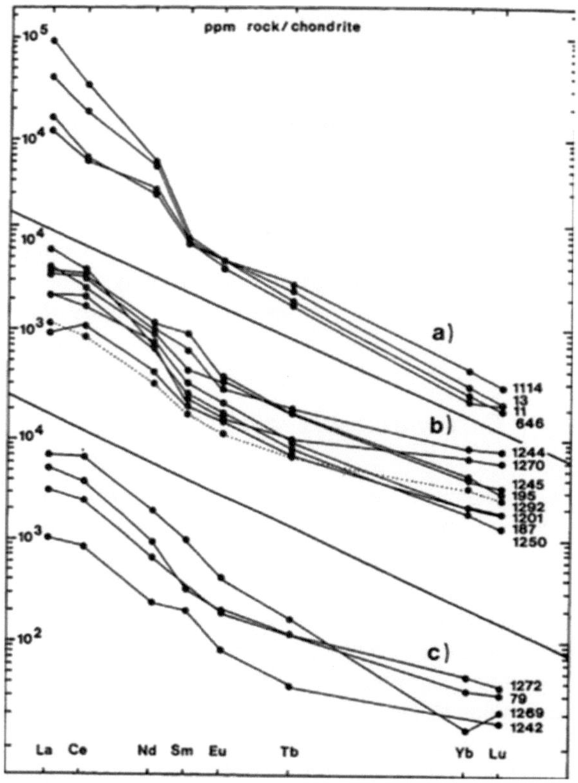

Fig. 1 Chondrite normalized distribution patterns: **a** ankeritic carbonatite, **b** sövite, and **c** alvikites [8] (Neutron Activation method in Hahn Meitner Institute Berlin, Germany)

results were published. Smith et al. [10] reported REE ore of 105 million tonnes averaging 3% REE_2O_3. Rai et al. [11] estimated 12 million tonnes of ore averaging 1.16% REE. While Krishnamurthy [2] reported an average of > 1% with some zones > 4% of REE. Since these estimates are based on the exploration carried out only in the easily accessible area in the western and south-western part of the sövite ring, in the vicinity of the ankeritic carbonatite plug, they cannot provide true REE potential of the Amba Dongar complex.

The main goal of the paper is to bring out earlier demarcated locations by the writer, not published hitherto, of the REE-rich zones all along the ring dike of sövite, ankeritic carbonatite, and carbonatite breccias and also to take a review of the exploration done so far. More emphasis is given to the eastern part of the carbonatite breccia rim and sövite ring dike as these have been completely ignored by all researchers who published papers on Amba Dongar. It is expected that this review and the suggestions made herein will serve as a guideline for further exploration.

2 Field Distribution of REE Zones

The assumption that REE-rich mineralized zones are restricted only in ankeritic carbonatite and in the vicinity of the ankeritic carbonatite is incorrect. There are appreciable amounts of REE minerals in sövite all over the ring dike, inner rim carbonatite breccia, and silicified carbonatite [6]. Most of these formations were well preserved when the writer mapped the area in 1967. There are certain parts where weathering is intense and soil samples from these weathered parts were also examined. These carbonatite units were systematically sampled and well-studied. Some important observations from such a study are given in the following sections.

2.1 *Carbonatite Breccia*

The carbonatitic activity began with the formation of a plug of carbonatite breccia forming a more than 1.5 km diameter plug in the central part of the ring structure. Outcrops of these breccias are also exposed outside the central ring structure. Taking into consideration all newly exposed breccia along road cuttings and breccia exposure in valleys it can be presumed that the original plug of breccia had occupied a much larger area than 1.5 km.

Detailed sampling of carbonatite breccia (Fig. 2a and b) was carried out and examined in detail. Typically, the best and least weathered exposures are found in valleys and also along hill slopes, and these are marked in Figs. 3, and 7a. Sövite fragments in carbonatite breccia vary in size from a few centimeters to almost a meter and these fragments enclose large apatite, barite, platy bastnäsite, and green monazite grains with irregular boundaries. Large crystals of apatite enclose pyrochlore. Occasionally carbonatite breccia is invaded by cross-cutting alvikite dikes Some outcrops of breccia analyzed total REE from 13,640 to 100,000 ppm Viladkar unpublished data.

2.2 *Sövite Ring Dike*

Sövite is heterogeneous in composition. However, it hosts a number of minerals including REE- and Nb-bearing minerals. A large number of samples from the entire ring dike of sövite were chosen for detailed investigation. Based on petrography and trace element contents (including LREE), areas of the potential REE and Nb rich are indicated in Figs. 3 and 4.

Volumetrically sövite constitutes the largest carbonatite mass in Amba Dongar diatreme. This multiphase intrusion is built by several phases with each phase having characteristic textural and mineralogical variation. The innermost part of the ring dike comprises a coarse-grained sövite (H, in Fig. 8) that gradually passes to finer-grained

Fig. 2 **a** Sampling of carbonatite breccia along the inner rim and two plugs of breccia outside inner rim. Most of the samples show presence of appreciable amount magmatic REE- minerals. **b** Thin section of breccia with sövite fragment showing presence of needles of bastnäsite (in oblong shaped area)

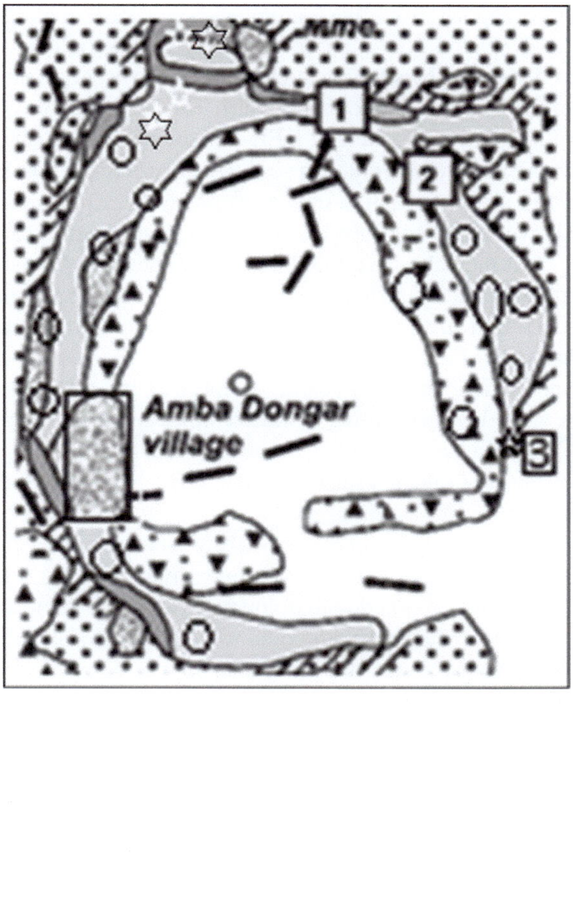

Fig. 3 REE-rich mineralized areas along sövite ring dike marked as oblong shaped, Nos. 1 and 2—sövite exposures rich in V, Nb and REE, No. 3—location of steep escarpment, sövite exposed in escarpment also shows enrichment in REE and Nb. Main ankeritic carbonatite plug is marked with dark line and this is the most explored are so far

Fig. 4 Eastern part of the sövite rind dike, circled areas indicate rich zones of REE and Nb mineralization. The black circle is contact between inner rim breccia and sövite with Nb and REE mineralization. This is good centre of rare earth mineralization and now easily accessible

sövite and finally to extremely fine-grained sövite in the outer part of the ring dike (G in Fig. 8). Both these samples show LREE content of different concentrations.

Exposures Nos. 1 and 2 (Fig. 3) are sövite rich in V, Nb, and REE. No. 3 in Fig. 3 and Fig. 4 shows the location of the steep escarpment that exposed the extreme end of the ring dike in the eastern part. This is the best exposure for the vertical sampling of sövite with pockets of REE-rich areas. Oblong-shaped area (Fig. 5) demarcate a swarm of alvikite dikes in sövite exposure in the eastern part of the ring dike rich in REE. The vertical line shows a fresh sövite exposure (for vertical sampling). Arrows show the existing footpath for climbing along the scarp.

It is then clear that the Rare Earth mineralized zones or pockets are located almost all over the ring dike of sövite while the richest and economically viable zones are found in the ankeritic carbonatite veins in sövite, ankeritic carbonatite plugs and sideritic carbonatites, and all along the contact between ankeritic carbonatite and sövite. The stars in Fig. 3 represent the areas where REE exposure was either removed during mining or buried under the mine dump and are now unapproachable. One such rich zone (with abundant green monazite) is located in the western part of the ring dike (in the valley leading to Mongra) and is totally covered with mine waste.

The eastern part of the sövite ring dike has been totally unexplored all these years. The N-S trending hill exposes fresh sövite outcrops with potential mineralization for exploration (marked with oblong shaped areas while the encircled area is the contact between sövite in the ring dike with the inner rim carbonatite breccia (Fig. 4).

Rich ore zones are also seen in the extreme southern part of the ring dike (Khandala-Kukudawa nallah (Fig. 6) though the approach is difficult from the southern part of the complex. However, this exposure can be easily approached from Kharvi, west of Amba Dongar. Additional possible areas in different parts of

Fig. 5 Vertical scarp of sövite in the far eastern part of the ring dike. Oblong shaped area demarcates swarm of alvikite dikes rich in REE. Vertical line shows a fresh sövite exposure (for vertical sampling). Arrows shows the existing footpath for climbing along the scarp

Amba Dongar are shown in Fig. 7. Figure 7b is very interesting in this respect as it represents a highly weathered part of ankeritic carbonatite which is very rich in LREE and P_2O_5. A well is located within ankeritic carbonatite exposure; however, a water sample is totally free from any trace element, F and P_2O_5, and is of good quality for drinking purposes.

In some parts of the sövite ring dike and plugs of ankeritic carbonatite (both in the south-western part and the mining area), one can see a thick blanket of weathered soil which is always rich in P_2O_5, Mn, Nb, and LREE. Though carbonatite breccia, sövite, and ankeritic carbonatite exposures contain rare earth minerals, uniformity in their distribution is lacking. Some of the REE-rich samples of the writer's collection are shown in Figs. 8 and 9.

In thin section, the majority of phase II ankeritic dikes and veins show medium-grained texture (0.8 to 4 mm) with ankerite and calcite usually containing inclusions of magnetite, apatite, pyrochlore, Columbite, zircon and REE minerals. Light-honey coloured pyrochlore occurs both as disseminated crystals, and in small aggregates. Total numbers of veins are innumerable; one such vein (Fig. 10a) was investigated in detail. Figure 10b, a BSE image shows the minerals present in this vein. Of these, most interesting is the cluster of REE-mineral synchesite. Analyses of these presented in Table 1.

Fig. 6 Another vertical scarp in the southernmost part of sövite ring dike exposing fresh sövite with good mineralization. This is easily accessible from Khandla village

2.3 REE Minerals

Previous studies of Viladkar [6] and Viladkar and Wimmenauer [7] established the presence of magmatic monazite, bastnäsite, cerite, barite, and thorite in thin sections of sövite as well as in ankeritic carbonatites. Doroshkevich et al. [9] analysed monazite, bastnäsite, parisite, synchesite, florencite, britholite, etc. from sövite samples close to the ankeritic carbonatite plug. BSE images of these minerals are exhibited in Fig. 11.

Recently, Magna et al. [12] reported a new REE mineral Wakefieldite along with Vanadinite in the sövites of the eastern part of the ring dike. Additionally, Viladkar and Sunder Raju (unpublished data) found Wakefieldite (La), Karelianite and (V_2O_3), Cooperite (PtS), in some sövites of Amba Dongar. Similarly, apatite shows a wide range of crystallization from cumulate to fine-grained sövite. The cumulate apatite in sövite carries an appreciable amount of REE and F.

Fig. 7 REE-rich zones in the different parts of carbonatite breccia ring and sövite ring: **a**—Carbonatite breccia exposure in the eastern part, **b**—A well within highly REE-enriched ankeritic carbonatite exposure, **c**—Another carbonatite breccia exposure in the eastern part (easily approachable), **d**—Part of sövite ring in the eastern part, **e**—Western part of the sövite ring dike, Sample No. 195 (G) in Fig. 7 comes from here

Fig. 8 **a**—REE-rich sövite sample, NW part of sövite ring; **b**—Sövite with veins of REE-rich ankeritic carbonatite, Northern part of ring dike near workshop; **c**—Sample from ankeritic carbonatite plug with a vein (black) rich in REE mineralization; **d**—Exposure of banded sövite and ankeritic carbonatite (in square). Ankeritic carbonatite exposed here is very highly enriched in REE, white veins contain apatite and barite (Sample No. 3, Viladkar, 1981); **e**—Sideritic carbonatite vein (dark), SW part of Amba Dongar. These contain more than 5% MnO and high total REE; **f**—Vein in sövite rich in REE minerals, Eastern part of ring dike of sövite, while descending HandevDongar; **g**—Outermost part ring dike, just north of Hill 2000, on way to Mongra; **h**—Innermost part of sövite ring, very coarse sövite, this is the earliest phase of sövite with coarse, euhedral calcite. It is invaded by REE bearing vein of ankeritic carbonatite

3 Discussion

There is a sharp increase in the production of REE from 75,500 tonnes in 2013 to 123,100 tonnes in 2016 [13, 14]. Future demand by 2026 will be linked to mainly the use of NdFeB magnets, electrical vehicles, wind turbines, etc. La and Ce will be in greater demand [15]. Of the total reserves of REE in the world (U.S. Geological Survey, in González-Álvarez et al. [16]) India shares only 6%.

Amba Dongar diatreme has good potential for REE and Nb, however, systematic investigation is needed. As discussed in an earlier section sövite and ankeritic

Fig. 9 **a**—Bore Hole samples: Centre of the Mining Area (Courtesy Soiltech Company, Pune. **b**—Northwest part of Mining Area (Courtesy GMDC). a: there are many such veins encountered in many boreholes and such veins are highly mineralized. b: this boreholes exhibits area of rich REE and pyrochlore mineralization

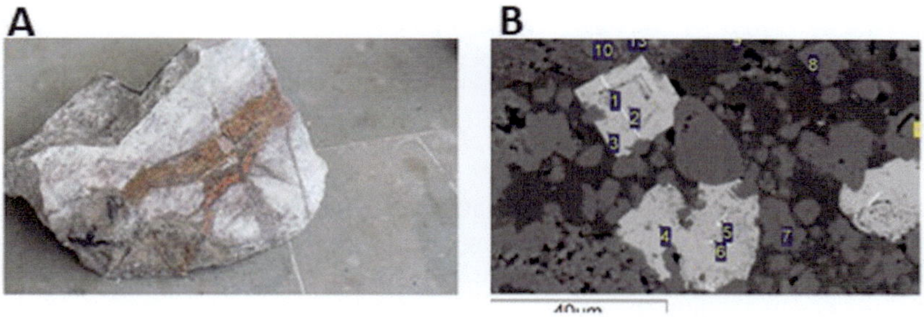

In B 1,2,3,4,5 and 6 are synchecite, 7, 8 and 10-apatite, 11 and 12 carbonates

Fig. 10 **A**—Ankeritic carbonatite vein in the sovite. **B**—BSE image-1, 2, 3, 4, 5 and 6 are synchecite, 7, 8, and 10 apatite, 11 and 12 carbonates

carbonatite show the presence of magmatic bastnäsite and monazite, suggesting that REE-minerals crystallized as primary minerals during the early stages of crystallization of carbonatite at Amba Dongar and their concentration went on increasing till end phase when ankeritic carbonatite formed and enrichment occurred in the hydrothermal stage. Hydrothermal re-mobilization of primary REE minerals has been discussed by several workers, e.g., Song et al. [17]; Trofanenko et al. [18]; Downes et al. [19].

Table 1 Synchecite from ankeritic carbonatite vein calculated for two of Ca, Sr and REE without Si and Al

CaO	9.64	17.74	17.61	17.57	14.08
SrO			1.00		
$Ce2O_3$	27.45	18.44	23.80	23.27	21.39
$La2O_3$	11.07	10.45	11.21	11.26	11.26
$Pr2O_3$	2.24	1.29	1.62	2.00	1.71
$Nd2O_3$	6.43	7.06	7.00	6.94	7.03
Tm					
Y	2.41	7.08	2.61	2.82	6.94
ThO	3.33				1.42
F	6.94	6.29	5.30	5.92	9.00
O = F	2.92	2.65	2.23	2.49	3.79
Total	69.51	68.35	70.15	69.78	72.83
La	0.28	0.21	0.22	0.23	0.24
Ce	0.68	0.37	0.47	0.47	0.46
Pr	0.06	0.03	0.03	0.04	0.04
Nd	0.16	0.14	0.14	0.14	0.15
Y	0.09	0.21	0.08	0.08	0.22
Th	0.05	0.00	0.00	0.00	0.02
Ca	0.70	1.05	1.03	1.04	0.88
Sr	0.00	0.00	0.03	0.00	0.00
F	1.48	1.09	0.91	1.03	1.66

Another mineral that is closely associated with REE minerals is barite which also occurs as a primary mineral in sövite and attains higher enrichment in ankeritic carbonatite (Fig. 11). The primary barite in sövite indicates that the primary calcio-carbonatite melt was rich in Ba. According to Néron et al. [20], presence of primary barite can be taken as a guide during exploration for REE-mineralization. The same is more or less true about the P_2O_5 content of the carbonatite melt. Apatite shows a wide range of crystallization from cumulate to later stages of crystallization. The apatite, in general, contains as much as 4% REE. The highest REE, however, is noticed in cumulate apatite (9.91% total REE and around 5% F). Britholite is quite abundant in sövites that are invaded by ankeritic carbonatite in different parts of the sövite ring dike.

Finally, the author would like to add that Amba Dongar is a very complex area in many aspects such as geology, structure, and compositional variations found in different carbonatites, and needs to be studied in extensive detail before arriving at any conclusions. The lateral extent and shape of the carbonatite intrusion need a detailed investigation. The mode of emplacement and the extent of carbonatite was discussed earlier [6]. The author has consistently argued that the central basalt cap

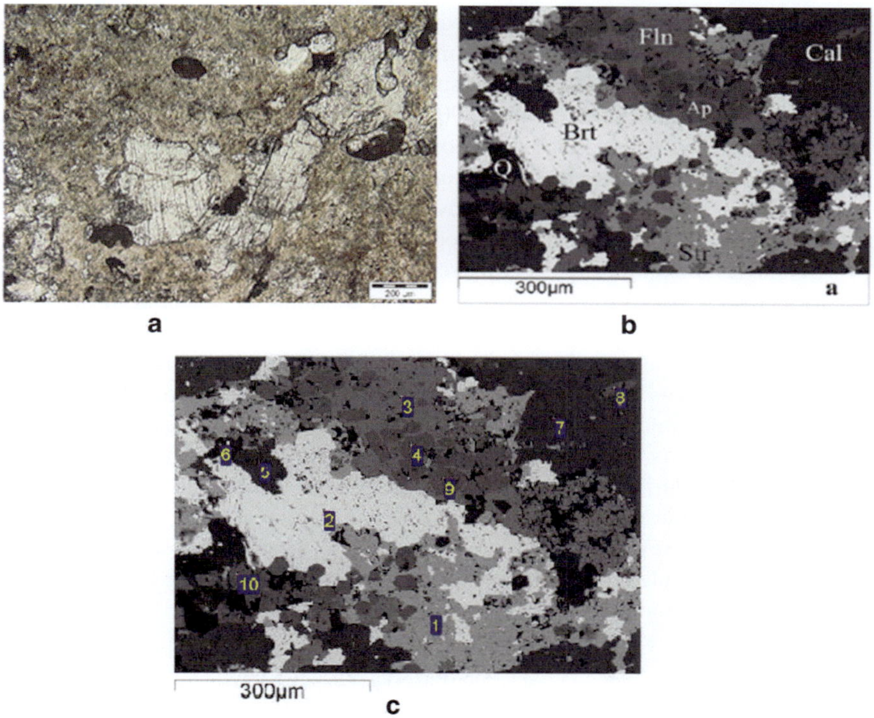

Fig. 11 Barite in sövite (**a**), and REE minerals associated with barite—**b** (BSE image), **c** Barite (2) and Strontianite (1)

over the original extensive carbonatite plugs is very thin and below it should be a massive carbonatite body. Initial shallow seismic investigations carried out by Shitole et al. [21] established the shape magmatic body and its lateral extent.

4 Summary, Conclusions and Suggestions

These suggestions are based on the field data obtained during several field seasons, the petrography of several carbonatite samples, and the geochemistry of samples since 1972.

(1) Amba Dongar is a very complex diatreme and structurally highly disturbed area. The diatreme is dissected by 37 valleys running radially to the Central Dome and many interesting carbonatite breccia and sövite outcrops are located in these valleys. These need to be explored. Some valleys, though appear steep, open out and can be approached easily.

(2) Amba Dongar needs to be remapped. The writer had mapped this area in 1967 and later researchers presented this map several times with slight changes. Now is the time to remap the area as exposures that were covered with thick vegetation (especially in valleys) are bald as 90% of vegetation cover has been removed. The increasing urbanization is responsible for encroachment on many parts of sövite and carbonatite breccia outcrops for agricultural fields. There are several carbonatite breccia and sövite outcrops with good potential for REE, Nb, and V. These need to be properly mapped.

(3) Geophysical exploration is highly recommended with an E-W extension from Moti Chikli (East) through Amba Dongar diatreme to Khandla in the west. Similarly, from the North beyond the Fluorite mine to Khandla-Kukudva Nalla in the South. This will definitely prove fruitful. In 2017, a group of students under the supervision of Prof. Sant carried out shallow seismic mapping and reached very interesting conclusions about the size, shape, and depth of the carbonatite body. In addition, some of the hidden carbonatite bodies could also be identified during aeromagnetic survey.

(4) Both the Atomic Minerals Directorate and the Geological Survey of India carried out drilling at several locations. On the basis of these results sub-surface map of the area should be prepared which will be useful and can be used as a guide for further exploration.

Acknowledgement I am grateful to Anna Doroshkevich, Geological Institute, Siberian Branch of the Russian Academy of Sciences, Sakhyanova Street 6a, 670047 Ulan-Ude, Russia, for microprobe analyses of synchesite.

References

1. Nagabhushnam B, Raju DS, Mundra KL, Rai SD, Purohit RK, Verma MB, Nanda LK (2018) LREE–Nb Mineralization in the South Western Part of Ambadongar Carbonatite Complex, Chhota Udepur District, Gujarat, India. Curr. Sci. 114, 1168–1170.
2. Krishnamurthy P (2019) Carbonatites of India. J. Geol. Soc. India. 94, 117–138.
3. Tirupathi PV (2020) Exploration for Rare metals and the Indian scenario. Geol. Soc. India, 95, 640.
4. Fosu BR, Ghosh P, Chew DM, Viladkar SG (2018) Composition and U—Pb ages of apatite in the Amba Dongar carbonatite–alkaline complex, India. Geological Journal, 1–17.
5. Sukheswala RN Udas GR (1963) Note on the carbonatite of Amba Dongar (Gujarat State) and its Economic Potentialities. Sci. Cult. 29, 563–568.
6. Viladkar SG (1981) The Carbonatites of Amba Dongar, Gujarat, India. Bull. Geol. Soc. Finland, 53, 17–28.
7. Viladkar SG, Wimmenauer W (1992) Geochemical and Petrological studies on the Amba Dongar carbonatites, (Gujarat, India). ChemErde, 52, 277–291.
8. Viladkar SG Dulski P (1986) Rare earth abundances in carbonatite, alkaline rocks and fenites of the Amba Dongar Complex, Gujarat, India. N. Jb. Miner. Mh. H-1, 37–48.
9. Doroshkevich AG, Viladkar SG, Ripp GS, Burtseva M (2009) Hydrothermal REE mineralization in the Amba Dongar carbonatite complex, Gujarat, India. Can Mineral. 47, 1105–1116

10. Smith MP, Moore K, Kavecsánszki D, Finch AA, Kynicky J Wall F (2016) From mantle to critical zone: A review of large and giant-sized deposits of the rare earth elements. Geosc. Front. 7, 315–334.
11. Rai SD, Mundra KL, Pandit PSC, Sahoo HS Verma MB (2018) REE and Nb mineral potentialities of Amba-Dongarcarbonatite complex and Kadipani fluorspar mine dumps. Expl.Res.Atomic Minerals, 27, 115–127
12. Magna T, Viladkar S, Rapprich V, Pour O, Hopp J (2020) Nb-V enriched sövites of the northeastern and eastern part of the Amba Dongar carbonatite ring dike-India: A reflection of post-emplacement hydrothermal overprint. Geochemistry, 80, 2–11.
13. Roskill (2016a) Lithium: Global industry, markets and outlook (13th ed.). London, UK: Roskill.
14. Roskill (2016b) Rare earths: Global industry, markets and outlook (16th ed.). London, UK: Roskill.
15. Goodenough KM, Wall F, Merriman D (2017) The Rare Earth Elements: Demand, Global Resources, and Challenges for Resourcing Future Generations. Natural Recourses Research. https://doi.org/10.1007/s11053-017-9336-5
16. González-Álvarez1 I, Stoppa F, Yang XY, Porwal A (2021) Introduction to the Special Issue, Insights on Carbonatites and their Mineral Exploration approach: A Challenge towards Resourcing Critical Metals. Ore Geology Review.
17. Song W, Xu C, Smith MP, Chakhmouradian AR, Brenna M, Kynicky J, Chen W, Yang YH, Deng M, Tang HY (2018) Genesis of the world's largest rare earth element deposit, Bayan Obo, China: Protracted mineralization evolution over ~1 b.y. Geology 46 (4), 323–326.
18. Trofanenko J, William-Jones AE, Simandi GJ, Migdisov AA (2016) The nature and origin of the REE mineralization in the Wicheeda carbonatite, British Columbia. Canada. Econ. Geol. 111 (1), 199–223.
19. Downes PJ, Demeney A, Czuppon G, Jaques AL, Verrall M, Sweetapple M, Adams D, McNaughton NJ, Gwalani LG, Griffins BJ (2014) Stable H-C-O isotope and trace element geochemistry of the Cummins Range Carbonatite Complex, Kimberley reion, Western Australia: implication for hydrothermal REE mineralization, carbonatite evolution and mentle source regions. Miner. Deposits, 49 (8) 905–932.
20. Néron Alexandre, Léo Paul Bédard, Damien Gaboury (2018) The Saint-Honoré Carbonatite REE Zone, Québec, Canada: Combined Magmatic and Hydrothermal Processes. Minerals 8 (9), 397. https://doi.org/10.3390/min8090397
21. Apurva Shitole, Dhananjay A. Sant, Imtiyaz A. Parvez, Govindan Rangarajan, Satish Patel, Viladkar SG, Murty ASN, Garima Kumari (2017) Shallow seismic studies along Ambadonger to Sinhada (longitude 74° 3'50"E) Transect, western India. International Seminar on Carbonatites-alkaline rocks and associated mineral deposits. Abstract Vol., Amba Dongar, India. Dec 8–11, 2017

Siwana Ring Complex, Western Rajasthan: A Potential REE Hub

Rajeev Bidwai

Abstract Rare earths (RE) with ever-increasing demand and with a global extensive resource of 120 million tonnes, play an important role in the green energy domain and advanced and latest technologies. Ever-increasing demand pushes for intense exploration activities to establish sizeable RE resources, known to be hosted in carbonatites, alkaline rocks, and placer deposits. India's RE reserves, mostly from Indian coastal domains are 6.9 million tonnes. Geological exploration and research have been focused on suitable geological domains including carbonatites and alkaline rocks (Rare earth Investing news: 2022). Neoproterozoic bimodal, volcano-plutonic, well-preserved collapsed caldera, Siwana Ring Complex (SRC), Barmer district, Western Rajasthan, India is a part of the third largest felsic Igneous Province called Malani Igneous Suite, in Trans Aravalli Block. It is associated with bimodal volcanism with peralkaline granite ring dyke and intruded by multi-phase granitic activity. Siwana caldera is structurally controlled by NE-SW, E-W, and NW–SE faults and fractures. Caldera margins, central caldera facies, and young felsic intrusive rocks with peralkaline characteristics hosting REE-U-Nb-Ag mineralization. Caldera margins have been widely explored and sizeable REE reserves have been established associated with microgranites, felsites, and tuffs. SRC is unique for having HREE-rich mineralization along with LREE. Few mineralized centers viz, Bhatikhera (Northern margin), Phulan (North northeastern margin), Ramaniya (Southern Margin), and Nal (southwest margin) hosts sizeable resources of average REE from 0.57–0.96% respectively. This paper presents the exploration details of the Siwana Ring Complex, its uniqueness, geological disposition, geochemical considerations, mineralogical attributes, structural limitations, and their bearing on multi-metal mineralization especially on rare earth mineralization.

Keywords Siwana ring complex · Caldera · RE mineralisation

R. Bidwai (✉)
Atomic Minerals Directorate for Exploration and Research (Retired), Indore, Madhya Pradesh, India
e-mail: bidwai_61@yahoo.com

K. Randive et al. (eds.), *Current Trends in Mineral-Based Products and Utilization of Wastes: Recent Studies from India*, Springer Proceedings in Earth and Environmental Sciences, https://doi.org/10.1007/978-3-031-50262-0_6

1 Introduction

Rare Earth Elements (REE) occur in a range from 150 to 220 parts per million (ppm) in various ore/accessory minerals in characteristic geological environments within Earth's crust. The concentration is effectively more as compared to certain base metals like Cu (55 ppm) and Zn (70 ppm). Ce and La are most abundant in RE mineralization with an average concentration of 63 ppm and 31 ppm respectively [2]. Rare earth elements with characteristic high density, melting point, conductivity, and thermal conductance are in high demand in emerging economies with advanced technologies and to be used in various industries.

Lanthanides comprising cerium, lanthanum, praseodymium, neodymium, promethium, europium, gadolinium and samarium (atomic numbers 57–64) are most abundant. Application of neodymium and praseodymium are most critical and used in permanent magnets, electronic industry, electric vehicles, medical equipment-based, wind turbines and data storage systems, aircraft engines, and studio lighting.

Heavy RE metals like dysprosium, yttrium, terbium, holmium, erbium, thulium, ytterbium, yttrium, and lutetium (atomic numbers 65–71) are less abundant as compared to lanthanides and finds applications in the development of clean energy technologies, hybrid cars, fiber optics and medical devices, the electronic industry, and permanent magnets. Indian REE deposits in variable concentrations and grades are hosted in carbonatites, alkaline/peralkaline felsic rocks including pegmatites, placer and beach sands, hydrothermal veins, enriched lateritic layers, and ion adsorption deposits.

Volcanogenic processes are responsible for U-RE-rich deposits in rocks. They occur close to volcanic centers and distal locations with leachable uranium associated with volcanic ash as the main source [3]. Hydrothermal fluids due to magmatic heat close to volcanic centers, leach uranium from source rocks and concentrate within fractures and vein fabric and also lake sediments. Caldera environments, especially along wall rock boundaries, form the ideal geological setting for fracture-related U-Th-RE mineralization. Further strata-bound type mineralization between acid-volcanic flows and disseminated type in the intrusive granite are also common [4].

Siwana caldera located within the Malani Igneous suite of Western Rajasthan is a unique geological setting in western India craton west of NE-SW trending rocks of Aravalli-Delhi Supergroup hosting multi-metal REE-Nb-Zr-Th-U-Ag mineralization associated peralkaline felsic volcanic and acidic intrusive (microgranites and felsites) with huge potential for REE deposit having significant HREE and high field strength elemental concentration.

2 Malani Igneous Suite (MIS)

The Malani igneous suite (MIS) lies in the northwest part of India, which is the third largest Igneous Rhyolite Province in the world after Whitsunday, East Australia, and Sierra Madre Occidental, Mexico. In India, the Malani Igneous Suite, in Western Rajasthan and the Bijli Rhyolite in Kotri-Dongargarh Belt in Central India are the two prominent felsic volcanic regions and are fascinating geological fraternity over decades for their disposition and economic potential with reference to REE, Rare metals, thorium uranium, Ag, and other related minerals [5].

Malani magmatism spread over 51,000 km^2 in western Rajasthan the Aravalli craton, is mostly covered by aeolian sands (Thar desert) and limited rock exposures occurring in various geomorphic structures. This geologic feature (MIS) extends from Jodhpur to the edge of the Thar desert over 232 km. The volcanic rocks are traced from Pokhran in the north to Sirohi district in the central part for 245 km along the N-S tract. Tosham Hill in Hissar district, Punjab represents the northeastern limit of Malani magmatism [6] (Fig. 1).

It overlies rocks of Delhi Supergroup and Erinpura granite (830 Ma) and is overlain by the Marwar group of rocks [7] exposing interlayered Basalt and Rhyolites over 31,000 km^2. Further an intrusive phase of Jalore, Siwana, and Malani Granite (18,500 km^2 (Fig. 1) and intruded by younger intrusive microgranite, felsites, and aplites.

Malani magmatism comprises volcanic rock with alternating felsic and mafic flows followed by massive plutons and sheets of peraluminous and peralkaline granite phase (Jalore, Siwana, and Barmer) and finally intrusion of basic and felsic dykes in bimodal volcanics and granites [8]. Rb–Sr isotope data (e.g., Crawford [9], Choudhary [10] that the Malani magmatism occurred for a duration of 100 Ma from 780 to 680 Ma [11].

3 Siwana Ring Complex, Barmer District, Western Rajasthan

Neoproterozoic Siwana Ring Complex (SRC) in MIS terrain is an elliptical, a well-preserved acid volcano-plutonic collapsed caldera, that lies between 72°10'-72° 35'/ 25°27'- 25°45' (45C 2, 6 and 10 toposheets of Survey of India) covering an area of ~ 700 km^2 (Fig. 2). SRC is well documented in geological records of research by geoscientists of governmental, private, and academic organizations covering, stratigraphic, geological, structural, geochemical, geochronological, metallogeny data and its evolution.

Elliptical collapsed Siwana caldera measuring 30 km × 25 m, trends ENE–WSW. It predominantly comprises basalt–rhyolite–pyroclastic flow sequence. This sequence is intruded by aegirine–riebeckite-rich granites along the sub-circular ring fracture. These rock units are intruded by peralkaline dykes and sills which are in agpaitic in nature. These rare earth elements (REE)-rich, felsite, microgranite, and

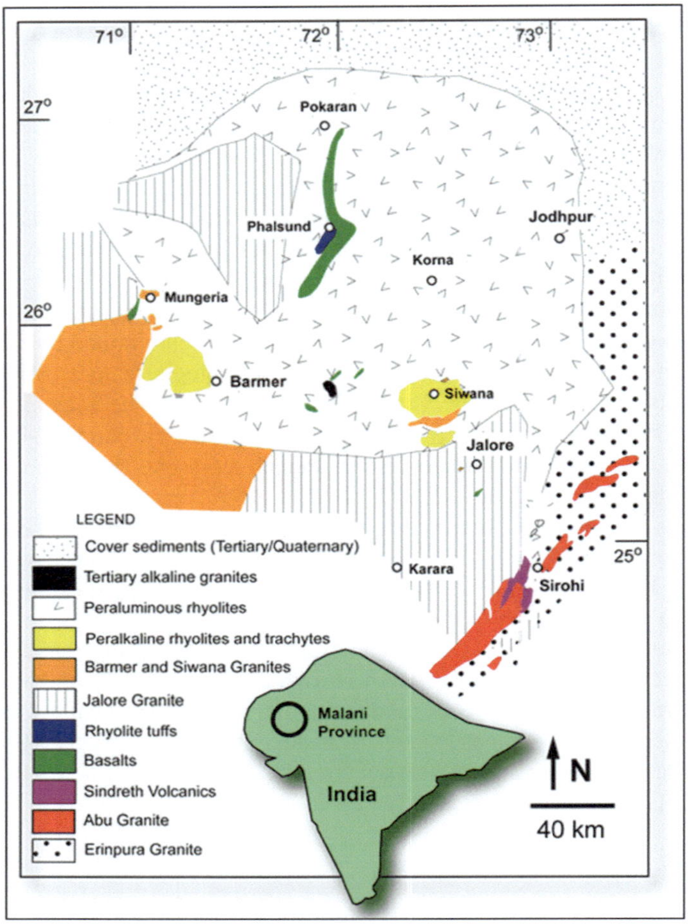

Fig. 1 Map of western Rajasthan indicating Malani magmatic rocks (modified from Bhushan [12])

aplite dykes of varied dimensions are very common along the margins of Siwana caldera [14].

The northern and northeastern margin of SRC is controlled by ENE-SWW trending major fault along with river Luni flows south-westerly. As a result, granites along ring fractures are not exposed in the northern margin. Basalt-Rhyolite-Trachyte-Dacite sequence interspersed with pyroclastic, ignimbrite-tuff flows in the northern margin is well exposed and well represented in present by a fault plane observed from Muthli in the southwest to Devra in the Northeast. Flow sequence in the northern margin shows fluctuating flow dips from 38° to 27° south-westerly and sub-horizontal in the center of the caldera in Siwana Fort Hill, Siwana. The travel of the viscous rhyolitic lava is observed along WSW-ENE for almost 6–8 kms due to the presence of ignimbrite and tuffaceous pyroclastic rocks. Lava channels and tubes

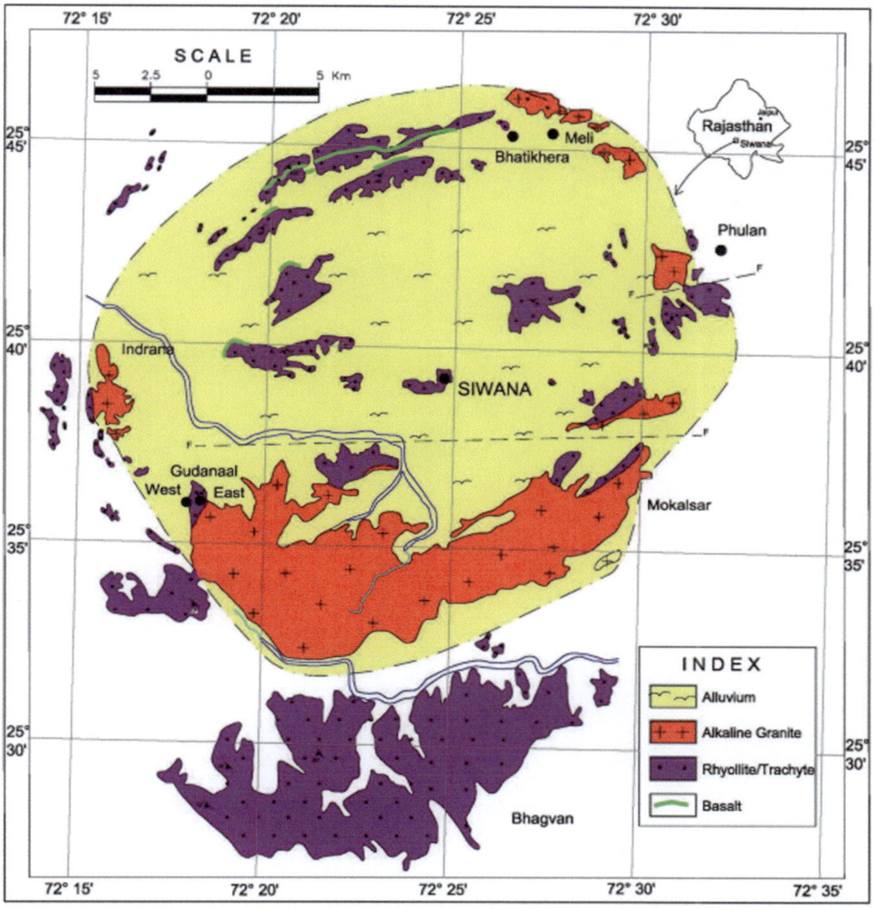

Fig. 2 Geological map of Siwana Ring Complex, Barmer district, Rajasthan (After Bhushan and Mohanty [13])

were identified in Sain ki Beri area in the northern margin as well as in the western and southern margin as eruption centers and travel conduits of the viscous rhyolitic lava flow [15]. Felsite and microgranite dykes at Bhatikhera for almost 3.0 km in strike length in the northern margin along basalt and rhyolite flows in the northern margin have been established as affluent REE resources. In addition, Phulan in the northeastern margin of SRC is a promising center for exploration with a number of parallel younger pyroclastic and micro granitic dykes and veins of a varied dimension of felsite and microgranite dykes along and across the basalt-rhyolite flow sequence were observed to be enriched in multi-metal concentration with predominant REE, and significant uranium, thorium, Zr, Nb and Ag (Fig. 3).

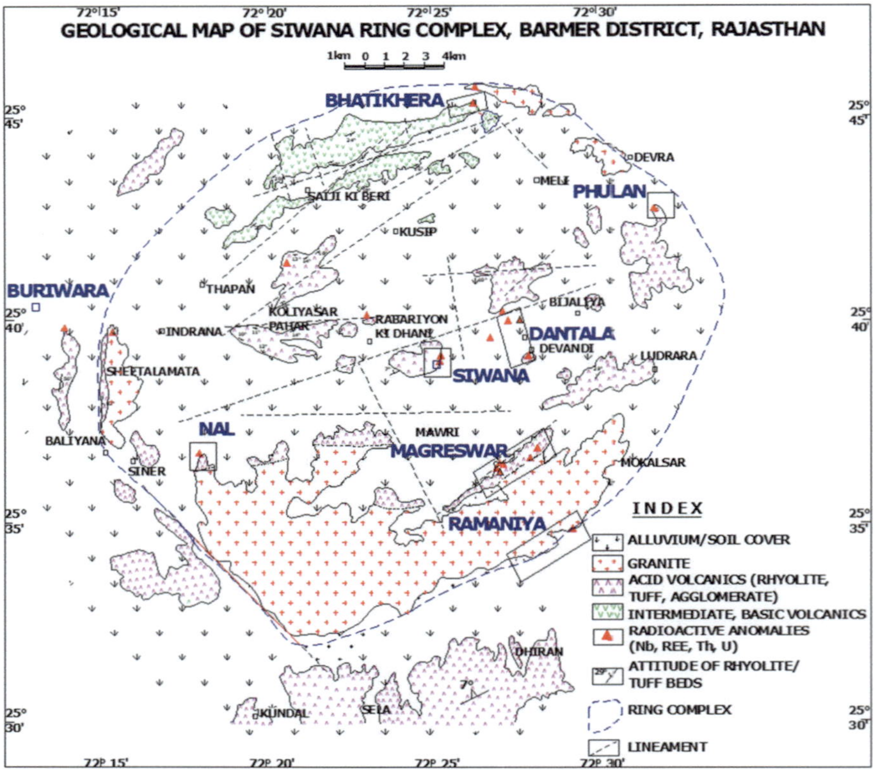

Fig. 3 Geological map of Siwana Ring Complex showing exploration for multimetal resources [5]

The southern margin of the SRC is well developed with a thick granite ring dyke observed within a basalt-rhyolite sequence and traversed by microgranite and felsite dykes varying in length a few meters to kilometers cutting across acid and mafic sequence. The sequence is traceable from Ramaniya in the east to Guda Nal in the west along the southern margin and a number of exploration centers have been established at Mokalsar—Ramaniya, Mawri-Magreshwar, Guda-Nal area for enriched REE resources along with significant Nb-Th-Zr-U concentrations (Fig. 3).

The western margin of SRC is also fertile with a distinct granite ring that is well developed intruding the basalt rhyolite sequence and the peralkaline granite and rhyolite. Younger intrusive and pyroclastic rocks are the target units for exploration for enriched REE and significant Nb, Zr, Th-U- Ag mineralization at Buriwara, Indrana-Sheetalamata temple area (Fig. 3) [5].

Central caldera facies are equally significant at Siwana Fort Hill, Bhim Ghoda Ka Pahar, Dantala, and Kerli ki Pahari wherein, younger sub horizontal rhyolitic flows have been observed intercalated with tuffs and pyroclastic rocks. Moat sediments have been mapped in Siwana Fort hill of ~ 30 m thickness between lower massive rhyolite and above columnar rhyolite flow. Dantala has emerged as a prolific REE exploration center for exploration wherein tuffs and felsites are the host rocks for enriched REE concentrations (Fig. 3).

4 Exploration for REE Resources

Exploration centers in volcanogenic caldera regions are broadly classified into three favorable locales for rare earth, rare metal, uranium and thorium, and silver concentration.

1. Caldera wall rock boundary (fracture controlled)
2. Between the rhyolitic flows or volcaniclastic bed (Stratabound)
3. Disseminated type in the intrusive granite

Examples known elsewhere are, Xiangshan Caldera, China, Long Valley, California, Valles Calera, New Mexico, Mcdermitt Caldera, Nevada, Canadian Cordillera, Marysvale Volcanic Field, Utah, Pocos de Caldas, Brazil (Fig. 4).

Potential source rocks are.

a. Intrusive granitoid with enriched Zr, REE, Th, and U by fractional crystallization and convection-driven thermos diffusion [4].
b. Volcanic and volcaniclastic sediments and ignimbrites.
c. Vitrophyre, pumice, and ash flow.

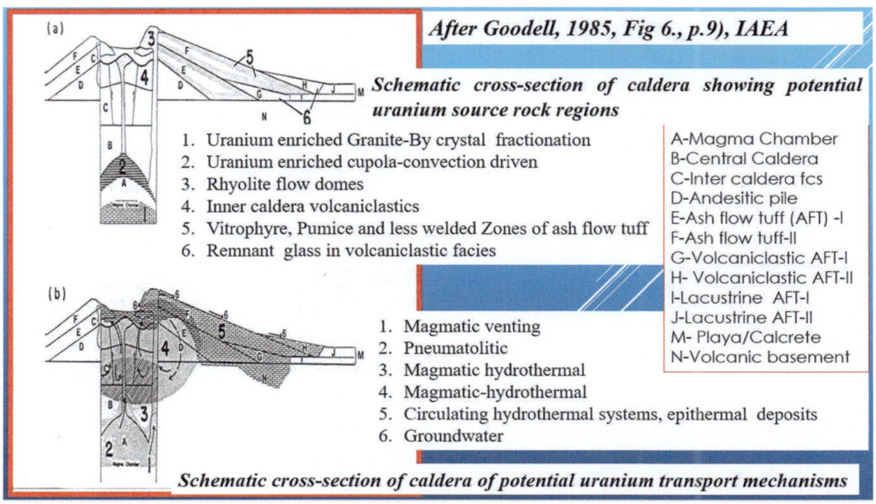

Fig. 4 Schematic cross-section of caldera showing potential source regions and transport mechanism

Siwana Ring Complex hosts multiple phases of magmatism comprising high silicic, low alumina and calcium, iron-rich bimodal volcanic suite intruded by peralkaline granite ring dyke and younger microgranite, granophyre, felsites, and aplites, rich in Y, Zr, Nb, LREE (La and Ce) present a diverse volcanogenic setting suitable for multi-metal mineralization along the margins of caldera at the contact with acid volcanic and granites; in addition to stratabound association with acid volcanic flows and Moat sediments within the caldera. Exploration efforts by the Atomic Minerals Directorate for Exploration and Research (AMD) over the years have established a volcanogenic environment for REE-Nb-Th-U-Ag rich mineralization hosted by acid volcanics, peralkaline granites, younger microgranites and felsites, aplites and moat sediments along the margins and within Siwana Caldera.

Accordingly, an exploration strategy has been adopted to target the margins of SRC and accordingly, Bhatikhera in the northern margin, Phulan in the northeastern Margin, Mokalsar-Ramaniya -Mawri-Magreshwar—Guda Nal area in southern margin and Buriwara-Sheetalamata in the western margin were developed for detailed exploration by Reconnaitory, and detail geological mapping, surface and subsurface geophysical survey, litho-hydro geochemical surveys for systematic sampling to ascertain distribution and dispersion of atomic minerals including REE resources (Fig. 3).

5 Characteristics of Granites and Rhyolites in SRC

Besides many variants of acid volcano-plutonic rocks two major rock units in SRC are rhyolites and granites. The granites, in general, are high silicic, low calcic type with low MgO, Fe_2O_3, and FeO with varying Na_2O and K_2O, peralkaline-meta-aluminous in nature with $Al/C + N + K < 1 \sim 1$ (0.51–1.02) and characterized by normative acmite (26.84–2.44 wt%) and enriched in varieties of zircons. The minor and trace elemental data indicates high Rb (1262–24 ppm, Av 206 ppm), and high Rb/Sr (Av 11.0) indicates the fractionated behaviour and their potential for accumulating rare metals. Depletion in Ba and Sr (Av 74 and 41 ppm) indicates fractionation of feldspar. Depletion of Sr and enrichment of Rb with low K/Rb ratio and high Rb/Sr ratio suggests derivation from highly fractionated melts. These are granular to porphyritic in texture with aegirine, riebeckite, epidote, and zircon are common [5].

Porphyritic to massive fine-grained rhyolites with phenocrysts of sanidine, quartz, riebeckite, aegirine, and hematite are common in a cryptocrystalline groundmass. The trachytes are relatively less silicic (Av 63%, n = 20, Rhyolite 72.24% 9.79, av. 72.24% n = 12) as compared to Rhyolites. These are highly peralkaline containing normative acmite up to 16.28 wt% and are porphyritic to massive in character. Enrichment of high field strength elements and high agpaitic ratios suggests peralkaline magma for the formation of rhyolites and trachytes.

Table 1 Mineral chemistry of Ishakawaite grains (n-11) (in %)

Y_2O_3	Nb_2O_3	Ta_2O_5	Fe_2O_3	TiO_2	CaO	UO_2	ThO_2
8.21	36.35	9.34	5.0	1.92	3.17	15.02	1.72

La_2O_3	Gd_2O_3	Dy_2O_3	Er_2O_3	Yb_2O_3	Total
0.04	1.02	0.96	0.81	0.97	87.0

Petrographic and mineralogical studies reveal a number of radioactive metamict phases responsible for radioactivity along with REE phases in acid volcanics and granites. The prominent REE phases include Cerianite (Ce^{+4} Th O_2), Synchisite (Ce, La $(CO_3)_2$ F, Monazite, LREE-Zr Silicate, Ca-LREE oxide, Ti-Ce –Oxide, Steenstrupine, Th-Zr-REE silicate which have been identified in Rhyolites by Raman Spectroscopy. Bastnaesite has also been reported in rhyolites. U-Samarskite or Ishakawaite is a Nb Ta phase of Y group rare earth containing 10–31% UO_2 [16], which has been identified in radioactive microgranite samples from Guranal (Southern margin) and Ludrara (Eastern margin). The mineral chemistry of Ishakawaite grain is tabulated below (Table 1).

The presence of minor fluorine and topaz, abundant aegirine, riebeckite, arfvedsonite, zircon-rich bimodal volcanic and peralkaline granitic rocks containing an anomalous concentration of RM-RE minerals besides U and Th suggest mobility of radio elements and rare earth elements in peralkaline magma. Its concentration along the hydrothermally altered metamict phases and favorable structural locales at the periphery and within the caldera along linear and radial fractures is envisaged. The volcanogenic environment in Siwana Ring Complex has similarities to the Streltsovka uranium district, in Russia and holds promise for multimetal mineralization.

6 REE-Rich Felsites/Microgranites

Felsite and microgranites are the younger intrusive as dykes in granite rings and volcanic flows and were known to contain high LREE, Nb, Zr, and Y in the southern margin near Mokalsar-Ramaniya belt (Jain et al. [17] and the references cited in). Spectrographic data of radioactive microgranites dykes of various dimension show high HFSE and LREE with Y (197–5377 ppm), Be (10–390 ppm), Nb (70- > 10,000 ppm), La (68–3922 ppm), Ce (249- > 10,000 ppm), Sn (34-472 ppm), Zr (449–4853 ppm) and Fersmite ($CaNb_2O_6$) has been reported for the first time [17]. Many Th-U-REE-rich radioactive anomalies along with higher silver concentration were located in the Siwana Ring Complex in various margins at the contact of granite and basalt-rhyolite contact intruded by younger intrusive of microgranite and felsites. Ramaniya (25°34′14.4″, 72°28′32.2″), Nal (25°36′29.8″, 72°17′51.8″), and Phulan (25°42′32.9″, 72°32′0″) at the periphery of SRC are favorable for structural vein type REE-Th-U mineralization with distinctly high LREE varying from

0.10–1.4% Ce, 0.15–0.55% La along with uranium up to 640 ppm, thorium up to 2.8% and Silver varying from 1.0–4.0 ppm with a maximum of 14 ppm associated with brecciated granite, felsites and quartz feldspathic rock overlying massive rhyolites. The significant exploration centers for REE are:

(a) **Bhatikhera Microgranite/felsite dyke:** A prominent microgranite/felsite dyke high La-Ce rich microgranite/felsite dyke located in the northern margin of SRC near Bhatikhera (25°45′20.0″, 72°26′33.0″, 45 C/6) extending for 1.4 km and width of 5-30 m trending E-W intruding the basalt-rhyolite sequence (Fig. 5). Minor offsets have been noted trending N-S to N20°W which may be the result of later faulting affecting the disposition of dyke. Ce up to 3.19% has been reported (Chemical Lab, Western Region AMD, Jaipur). Silver in grab samples has also been analyzed up to 15 ppm [14].

These dyke rocks are greyish to greenish-white fine-grained rocks (Fig. 6a). These are composed of quartz, feldspar, and aegirine. Peralkalinity represented by the presence of aegirine shows a rosette structure. Intergrowth texture of quartz and aegirine along with feldspar and aegirine (Fig. 6b) is common. For the first time significant silver values of 1 to 15 ppm have been reported in microgranite dykes intruding the basalt-rhyolite flow sequence [14].

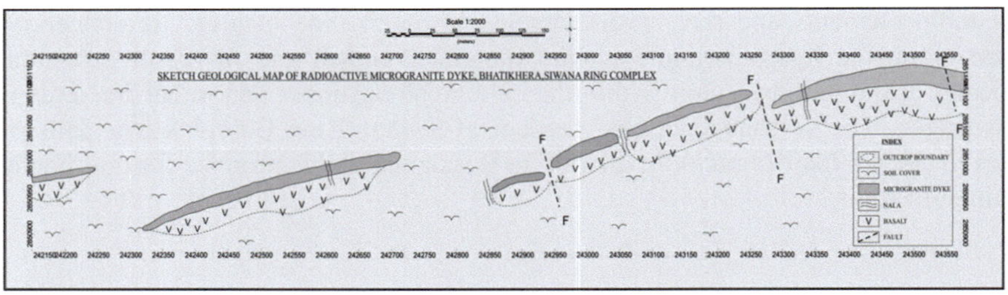

Fig. 5 Microgranite/Felsite dykes in Bhatikhera, Northern Margin, SRC

Fig. 6 **a** Field photograph of Microgranite/Felsite dyke at Bhatikhera **b** Intergrowth texture of quartz (Q), feldspar (F) and aegirine (A). TL, 2N, air

Chemical analysis of microgranites/felsites from Bhatikhera (n-4) indicates that samples (n = 4) are high SiO_2 (72.81–70.65), Fe_2O_3 (8.72–8.28%), Na_2O (5.57–3.72%), low to moderate Al_2O_3 (7.63–6.01%) and low FeO (0.72–0.50%). It also contains high field strength (HFS) elements like Nb (649–629 ppm), Y (1472–1466 ppm), and Zr (12,383–12,120 ppm) and also the zinc (1338–846 ppm). LREE's especially La, 942–842 ppm) Ce up to 3.19% have been analysed. Peralkalinity is represented by high zircon, aegirine, riebeckite, and alkali-alumina ratio, high zircon, and also the presence of aegirine and riebeckite indicated peralkalinity in the rock. Two significant lodes with moderate tonnage and grades of total REO and Nb_2O_5 have been established with remarkable HREE concentration. The presence of relatively high HREE makes Siwana a unique RE deposit.

(b) **Felsite Dyke in Phulan Area in North-Eastern Margin and Sheetalamata Area, Western Margin, SRC**

At **Phulan** (25°42′32.9″, 72°32′0″), numbers of N-S trending felsite dykes measuring 1 m-10 m in width and 10 to 800 m strike length are occurring as intrusive into basalt-rhyolite sequence. These dykes show high-order radioactivity with uranium 0.012-0.064% U_3O_8 and ThO_2 0.026- 0.29%. The contact of granite and basalt-rhyolite is highly structurally disturbed and these dykes trending NW–SE are located at the contact. Geochemically, it is characteristically low in alumina FeO, CaO, and high silicic and high Na_2O. Both HREE and LREE are high as compared to the rest of the area as in most of the places, LREE is higher than HREE. Here (n = 4) LREE with La (0.55–0.44%), Ce 1.4–2.08%) and HREE with Gd (1478–674 ppm), Dy (1354–592 ppm), Er (678–314 ppm), Yb (454–232 ppm) are high. Besides, the HFS elements show higher concentration with Y (0.8–0.35%, Zr (1.41–0.73%), Nb (576–299 ppm). At Sheetalamata hill (25°39′52″, 72°15′12″, 45 C/6) the radioactivity associated with granites in the eastern flank in the lower contours extending for ~ 1 km and are traversed by radioactive felsite dykes extending for 500 m × 1-5 m. In the western flank, rhyolites and basalts are well exposed. A mylonitized and cataclasites-bearing shear zone is observed at the contact of granite and rhyolite and radioactivity is observed for ~ I km. These granites are radioactive and have been analyzed (0.01–0.016% U_3O_8 and 0.01–0.048% ThO_2, n = 10). Felsite dyke has analysed upto 0.069% U_3O_8 and 0.26% ThO_2. Felsite veins bearing granites are moderately Silicic (66–70.2%), low calcic (0.26–0.56%) high alkali (Na_2O + K_20- 5.48–8.65%), and 0.15–0.47% F. In these samples (n = 5), LREE with 422–3874 ppm La, 9556–8054 ppm Ce, and 520–3622 Nd are high and the HFS elements like Y (846–1532 ppm), Zr (470–4470 ppm) are also significant.

Felsite from Phulan contains a few minute grains of **cerianite** (Fig. 7). Cerianite grains are fine-sized (~0.05 mm in size), euhedral, colorless, and isotropic. Orange-coloured radiation haloes have formed around cerianite. Altered variants acquired pseudo-tetragonal prismatic shapes. They are orange-yellow, fibrous, and birefringent. Dispersed tracks have been recorded between radioactive and altered phases indicating the mobility of uranium (Fig. 7a and b).

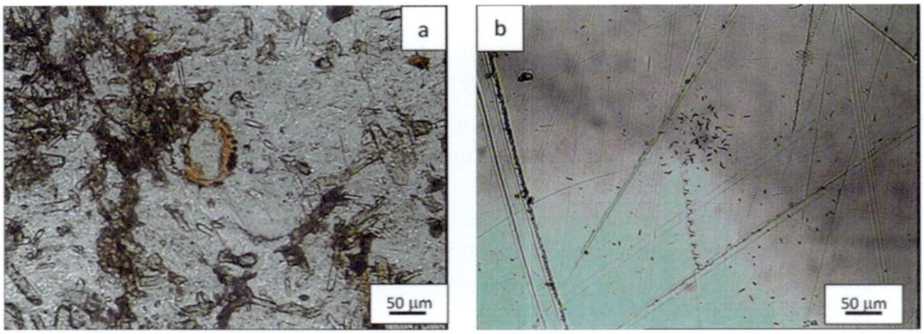

Fig. 7 Euhedral cerianite (C) with orange-colored halo (**a**) alpha tracks (**b**)

(c) REE Rich Exploration Centres in Southern Margin, SRC

In the southern margin, the granite ring is well developed and shows sharp contact rhyolite on the northern and southern flanks. Promising radioactive mineralization has been reported at many places along the southern margin, out of which two areas are discussed having the possibility of volcanogenic mineralization (Fig. 8).

1. **Ramaniya:** In the south-eastern margin located 1.5 km north of Ramaniya village ((25°34′14.4″, 72°28′32.2″) the outer rhyolite is very thin and limited in exposure and shows faulted contact with well-developed granite and traversed by a number of microgranites granites reported earlier [17]. Between the massive rhyolite of

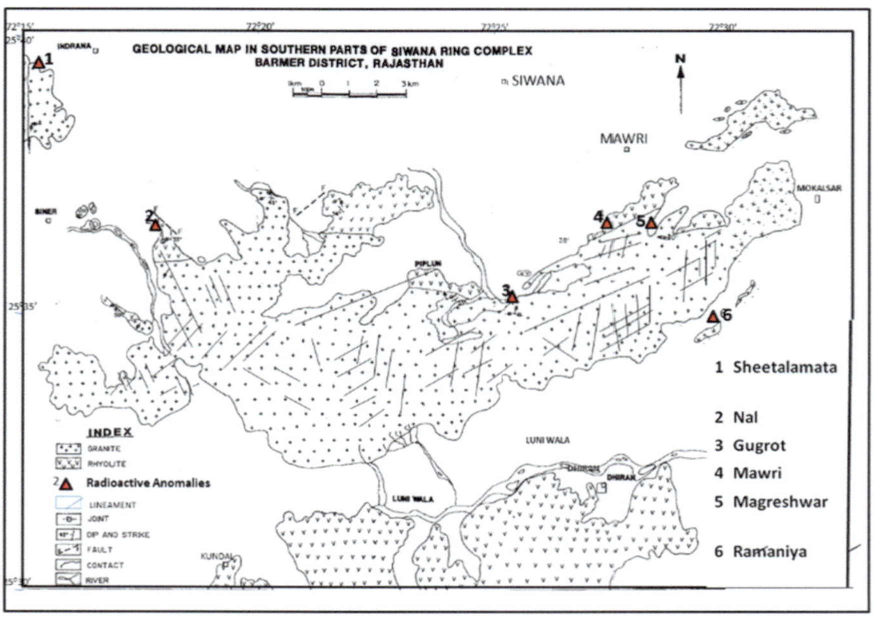

Fig. 8 Geological map of the Southern margin of the Siwana Ring Complex

Fig. 9 Ramaniya radioactive hill in the south-eastern margin of Siwana Ring Complex, Barmer district, Rajasthan

the outer volcanic sequence and the uppermost porphyritic rhyolite, there is a small hill having a dimension of 70 m x 30 m trending N70°E-S70°W comprised brecciated, sheared, and mylonitized granitoid traversed by silica and micro-granite veins (Fig. 9). The entire hill is radioactive and a magnetic survey suggests its continuity in the strike by 300 m beneath Nala sand and sand dunes cover. Radiometric analysis of grab samples shows 0.01–0.029%U_3O_8 and 0.018–0.12%ThO_2 (n = 14). LREE are relatively high with La (322–328 ppm), Ce (748–808 ppm) and HFS elements like Y (438–1110 ppm), Zr (1526–23,800 ppm) and Nb (122–198 ppm) are also high.

Medium to coarse-grained **aegirine-riebeckite granite** showing Hypidiomorphic texture in general. Quartz and alkali feldspar are the essential minerals while aegirine and Na-amphibole (riebeckite & arfvedsonite) are important accessory constituents. Euhedral prismatic to longitudinal crystals Aegirine and Na amphibole (riebeckite-arfvedsonite) are fairly abundant. The radioactive brecciated rocks are a composite rock comprising aegirine and riebeckite-arfvedsonite occur in all shapes (from prismatic, longitudinal to fibrous) and sizes and are often intermingled. Granular aggregates of sphene, fibrous clay material, and lumps of hydrous iron oxide are the alteration products. Ilmenite is the main ore mineral. Numerous iron-oxide-filled fractures randomly traverse the rock. Few radioactive phases confined within granitoids are (a) Heterogeneous (brown-yellowish-purple-turbid-colourless) altered phase (of about 0.20 mm size) after some primary radioactive mineral is present within granitoids. (b) Hydrous Fe-oxide and comprised of U-Ti-Fe and clay materials. (c) Small, euhedral, metamict phase occur as inclusions in aegirine /riebeckite/ arfvedsonite. Strong pleochroic haloes are formed around them. (d) Discrete low-density tracks correspond to this metamict phase. Irregular lumps/granular aggregates of earthy/turbid anatase/leucoxene have yielded low-density alpha tracks. (e) Anatase/leucoxene are sometimes smeared or bordered by lumps of ferruginous material. (f) Very dispersed and feeble alpha tracks are contributed by uranium adsorbed with clay or ferruginous material. Discrete, about 1 mm long, euhedral grains of topaz are generally rimmed by altered micaceous material (Fig. 10). Topaz crystals are internally fractured and

Fig. 10 Euhedral and fractured topaz (T) rimmed by Micaceous material, fractured quartzo-feldspathic vein

found in fractured riebeckite bearing quartzo-feldspathic vein within traversing the granitoid. The presence of topaz is a significant factor for volcanogenic radioactive mineralization.

2. **Nal area**: In the southwestern margin near Nal area 25°16′2.7″, 72°19′20.8″ (45C/6), the massive rhyolites are capped by Quartzo feldspathic rock observed for ~ 1 km strike length and 10- 50 m in width occurring on the eastern, western flank and also in the valley part of the main hill containing massive rhyolite (Fig. 11). High-order radioactivity is observed in the capped quartzo feldspathic rock pat. The underlying massive rhyolite is not radioactive. The radiometric assay of grab samples assayed 0.01–0.023%U_3O_8 and 0.016–0.17% ThO_2 (n = 21). These are relatively less silicic (61.0–70.1%), High Fe_2O_3, (6.65–18.67%), Low FeO (0.07–1.54%), Low to moderate alumina, (5.43–8.91%), high alkali (7.3–13.1% (Na_2O + K_2O), with moderate fluorine (0.15- 0.38%). High HFS elements Y (1297- 39,454 ppm), Zr (1474–10,090 ppm), Nb (184–484 ppm), and high LREE La (134–7634 ppm), Ce (312–17,870 ppm) are also characteristic feature. Quartz, feldspars, aegirine, and epidote are the main mineral constituents. Quartzo-feldspathic part is coarse-grained (0.50–1.2 mm) comprising interlocking quartz and feldspar. Quartz usually has irregular boundaries, is non-undulose, and sometimes encloses feldspar and aegirine grains. The coarse feldspars are perthitised with patchy textures and are highly fractured. Felsitic part is represented by randomly oriented fine laths of alkali feldspar and aegirine. Aegirine occurs as randomly oriented longitudinal/columnar prisms. Microgranite dykes trending N30-50°W-S30-50°E demarcated up to 950 m in length with upto 8 m thickness.

Fig. 11 Radioactive Quartzo feldspathic rock overlying massive rhyolite at Nal area

A number of radioactive phases are identified (a) Pseudo-tetragonal **amorphous gummite (?)** outlined by hydrous iron-oxide (b) Skeleton allanite with corresponding low-density feeble tracks in quartzo-feldspathic part (c) Tetragonal metamict prism of thorite/ uranothorite (d) Fibrous clayey material bordered by radiating needles of hydrous Fe-oxide.

(d) **Central Caldera related**

Dantala, (Central Caldera, SRC): The Dantala area falls in the central caldera of Siwana Ring Complex and is about 3 km north of Siwana. Two outcrops of radioactive REE-bearing tuff having dimensions of 500 m × 50 to 200 m and 150 m × 30 to 50 m have been recorded during the early part of the geological survey The radioactive tuffs are altered (ferruginised, silicified, and argillised) brecciated tuffs. These REE-bearing outcrops occur within a soil-covered area covering 1 km in length and 500 m in width with an area of 0.5 sq km. Boreholes were drilled at wide grids of 500 m × 500 m. The average grade of 0.523% REO with an average thickness of 32.2 m is observed contained in 15.5 million tonnes of ore.

Besides a number of radioactive anomalies have been located associated with massive rhyolite, tuffs, and moat sediments within Siwana caldera. The fractured massive rhyolite at Mawri- Asadra-Magreswar sector extending for 800 m × 5–30 m in the southern margin (figure-8) having uranium upto 0.047% U_3O_8 and negligible thorium and Kusip in the northern margin extending for 700 m × 1–10 m, thoriferous in nature, associated with massive rhyolite at the contact of basal basalt are prominent. In the central caldera in Siwana Fort Hill radioactivity associated with 30 m thick moat sediments showed up to m 200 ppm U_3O_8 with no thorium occurring between the upper youngest flow columnar rhyolite and basal massive and porphyritic rhyolite. Th: U ratio is 1:1. Radioactivity-associated hematised and brecciated tuff at Dantala and Devandi in the central caldera are also significant and shall contribute to establishing a volcanogenic model of radioactive and REE mineralization in the periphery and also in the central caldera facies. Thus, Massive rhyolites, Tuffs, felsite/microgranites, and granites are the prominent host rocks containing REE-rich Th-U-Ag mineralization.

7 Discussion

The compound annual growth rate of the rare earth elements market shall be more than 4% during the next five years. High-rising green technology demands from emerging economies indicate dependability on the RE elements and their high-grade concentrations. Inconsistent availability of rare earth elements may restrict growth. The increasing scandium usage in Aerospace applications demands scandium and gives ample scope for its need and growth.

Lanthanides with lower atomic numbers of 57–64 are significantly used for permanent magnets, electronic industry, electric vehicles, medical equipment, wind turbines and data storage systems (Nd), aircraft engines, studio lighting, and related areas of the film industry (Pr). Heavy rare earth metals of atomic numbers 65–71 are less common. Dysprosium, yttrium, and terbium are very significant for the development of clean energy technologies, hybrid cars, fiber optics and medical devices, the electronic industry, and permanent magnets.

Rare Earth mineralization is confined mostly in carbonatite contains 50% carbonate minerals as the major host. Globally major REE deposits have been reported in carbonatites (Bayan Obo, Mongolia, Mountain Pass mine, California [18]. Pegmatites and peralkaline rhyolites are established RE sources with high field strength elements including Nb and Zr and have been well-established in Egypt, Canada, and Greenland [19–21]. Indian rare earth deposits in variable concentrations and grades are hosted in carbonatites, alkaline/peralkaline felsic rocks including pegmatites, placer and beach sands, hydrothermal veins, enriched lateritic layers, and ion adsorption deposits. Carbonatites an alkaline suite of rocks are the primary targets for REE-rich resources. LREE are concentrated in higher proportion as compared to HREE globally and India is no exception. However, the inland placers, carbonatites, pegmatites, and alkaline-peralkaline suite of rocks have been established in Sevattur, Rasimalai, Kanigiri, Marlagalla, and many more in Tamil Nadu, Alkali suite of rocks in Andhra Pradesh, Assam, Meghalaya, Rajasthan, Gujarat in Indian perspective.

Malani Igneous Suite, a large felsic rhyolite province, west of the Aravalli-Delhi Fold Belt, Western Rajasthan, is the third largest in the world. It has huge potential for REE mineralization associated with the third magmatic intrusive phase in the form of microgranite, felsites, and aplites. Siwana Ring Complex in Barmer district, a collapsed cauldron structure of 750 km^2 has emerged as a potential REE hub for the LREE, HREE, and Nb resources. The acid volcano-plutonic caldera of SRC has been nicely documented on records by researchers from government, academic, and private entrepreneurs in all dimensions of geological perspective regarding its disposition in Indian and global perspective, geochemical and petro-mineralogical characterization of each rock unit of acid and mafic flows, volcanological perspective and metallogeny. A tertiary alkaline suite of rocks of Sarnu-Dandali has emerged as a big REE resource with REE deposits rich in LREE identified at Kamthai. A similar environment exists in Mer-Mundwara for similar REE resources.

Geological settings related to the caldera and its environs are ideal for different styles of uranium deposits in the magma chamber, rift margins, stratabound flows, domes, and various ash and volcaniclastic and lacustrine sediments-oriented mineralization (Fig. 12). A geological model has been envisaged for uranium and associated elemental concentration in volcanic rock, from far to close proximity to volcanic caldera. Such a model has been observed in well-known metals, non-metals, and metalloids as given in Fig. 12. A similar environment exists in Western Rajasthan and with reference to the Siwana caldera possible atomic mineral resources including U, REE, and rare metals, can be explored. The presence of a huge oil–gas reserve in the Barmer-Jaisalmer basin enhances confidence in the geological model.

Atomic Minerals Directorate for Exploration and Research, the exploration agency in the country has focussed on this geological environment for volcanogenic uranium potential and REE potential in the last seven decades. With the analogy with the numbers of caldera systems in time and space, the Siwana caldera holds big potential for multimetal mineralization. Since the 1950's importance of REE has been ascertained. However, with the increasing demand over the years for REE resources in the industry, technology, and many sectors of day-to-day life including agriculture, medicine, and many more, besides high-end technologies, the focus has been reoriented globally. Otherwise, the REE were the geochemical indicators in petrogenesis for interpretation and establishing genesis for metal deposits in earth dynamic systems.

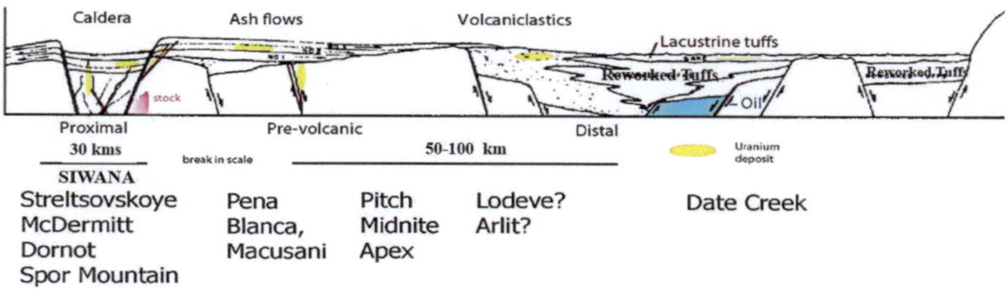

Fig. 12 Schematic cross-section of an idealized volcanic system with uranium-mineralized zones or deposits (Modified after Sherborne et al. [22])

A bimodal volcanic suite of rocks of Siwana caldera intruded by alkaline to peralkaline granite ring dyke and both these units intruded by younger microgranite, aplites, granophyre, and felsites are an ideal site for volcanogenic REE- uranium, Niobium and a minor amount of silver multimetal mineralization.

Reoriented rigorous exploration since 2010, has brought to light a number of exploration centers in the Siwana caldera for REE resources with special mention of high HREE which is unique to these mineralized centers. Generally, 90% of REE deposits are LREE (La-Ce-Nd rich). Thus, concentrated efforts in SRC by AMD presently and in the future at exploration centers at margins and central caldera associated with felsites, microgranites, brecciated granitoid, tuffs, and associated pyroclastic rocks would establish Siwana Ring Complex a REE hub with sizeable resources of REE and Nb. Further, volcanogenic uranium along with associated metals would enhance the potential and importance of the Siwana Ring Complex.

References

1. Rare earth Investing news: (2022): https://investingnews.com/daily/resource-investing/critical-metals-investing/rare-earth-investing/rare-earth-reserves-country/
2. Keith R Long, Bradley S van Gosen, Nora K Foley, Daniel Cordier (2010): The Principal Rare Earth Elements Deposits of the United States, USGS Scientific Investigations Report 5220
3. Nash JT (2010): Volcanogenic uranium deposits: geology, geochemical processes, and criteria for resource assessment: U.S. Geological Survey Open-File Report 2010–1001, 110 p.
4. Goodell PJ (1985). Classification and model of uranium deposits in volcanic environments in International Atomic Energy Agency, Uranium deposits in volcanic rocks: International Atomic Energy Agency Panel Proceedings of a Technical Committee Meeting, El Paso Texas April 2–5, 1984, 1–17.
5. Bidwai R, Srinivasan S, Nanda LK, Atanu Banerjee, Yadav OP, Rai AK Parihar PS (2015): Anomalous rare earth abundances in Siwana Ring Complex, Barmer district, Western Rajasthan. Geological Survey of India Special Publication 101. Recent developments in Metallogeny and Mineral Exploration in Rajasthan, 163–176.
6. Mukherjee AB, Roy A (1981): Cooling conditions of high-level Precambrian granite of Siwana: Evidence of experimental melting behaviour and the sodic amphibole reaction relation. Indian Journal of Earth Sciences. 8 (2), 99–108.
7. Roy AB, Jakhar SR (2002): Geology of Rajasthan (Northwest India) Precambrian to recent: Jodhpur, India, Scientific Publishers, 421 p.
8. Kochhar N (1996): A-type Malani magmatism, North-western Peninsular India, In: Mineralisation in the Western Indian Craton. M Deb and D M Banerjee (Eds). Proceedings National Seminar, University of Delhi, 10–15.
9. Crawford AR (1975): Rb Sr age determination for the Mt. Abu granite and related rocks of Gujarat: Journal Geological Society of India, 16(1) p.73.
10. Chaudhary AK, Gopalan K Sastry A (1984): Present status of the geochronology of the Precambrian rocks of Rajasthan. Tectonophysics,131–140.
11. Rathore SS, Venkatesan TR, Srivastava RK, (1999): Rb-Sr isotope dating of Neoproterozoic (Malani Group) magmatism from south-west Rajasthan, India: Evidence of younger Pan-African thermal event 40Ar39Ar studies: Gondwana Research, 2, 271–281.
12. Bhushan SK (1995): Late Proterozoic continental growth: Implications from geochemistry of acid magmatic events of west Indian Craton, Rajasthan. Geological Society of India Memoir., 34, 339–355.

13. Bhushan SK, Mohanty M (1988): Mechanics of intrusion and geochemistry of alkaline granites from Siwana, Barmer district, Rajasthan. Indian J. Earth Sci., 15(2), 103–115B
14. Bidwai R, Srinivasan S, Nanda, LK, Banerjee Atanu, Bangroo PN, Rai AK, Parihar PS (2014): Anomalous silver concentration in volcano-plutonic rocks of Siwana Ring Complex, Western Rajasthan. Current Science, 106(2), 159–162.
15. Bidwai R, Krishnamurthy P (1997): Possible lava channels and tubes from the Siwana Caldera Sequence, Siwana, Barmer District, Rajasthan. Jour. Geol. Soc. India, 50(3), 303–305.
16. Nilssen B (1970): Samarskite: Chemical composition, formula and crystalline phase produced by heating, Norsk geologisk tidsskrift, 50, 357–373.
17. Jain RB, Miglani TS, Kumar S, Swarnkar BM, Singh R (1996): Rare metal and rare earth rich peralkaline, agpaitic granitoid dykes of Siwana Ring Complex, district Barmer, Rajasthan, Current Science, 70, 854–858.
18. Jackson WD, Christiansen G (1993): International Strategic Minerals Summary Report – Rare - earth Oxides: USGS circular 930 N.
19. Saleh GM (2006): Geologic relationships and mineralization of peralkaline/alkaline granite - syenite of the Zargat Na'am ring complex, South-eastern Desert, Egypt: Chinese Journal of Geochemistry, 25 (2)
20. Salvi S, Williams - Jones AE (2006): Alteration, HFSE mineralization and hydrocarbon formation in peralkaline igneous systems: Insights from the Strange Lake Pluton, Canada: Lithos, 3 (40).
21. Williams-Jones AE, Migdisov AA, Samson IM (2012): Hydrothermal Mobilization of the Rare Earth Elements—a Tale of "Ceria" and "Yttria": elements, 8, 355–360.
22. Sherborne JE, Jr Buckovic WA, Dewitt DB, Hellinger TS, Pavlak SJ (1979): Major uranium discovery in volcaniclastic sediments, Basin and Range Province, Yavapai County, Arizona: American Association of Petroleum Geologists Bulletin. 63, 621–646.

Application of EDS Based Automated Mineralogic System to Investigate Process Efficiency of a Processing Plant of Iron Ore Mine from Chhattisgarh, Central India

Basant Rath, P. V. S. Raju, I. Panduranga Reddy, S. K. Chaurasiya, and Vibhuti Roshan

Abstract Iron ore slimes from the processing plant of Kirandul were characterized by optical microscopic techniques and Automated mineralogical techniques to understand the distribution of iron-bearing phases and their recoverability. The study indicates that the dominant mineral phase is Limonite with minor Gibbsite, porous goethite, low iron limonite, and Goethite. Quartz, Magnetite, and porous Hematite are present in trace proportions. The iron content in the sample is 44.76% of the sample and is distributed in Limonite (55%), Porous Goethite (10.15%), and Goethite (5%). The variety of magnetite and hematite only contribute 0.37% Fe or 0.17% of the total Fe in this sample is found in hematite and magnetite. Limonite is the dominant phase in the tailings sample with 83% of the limonite occurring in liberated form. Magnetite-rich particles are in the liberated state. Monomineralic Limonite is constituted by Limonite particles that are > 95% Limonite, which account for 35% of the sample and almost 97% of all the Limonite in the sample. Around 37% of all Fe in this sample is found in the monomineralic limonite particle type. Around 80% of the P in the sample (0.02%) is found in Limonite. Around 25.82% Si is found in the sample and this is distributed dominantly between quartz (34%), Limonite (23%), and low Fe Limonite (18%). Al concentration is 2.67% of the sample and is dominantly found in Gibbsite (35%), Limonite (36%), and low Fe Limonite (11%). The study indicates that the process is functioning well with a low amount of valuable hematite or magnetite present. Any occurrence of magnetite, hematite, porous magnetite, or porous hematite is typically found in a middling or locked occurrence at ultra-fine size.

B. Rath (✉) · S. K. Chaurasiya · V. Roshan
Research and Development Centre, NMDC Ltd, Hyderabad 500007, India
e-mail: basant_rath@nmdc.co.in

P. V. S. Raju
CSIR-National Geophysical Research Institute, Hyderabad 500007, India

I. Panduranga Reddy
Sr. Professor, Department of Geology, Osmania University, Hyderabad 500007, India

K. Randive et al. (eds.), *Current Trends in Mineral-Based Products and Utilization of Wastes: Recent Studies from India*, Springer Proceedings in Earth and Environmental Sciences, https://doi.org/10.1007/978-3-031-50262-0_7

Keywords Automated mineralogy · Processing of Ore · Monomineralic · Liberation · Interlocking

1 Introduction

Iron ore occurs in different parts of the world and is mainly formed from the Banded iron formation (BIF) which mainly occurs in the shield part of the world, particularly from the Precambrian time. These iron formations were mainly associated with oxide facies only [1].

Bailadila iron ore group is the biggest iron-bearing deposit in the Bastar craton which is around 40 km in length and around 1 km wide. Owing to the highly resistant minerals association and less affected by weathering, there are various BIF-hosted hills are formed which are numbered 14 deposits starting from north to the south of the Bailadila hill [2].

Indian iron ore is considered high in grade but at the same time, it is very complex in nature in terms of structural deformation and association of other gangue minerals. Major ore minerals observed in Indian iron ore are Hematite, Goethite, Limonite, Magnetite, Martite, etc. whereas; gangue minerals observed are Kaolinite, Gibbsite, and Quartz. Geological structures get changed slightly as new benches are open due to continuous mining and the grade has been slightly affected by this process. The Fe% is going down at the same time P and silica percentages are increasing in the deeper benches. The average chemical assay of these iron ores is around 45–60% Fe, 5–8% SiO_2, and 3–10% Al_2O_3. Most of the siliceous and aluminous minerals occur as quartz, kaolinite, and gibbsite with traces of illite as well [3]. Further, initially, the lump-fines ratio was around 70:30 whereas in the present scenario, it has become nearly reverse. These huge amounts of slimes (<150 mm) are separated from Run of Mines (ROM) and stored separately as they cannot be used directly in a blast furnace for steel making. Handling these slimes becomes problematic over a period of time in terms of environmental issues, space constraints, etc. This paper deals with the characterization of a low iron content sample that has relatively higher silica and alumina. This study helps to understand the liberation parameters of the selected sample for modification or decision-making in the downstream mineral beneficiation process.

1.1 Details of Processing Plant

Generally, the Run of mines (ROM) is crushed to −1 m size then it goes to a primary crusher where the ore is crushed to −300 mm size (Fig. 1). After this, it is a scalping screen with a 150 mm opening where more than 80% of −150 mm ore is screened and the remaining 20% of + 150 mm size ore is taken to a secondary crusher and screened at −150 mm. The ultimate crushed product received is −150 mm. This product is

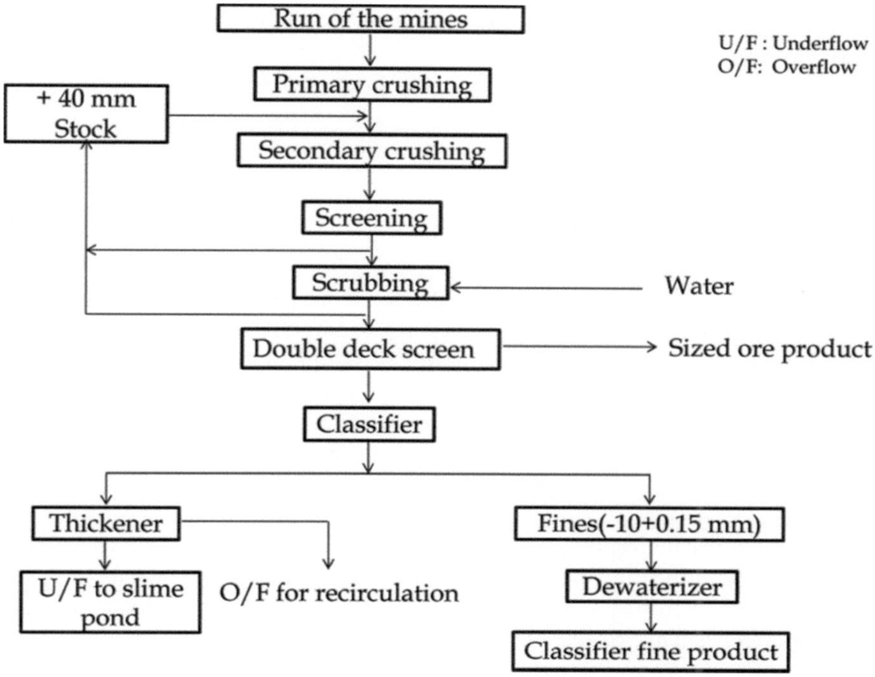

Fig. 1 General flow sheet used for fines/slimes generation in iron ore mine

screened at −150 mm + 10 mm (Lump), −10 mm + 100 mesh (Fines), and (−40 mm + 6 mm) as natural pellet under dry and wet operation as per requirement. The lump ore i.e. −150 mm + 10 mm is then processed at the tertiary crushing plant and under the vibratory screen to get calibrated lump ore (−40 mm + 6 mm) and fines (−6 mm). At this stage mostly the lump ore shows around Fe−66%, SiO_2−1.5%, Al_2O_3−1.3% and fines ore shows Fe−65%, SiO_2−2% and Al_2O_3−2%.

When CLO (−40 mm to + 06 mm) is to be produced, the NPO size −40 to + 10 mm (13.23%) screened from secondary screens is mixed with the CLO (20.16%) produced after tertiary crushing and the total CLO will be 33.39% of grade Fe%−66.67, SiO_2%−1.53 and Al_2O_3%−1.40. In the case of Lump Ore (−150 mm to + 10 mm) production, the screened NPO (13.23%) shall be mixed with the −150 to + 40 mm fraction of lump (29.34%) received after the primary screening. The combined grade and weight percentage of the total lump shall be Fe%−66.53, SiO_2%−1.60 & Al_2O_3%−1.48, and 42.57% respectively.

The Fines (−06 mm) generated after tertiary crushing (09.18%) of grade Fe%−65.20, Si%−2.16 & Al_2O_3%−2.1 0 shall be mixed with the Fines generated in the second stage of processing i.e. 40.20% and the total Fines (49.38%) size −10 mm to + 100 mesh shall be of grade Fe%−64.70, SiO_2%−2.44 & Al_2O_3%−2.53.

The slimes generated after this process varies in size and accordingly in chemical assay as well i.e. Fe from 44–59%, Al_2O_3 1–5%, SiO_2 2–25%, and LOI 1–5%. The tailing samples in studied in various sizes and here discussed the highest silica-bearing sample.

2 Experimental Results

The 'As Received Slime' sample was collected from the slime pond of Kirandul iron ore mines studies in detail using various mineral characterization tools in the laboratory for detailed mineralogical and micro-morphological characterization of the phases present in the slime sample. The 'as received' samples were less than 20 mm in size and more than 90% of particles were less than 65 mesh (212 microns) with 3.85 specific gravity 2.3 ton/m^3 bulk density. These samples were investigated using a Leica light microscope and a Zeiss Electron microscope (Automated Mineral Analyzer).

Several representative slime samples were taken for cold mounting using resin and hardener and then lapping /polishing was done. After proper polishing of the sample, optical microscopic study, and electron microscopic study in combination with Energy Dispersive Spectroscopy was done. The ultrafine particles were taken for secondary electron imaging and the surface morphology of the samples was analyzed.

Zeiss Sigma 300 VP Gemini hardware and Mineralogic software were used in an automated mineralogy system, where with the help of software application customization a recipe was prepared to get detailed mineralogical information such as elemental deportment, modal analysis, and mineral liberation for ores and rocks. A properly selected sample was taken and polished sections were prepared by mixing with resin and hardener after which it was polished using different grades of diamond paste. Graphite powder was added to de-agglomerate the ultrafine particles. The polished block was taken and scanned using Mineralogic software. Following this, the detailed modal mineralogy and liberation parameters of the iron-bearing phases were determined.

The study indicates that the dominant mineral phase is Limonite with minor Gibbsite, porous goethite, low iron limonite, and Goethite. Quartz, Magnetite, and porous Haematite are present in trace proportions. The samples show around 45% Fe, 26% SiO_2, 3.5% Al_2O_3 and 4.6% LOI. The iron content present in the sample is distributed in Limonite (55%), Porous Goethite (10.15%), and Goethite (5%). The variety of magnetite and hematite only contributes 0.37% Fe or 0.17% of the total Fe in this sample is found in hematite and magnetite (Table 1). The scanned detail of various minerals' 2-D distribution in the slime sample is shown in Fig. 4.3.

Table 1 Mineral modal analysis of iron ore slimes by automated mineral analyzer

Minerals	Number	Area %	Weight %	Grain size (μm)	Average composition	Assigned composition
Limonite	39,435	54.36	55.32	39.6	Fe 45.73: O 24.15: Al 1.93	Fe 52.26: O 44.91: Al 1.93
Gibbsite	2775	3.83	4.75	42.1	O 28.76: Al 21.65: Si 0.8	O 61.53: Al 34.59: Si 0.8
Porous Goethite	6849	9.44	10.15	58.54	Fe 57.24: O 20.47: Al 0.09: Si 0.05	Fe 62.85: O 36.01: Al 0.09: Si 0.05
Limonite (Low Fe 25–35%)	3890	5.36	4.8	34.07	Fe 31.78: O 22.3: Al 6.75	Fe 30: O 22.3: Al 6.75
Goethite	4302	5.93	4.98	27.77	Fe 57.74: O 19.66: Al 0.34: Si 0.17	Fe 62.85: O 36.01: Al 0.34: Si 0.17
Quartz	14,620	20.15	20.56	29.61	Si 32.34: O 27.2: Al 0.2	Si 46.74: O 53.26: Al 0.2
Hematite	187	0.26	0.19	31.22	Fe 66.66: O 17.39: Al 0.2	Fe 69.94: O 30.06: Al 0.32
Porous Hematite	3	0.00	0.16	226.65	Fe 65.88: O 13.52: Al 0.8	Fe 69.94: O 30.06: Al 0.8
Magnetite	110	0.15	0.14	36.65	Fe 73.55: O 15.69: Al 0.8	Fe 72.36: O 27.64: Al 0.14
Aluminogoethite	295	0.41	0.07	19.27	Fe 47.91: O 20.28: Al 1.78	Fe 47.91: O 20.28: Al 1.78
Siderophyllite	43	0.06	0.04	31.6	O 23.2: Al 12.67: Si 10.97: Fe 10.72: K 2.57: Cl 0.16	O 34.65: Al 16.69: Si 11.58: Fe 23.03: K 8.06: Cl 0.16
Kaolinite	22	0.03	0.02	34.97	O 24.64: Si 18.99: Al 18.48	O 55.78: Si 21.76: Al 20.9
Porous Magnetite	4	0.01	0.02	66.69	Fe 70.53: O 14.87: Al 0.02	Fe 72.36: O 27.64: Al 0.02
Calcite	13	0.02	0	4.03	O 34.: Ca 34.45: Fe 0.75: Mg 0.28	O 47.96: Ca 40.04: Fe 0.75: Mg 0.28

2.1 Liberation Studies

The liberation studies indicate that Limonite is the dominant phase in the tailings sample with 83% of the limonite occurring in liberated form, the next phase being Alumino goethite which is liberated to the extent of 27%. Hematite and Magnetite are mostly locked with only 7% and 4% respectively occurring in liberated form (Table 2).

The cumulative liberation Yield graphs in Figs. 2 and 3 also confirm.

Limonite is the dominant phase in the tailings (53%) and in general, has a low contact association with most phases indicating its occurrence as liberated and monomineralic particles. Magnetite found in the tailings displays around 11% association with porous goethite and around 33% association with porous limonite. Around 40% of the magnetite is associated with background (resin) which can be used to indicate and support the reasonable liberation of magnetite and that its perimeter is exposed.

Based on these data, it has been concluded that the main iron-bearing mineral is Limonite out of which around 83% of grains are in a liberated state. Around 79% of alumina-bearing goethite are in a liberated state and the major silica-bearing gangue mineral is Quartz and around 89% of quartz particles are in a liberated state (Table 2). Keeping these in view, it can be concluded that the ore can be beneficiated and the grade can be enriched without adding many operation processes.

Limonite shows 83% liberation and this is also supported by a steep curve showing a high amount of perimeter exposure. A steep line of porous hematite displays the middling of this phase.

3 Results and Conclusion

The study indicates that the dominant mineral phase is Limonite with minor Gibbsite, porous goethite, low iron limonite, and Goethite. Quartz, Magnetite, and porous Haematite are present in trace proportions. The iron content in the sample is 44.76% of the sample and is distributed in Limonite (55%), Porous Goethite (10.15%), and Goethite (5%). The variety of magnetite and hematite only contribute 0.37% Fe or 0.17% of the total Fe in this sample is found in hematite and magnetite. Limonite is the dominant phase in the tailings sample with 83% of the limonite occurring in liberated form. The study indicates that the process is functioning well with a low amount of valuable hematite or magnetite present. Any occurrence of magnetite, hematite, porous magnetite, or porous hematite is typically found in a middling or locked occurrence at ultra-fine (less than 20 microns) size. Removal of siliceous minerals i.e. quartz is possible through applying magnetic separation or spiral concentration or by adopting a combination of both i.e. spiral concentration followed by magnetic separation. The mineral modal analysis and liberation studies indicate that the ore can be enriched up to 65% Fe, 2–3% Si, and 1–2% alumina by physical beneficiation method.

Table 2 Liberation parameters of iron ore slimes obtained by automated mineral analyzer is as follows-

	Liberated	Middling	Locked	< 10%	< 20%	< 30%	< 40%	< 50%	< 60%	< 70%	< 80%	< 90%	< 100%	1
Limonite	83.14	12.78	4.08	1.52	1.46	1.1	1.04	1	1.63	1.52	2.19	5.39	71.63	11.52
Aluminogoethite	27.29	37.15	35.55	19.05	12.61	3.89	0	6.04	12.27	18.05	0	0.79	18.41	8.88
Porous magnetite	0	17.9	82.1	0	0	82.1	0	0	2.98	0	14.93	0	0	0
Porous hematite	0	100	0	0	0	0	0	0	59.79	40.21	0	0	0	0
Siderophyllite	73.11	4.55	22.34	11.66	7	3.68	2.1	0	0.56	0	1.89	0	0.81	72.3
Kaolinite	100	0	0	0	0	0	0	0	0	0	0	0	7.8	92.2
Alumogoethite	79.16	6.6	14.25	11.28	2.31	0.67	1.06	0.49	1.38	0.73	1.01	1.92	6.4	72.76
Hematite	6.93	32.97	60.09	47.62	10.9	1.58	14.32	6.08	3.58	4.35	1.58	3.05	3.15	3.78
Magnetite	3.85	32.26	63.89	37.14	8.41	18.34	5.46	9.73	7.24	3.16	2.23	4.45	2.07	1.78
Goethite	15.82	40.21	43.97	22.84	14.43	6.71	5.24	6.53	4.54	7.76	8.2	7.93	11.89	3.93
Limonite (Low Fe–25–35%)	46.6	39.49	13.91	7.24	2.98	3.69	3.27	3.84	2.81	8.24	6.09	15.24	39.83	6.77
Porus Goethite	0.97	77.15	21.88	5.5	7.05	9.33	11.89	17.75	12.77	15.31	13.46	5.97	0.95	0.03
Gibbsite	66.19	26.94	6.87	2.52	2.67	1.68	1.58	2.04	2.08	2.08	7.67	11.5	38.04	28.15
Quartz	89.33	7.96	2.71	1.98	0.09	0.64	1.19	0.97	0.93	0.14	0.73	4.01	7.05	82.28
Calcite	91.27	8.73	0	0	0	0	0	8.73	0	0	0	0	0	91.27

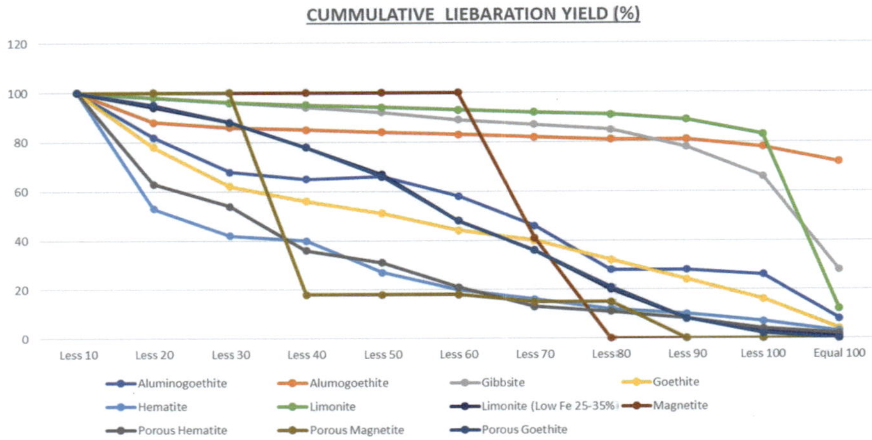

Fig. 2 Cumulative liberation parameters of iron ore slimes by automated mineral analyzer

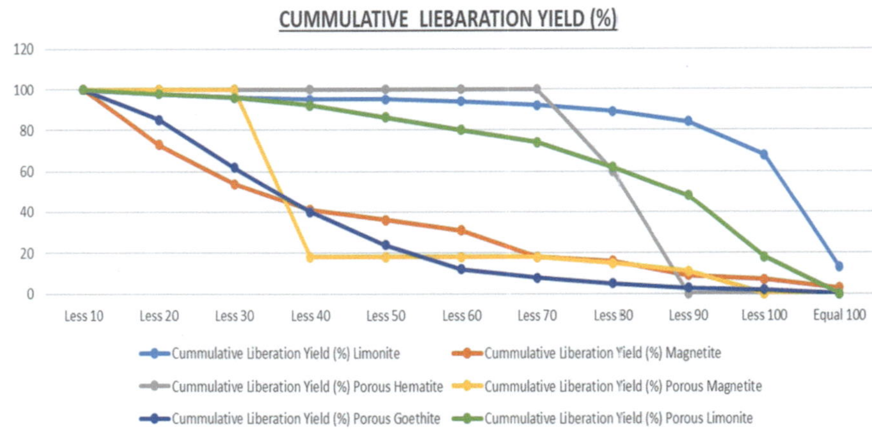

Fig. 3 Cumulative liberation yield graph of iron ore slimes by automated mineral analyzer

4 Photomicrographs

Ore microscopy was done with a Leica DM6M microscope with an LED natural light source (Fig. 4a and b) and the MEIJI stereo zoom microscope was used for 3D imaging and image analysis of heavy media separation test products (Fig. 5a and b). A few selected Back Scattered Electron (BSE) images are also presented here in Fig. 6a–f.

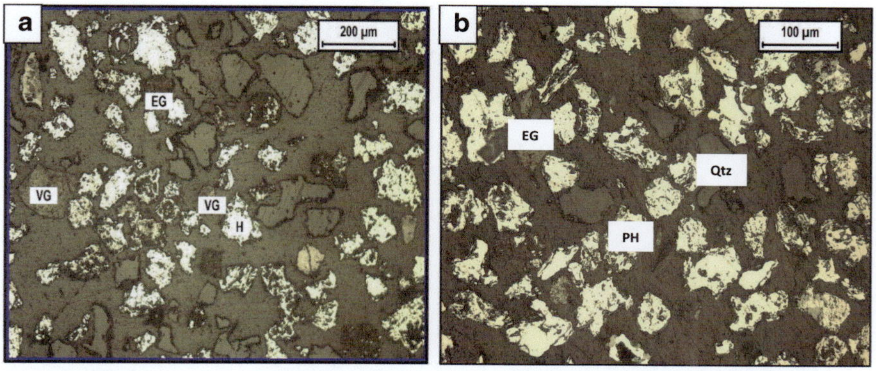

Fig. 4 **a** and **b** Photomicrographs of light microscopy indicate about distribution of Hematite (Hem), Porous Hematite (PH), Vitreous Goethite (VG), Earthy Goethite (EG), and Liberated Quartz (Qtz). Porous hematite (PH) is shown in images (**a**) and (**b**). Image A-Reflected Light_10x. image B-Reflected Light_20x

Fig. 5 **a** and **b** Stereomicroscopic photographs of the tailing sample indicate about distribution of ematite (Hem), Vitreous Goethite (VG), Earthy Goethite (EG), and Quartz (Qtz) in image C. Image D shows all liberated Quartz (Qtz) particles after processing it in heavy media separation test. Image A-Incident Light white field_1x. Image B-Incident Light Dark filed_2.5x

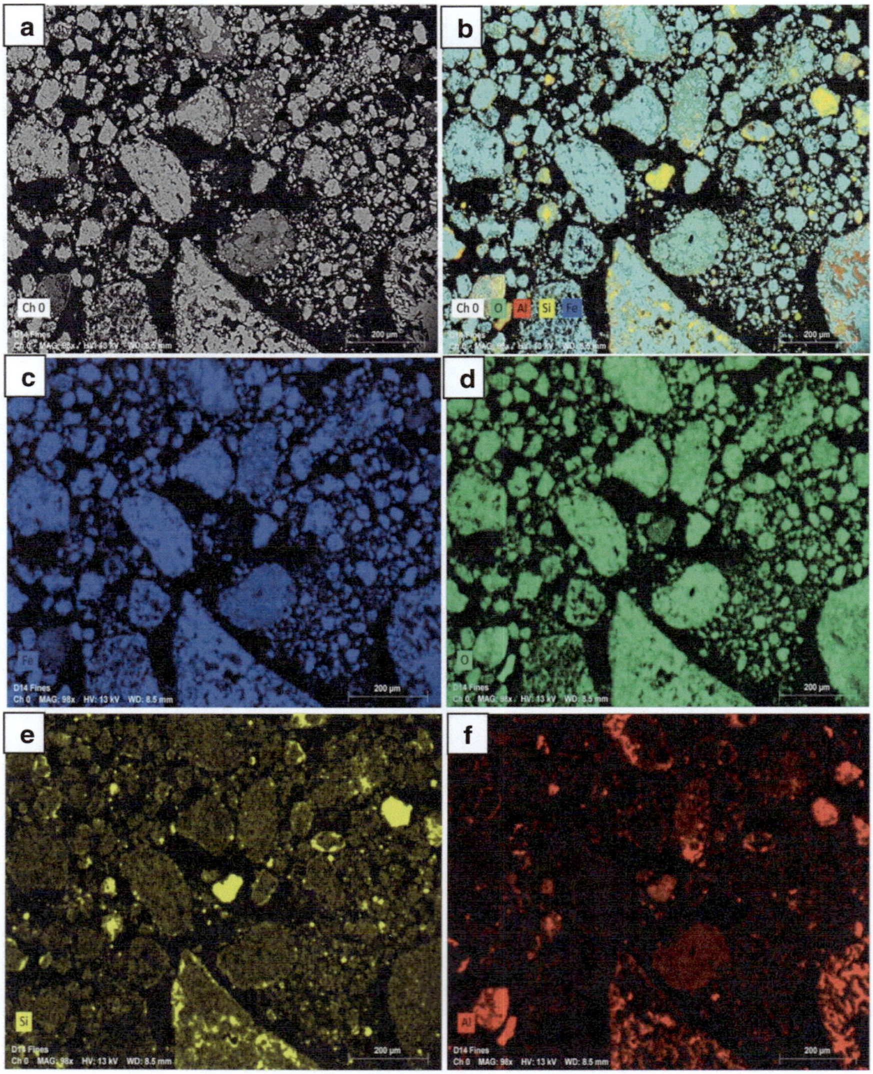

Fig. 6 The above photomicrographs show Back Scattered Electron (BSE) image E and its corresponding elemental map in F which shows the distribution of various elements i.e. Fe, Si, Al, and O. Image G shows an elemental map of Fe, image H shows an elemental map of O, images I shows an elemental map of Si and image J shows an elemental map of Al

Acknowledgements The authors would like to thank the management of NMDC Limited for approving the work done to present and publish the paper. The authors also wish to acknowledge the suggestions received from various departments of the R&D Centre, NMDC Limited during the work.

References

1. Chakraborthy KL, Tapan Majumdar (1986) Geological Aspects of the Banded Iron Formation of Bihar and Orissa, Jour. Geol. Soc. India, 28, 109 to 133.
2. Crookshank H (1938) The iron ores of the Bailadila Range, Bastar State. Trans. Min. Geol. Met. Inst, Ind., 34.
3. Roy Subrata, Das A (2008) Characterization and processing of low-grade iron ore slime from the Jilling area of India. Mineral Processing & Extractive Metallurgy Review. 29. 213-231. https://doi.org/10.1080/08827500801997886

Flowability Characteristics of Iron Ore and Their Influence on Material Handling System Design Parameters

T. V. S. Subrahmanyam, Somiran Mandal, Bhagwan Singh, and S. K. Chaurasiya

Abstract The total production of iron ore in India was 246 million tonnes in 2019–20 and 188.6 million tonnes (estimated) in 2020–21, according to the annual report 2020–21 of the Ministry of Mines. Odisha state is the leading producer of iron ore accounting for nearly 59.64% of total production followed by Chhattisgarh (14.11%), Karnataka (12.76%), Jharkhand (10.93%) and remaining (2.56%) production reported from Andhra Pradesh, Madhya Pradesh, Maharashtra and Rajasthan. The production of millions of tonnes of iron ore involves intensive material handling operations including storage and conveying the Run-of-Mine (ROM) and finished product. During the handling of the iron ore in interplant processing, the ore is stored in stockpiles or silos, and transported by means of a series of conveyors that is accomplished through transfer chutes. The flawless movement of iron ore during the entire process plays a key role in realizing the highest levels of productivity and plant capacity utilization. However, iron ore flowability problems like plugging/choking of transfer chutes, and arching/bridging/ratholing in silo/stockpiles are often encountered especially while handling wet and sticky ores during monsoon and post-monsoon seasons. This necessitates having a scientific understanding of the flowability characteristics of various grades of iron ore simulating the field conditions. The material handling systems are required to be designed and tailor-made based on the flowability characteristics established by standard test methods simulating field conditions. The flow properties are dependent on particle size distribution, particle shape, chemical composition, moisture content, and temperature. The present study involves the evaluation of flow properties of different grades of iron ore samples collected from the Odisha-Jharkhand region and the calculation of design parameters for material storage and handling systems, like silo/stockpiles and transfer chutes. The influence of flowability characteristics on various material handling system design parameters was analyzed.

Keywords Iron ore · Flowability · Shear test · Silo · Hopper · Transfer chute · Jamming · Arching · Rat holing · Moisture

T. V. S. Subrahmanyam (✉) · S. Mandal · B. Singh · S. K. Chaurasiya
R&D Centre, NMDC Limited, Hyderabad, India
e-mail: tvssubrahmanyam@nmdc.co.in

K. Randive et al. (eds.), *Current Trends in Mineral-Based Products and Utilization of Wastes: Recent Studies from India*, Springer Proceedings in Earth and Environmental Sciences, https://doi.org/10.1007/978-3-031-50262-0_8

1 Introduction

The production of millions of tonnes of iron ore in India annually involves intensive material handling operations including storage and conveying the Run-of-Mine (ROM) and finished product. During the handling of the iron ore in interplant processing, it is stored in stockpiles or silos, and transported by means of a series of conveyors that is accomplished through transfer chutes. The flawless movement of iron ore during the entire process plays a key role in realizing the highest levels of productivity and plant capacity utilization. However, iron ore flowability problems like plugging/choking of transfer chutes, and arching/bridging/ratholing in silo/stockpiles are often encountered especially while handling wet and sticky ores during monsoon and post-monsoon seasons. This necessitates a scientific understanding of the flowability characteristics of various grades of iron ore simulating the field conditions. The material handling systems are required to be designed and tailor-made based on the flowability characteristics established by standard test methods simulating field conditions. The flow properties are dependent on particle size distribution, particle shape, chemical composition, moisture content, and temperature. In view of the above, this study was carried out with the aim of understanding the flow properties of different grades of iron ores and their influence on the material handling system design parameters.

2 Materials and Methods Selected

Seven different grades of iron ore samples (designated as S. No. 1 to 7 henceforth) from mines located in the Jharkhand and Odisha (Zone-I) belt in India were selected for the present study (Table 1).

Table 1 Iron ore samples considered for the study from the Jharkhand-Odisha Belt

S. No.	State	Source/Mines
1	Jharkhand	Gua Mines
2	Jharkhand	Gua Mines
3	Jharkhand	Gua Mines
4	Jharkhand	Noamundi Mines
5	Jharkhand	Noamundi Mines
6	Jharkhand	Noamundi Mines
7	Odisha	Joda Mines

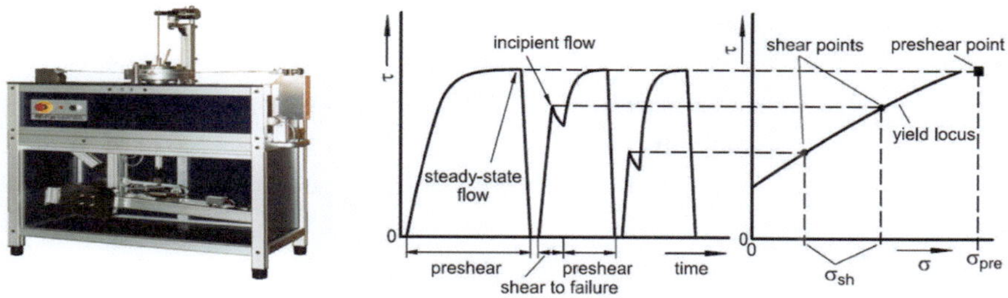

Fig. 1 Ring shear tester and shearing procedure [1]

3 Experimental Procedure

Shear testing

Shear tests were conducted at different moistures using a Ring Shear Tester [1] (Fig. 1). The standard shear cell is homogenously filled with the iron ore sample to be tested and subjected to shearing. Similarly, wall friction tests were also conducted at different moistures against liner specimens. The yield locus [a plot of consolidation stress (σ) versus shear stress (τ)], wall yield locus [a plot of consolidation stress (σ) versus shear stress (τ)], and flow function [2] [a plot of major consolidation stress versus unconfined yield stress] were generated.

4 Results and Discussion

4.1 Size and Chemical Analysis

The size analysis is determined by using a standard set of test sieves and the average particle size (D50 & D80) is established (Table 2) for the samples considered for study.

Table 2 Size distribution of tested ore

S. No.	D80 (mm)	D50 (mm)	% of 1 mm in Tested Ore
1	1.49	0.34	72.22
2	1.16	0.14	77.40
3	2.68	1.10	50.22
4	2.02	0.36	64.39
5	1.39	0.11	75.39
6	1.29	0.21	77.19
7	2.49	0.93	52.00

Table 3 Chemical analysis

S. No.	Fe	SiO_2	Al_2O_3	LOI
1	58.60	4.72	4.95	6.09
2	65.00	3.56	1.12	1.77
3	59.20	3.48	4.64	6.23
4	63.60	1.84	2.51	3.19
5	65.40	1.05	2.09	2.36
6	65.60	2.04	1.90	2.22
7	66.00	1.48	1.59	2.33

The chemical analysis of samples is established by the wet classical method and presented in Table 3. The presence of Alumina affects the flow properties significantly.

4.2 *Moisture Content and Bulk Density*

Apart from particle size distribution, the moisture content of ore significantly influences the flow properties. The moisture content, saturation moisture limit (SML), and critical moisture were established and presented in Table 4. The critical moisture is the moisture content of ore where it exhibits the highest cohesive strength which is drawn from shear test data. The bulk density is also measured at critical moisture during the shear testing and presented in Table 4.

Table 4 Moisture content

S. No.	Moisture (%)	Saturation moisture limit (%)	Critical moisture (%)	Bulk density (@Critical moisture) (kg/m^3)
1	1.1	18.8	14.1	1672
2	0.35	18.3	13.7	1691
3	1.07	19.0	12.3	1711
4	1.1	17.0	12.7	1750
5	0.8	17.8	11.5	1608
6	0.7	13.2	9.9	1947
7	0.7	13.8	10.3	1882

Table 5 Flow properties—flow function and effective angle of friction

S. No.	Flow function (FFC) (Average)	Effective angle of friction, Φe (PhiE), Deg (Average)
1	1.50	60.50
2	1.66	60.20
3	2.03	54.70
4	1.61	60.33
5	1.54	58.90
6	1.60	60.10
7	1.79	58.60

4.3 Flow Properties

The data generated by the standard shear testing procedure is analyzed by plotting yield loci and constructing Mohr stress circles. The resultant average flow characteristics across various moisture levels are presented in Table 5. The effective angle of internal friction is a measure of particle friction against itself. The flow function (ratio of consolidation stress ($\sigma 1$) to unconfined yield strength (σc)) is a measure of iron ore flowability. The larger the flow function average (ffc), the better the flowability.

4.4 Design Parameters of Hopper/Silo

The average mass flow hopper design parameters like minimum hopper slope for mass flow (ϴc & ϴp for cone and wedge shape hoppers) and minimum outlet dimension to prevent cohesive arching at critical moisture for instantaneous conditions (Dc instantaneous) are calculated using the flow parameters [3] and presented in Table 6. The kinematic angle of wall/liner-friction of a Q&T plate of surface roughness (Average) of 2.5 microns was considered for calculating the mass flow hopper slopes at a typical outlet dimension of 1.0 m [4].

Table 6 Critical design parameter-hopper slope and opening

S. No.	Avg ϴc	Avg ϴp	Dc (Inst)
1	82.0	71.3	0.26
2	76.1	66.3	0.58
3	78.3	68.1	0.36
4	83.8	71.7	0.42
5	81.9	70.7	0.44
6	79.7	68.2	0.42
7	83.2	71.5	0.28

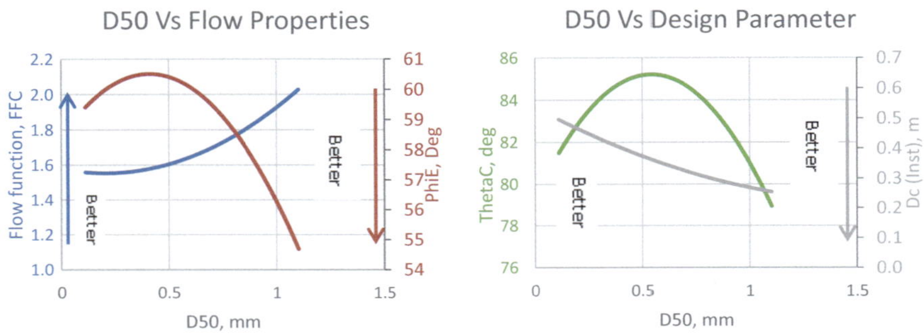

Fig. 2 Variation flow properties and design parameters with D50

4.5 Effect of Particle Size on Flow Properties and Design Parameters

The particle size has a definite effect on the flow properties and design parameters. With the increase in particle size, the flowability (FFC) has increased and the angle of internal friction decreased [5]. A reduction in minimum hopper opening dimension was observed with an increase in particle size. The minimum mass flow hopper slope (Theta C, ϴc) of the conical hopper increased initially and then registered a decrease. This might be due to the influence of other variables like particle shape and liner surface properties (Fig. 2).

4.6 Effect of Saturation Moisture Level on Flow Properties and Design Parameters

As the SML (Moisture absorbing capacity) increases, the flowability (flow function FFC) decreases and the effective angle of friction (PhiE) also increases as expected. Consequently, the design parameters also registered an increased value with SML up to certain levels. The influence of other parameters like size and chemical constituents altered the trend after a certain level (Fig. 3).

4.7 Effect of Alumina on Flow Properties and Design Parameters

As the alumina content increases the flowability (flow function (FFC) decreases and the effective angle of friction (PhiE) increases. Consequently, the design parameters also call for high slope and outlet dimensions with an increase in Alumina content

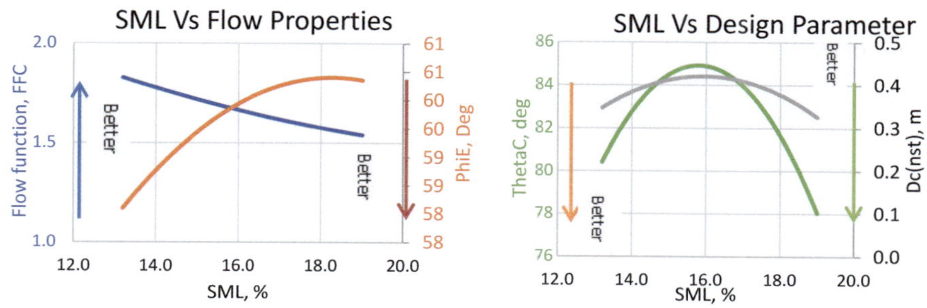

Fig. 3 Variation of flow properties and design parameters with SML

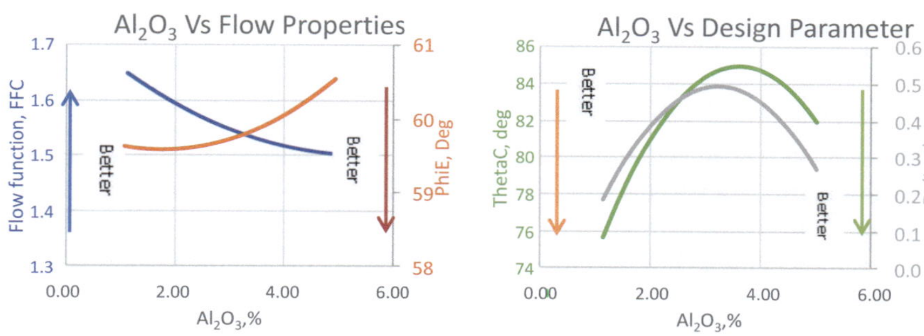

Fig. 4 Variation of flow properties and design parameters with alumina

up to 4.5%. The influence of other parameters like size and chemical constituents altered the trend after a certain level (Fig. 4).

4.8 *Effect of Alumina/Silica Ratio on Flow Properties and Design Parameters*

The increase in the alumina-to-silica ratio resulted in a decrease in flowability (flow function FFC) and an increase in the effective angle of friction (PhiE). Consequently, the design parameters also call for high slope and outlet dimensions with an increase in the Alumina/Silica ratio. The influence of other parameters like size and chemical constituents altered the trend after a certain level (Fig. 5).

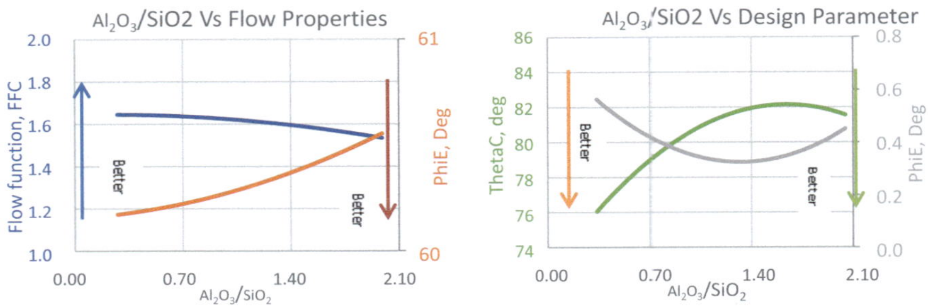

Fig. 5 Variation of flow properties and design parameters with alumina/silica ratio

5 Conclusions

The following conclusions are drawn from the study:

- The particle size, moisture content, and chemical composition of iron ores have a significant influence on the flow properties and consequently the material handling system (silos/hoppers/stockpiles) design parameters.
- Iron ore (bulk solids in general) is complex in nature to be categorized in general for its flowability as it depends on the combined influence of many parameters.
- It is advisable to shear test each sample individually simulating the field conditions and tailor-make the systems like hoppers, silos, stockpiles, and transfer chutes. This helps in trouble-free material handling operations without flow-related problems and enhances the productivity of processing plants.

Acknowledgements The authors are grateful to the management of NMDC for their support in the execution of this research project.

References

1. ASTM D 6773–08: Standard Test Method for Bulk Solids using Schulze Ring Shear Tester
2. Annual Report 2020–21, Ministry of Mines, Government of India
3. A.W. Jenike (1964): Storage and flow of solids, *Bulletin* 123, Engineering experiment station, University of Utah, Salt Lake City, UT, 1964
4. Dietmar Schulze, Powders and Bulk Solids, Behavior, Characterization, Storage and Flow.
5. Ganesan V, Rosentrater KA, Muthukumarappan K (2008) Flowability and handling characteristics of bulk solids and powders – a review with implications for DDGS, Biosystems Engineering.

Fine Size Dry Iron Ore Beneficiation Using Thin Bed Air Fluidized Dry Separator

Ganesh Chalavadi, Ranjeet K. Singh, and Monica Sahoo

Abstract Iron ore beneficiation is generally done through a wet processing route. The areas of water scarcity and owning to issues like slurry generation make wet processing route application a tougher one. At different stages of wet beneficiation, nearly 1 m^3 of water is required for processing one tonne of iron ore. Processing the iron ore through a dry processing route may be considered an alternative method where the air is used as fluid media instead of water, which reduces the slurry generation. In the present study, iron ore feed with a size range of − 1 mm to + 0.1 mm with an assay of 58.28% Fe is subjected to dry processing on a thin/shallow bed air fluidized dry separator (TBAFDS). This unit is a gravity-based separator that utilizes the difference in minimum fluidization velocities of heavier and lighter particles. Airflow rate, frequency of deck vibration, and side tilt are three parameters that were considered for the experimentation. Iron ore processed in a single stage, nearly 4.5 to 5% enhancement of Fe content was achieved in heavy stream products of TBAFDS. The middling and light stream of the first stage is processed further in multistage on TBAFDS to increase the yield% of high Fe content product. At the end of the third stage, a product stream of 62.25% Fe content with a yield of 53.56% was achieved.

Keywords Dry beneficiation · Fluidization · Iron ore · Minimum fluidization velocity · Frictional force

---

G. Chalavadi (✉)
Academy of Scientific and Innovative Research (AcSIR), Chennai, India
e-mail: gc@nmlindia.org

G. Chalavadi · R. K. Singh · M. Sahoo
CSIR-National Metallurgical Laboratory, Jamshedpur 831007, India
e-mail: kranjeet@nmlindia.org

M. Sahoo
e-mail: monu@nmlindia.org

K. Randive et al. (eds.), *Current Trends in Mineral-Based Products and Utilization of Wastes: Recent Studies from India*, Springer Proceedings in Earth and Environmental Sciences, https://doi.org/10.1007/978-3-031-50262-0_9

1 Introduction

Work done in dry beneficiation iron ore is limited to magnetic separation and dense media fluidization. Pre-concentration of lean hematite and limonite ores using a dry high-intensity magnetic separator was reported by Zhang [1]. Chen [2] reported iron impurities' dry separation performance using a novel dry vibrating high gradient magnetic separation method. He [3] employed a dry density-based fluidized bed separator utilizing a dense medium with a density range of 5–4.0 g/cc with the objective to beneficiate coarser particulate iron ore. Oshitani [4–6] reported the utilization of a dense medium fluidized bed for the dry separation of lump iron ores, as well as the implementation of a fluidized bed separator for the dry separation of fine-size iron ore treatment. Lump iron ores dry separation using dry dense medium gas-fluidized bed was reported by Oshitani [4–6]. Tripathy [7, 8] conducted a comprehensive study on the effective separation of hematite fines using a dry high-intensity magnetic separator. The study also involved the utilization of response surface methodology to evaluate the performance of a specialized dry high-intensity induced roll magnetic separator specifically designed for hematite fines. Dense media fluidization and dry magnetic separation have inherent limitations. The efficiency of magnetic separation highly depends upon the particle size, particle position exposure to the magnetic field, and magnetic susceptibility of associated minerals. On the other hand, in dense media air fluidized separator, it is difficult to maintain and stabilize the suspension media. The fluidization technique (without any media suspended) can overcome all the mentioned limitations as it is not a selective process, and higher throughput rate and bottlenecks due to suspension media are avoided. Chalavadi and Singh [9] established the feasibility of separation of iron ore fines ($-$ 1 mm to $+$ 0.1 mm) on TBAFDS which uses fluidization principles. In this paper, the physical dry processing of iron ore in multi-stages on TBAFDS is discussed.

2 Material Characterization

For the characterization study under an optical microscope, using reflected light, polished sections were prepared from the representative head sample using epoxy, resin, and hardener. From the detailed mineralogical study of the given iron ore head sample, hematite is observed to be present as the major iron ore mineral. It is present in the form of massive, microplate, subhedral to anhedral as well as interlocking characteristics. Apart from hematite, goethite is also present as the chief mineral here. Both vitreous and ochreous goethite are viewed to be present (Fig. 1a, b). Most of them are found to be replacing hematite (Fig. 1a) and likewise, the hematite is viewed to be replacing goethite. The texture of goethite varies from colloform to radiating to fibrous or drusy. The presence of vitreous goethite in the present sample depicts the unceasing dehydroxidation process there. Some relict martite grains have seemed to be present here (Fig. 1d) thereby illustrating the oxidation of the magnetite. The observed mesh or trellis texture in the martite represents the continuous weathering and leaching there. Overall, the hematite and martite are viewed to be present as interlocked grains with goethite and clay minerals. Incipient weathering, solution, leaching, and re-precipitation lead to the formation of cavities in the given sample which is observed to be filled by alumino-silicate minerals like clay and quartz (Fig. 1c). Most of the clay minerals are found to be interlocked with ochreous goethite as well where they occupied the cavity or pore spaces and goethite mineralization is constrained to the pore boundaries.

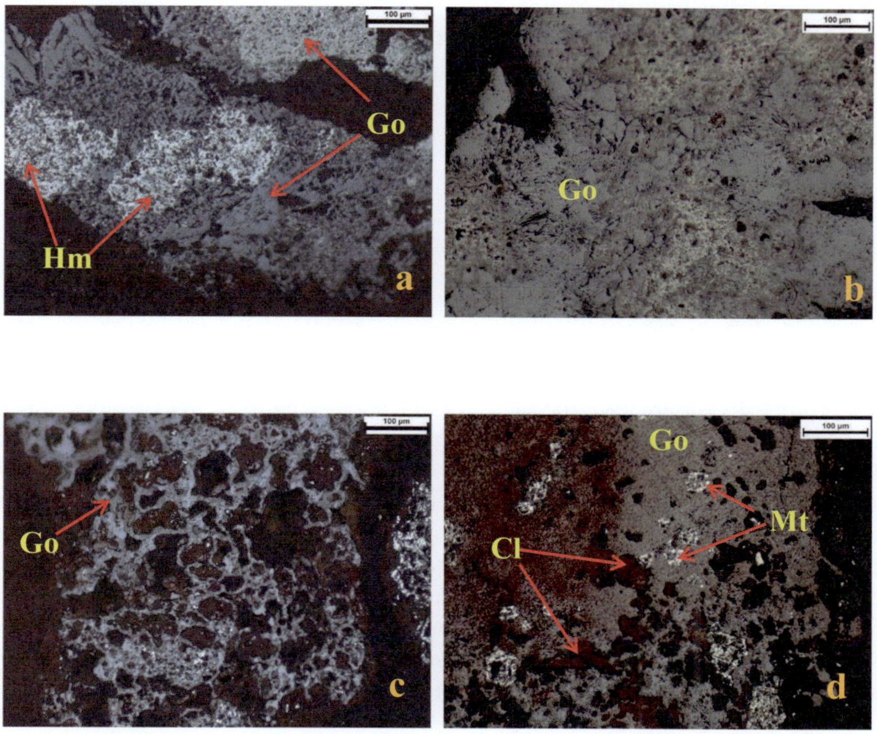

Fig. 1 Photomicrographs representing the mineralogical and textural characteristics of the iron ore and interlocked gangue minerals. **a** ochreous and vitreous goethite (Go) and hematite (Hm) are getting replaced by ochreous goethite; **b** ochreous and vitreous goethite (Go) showing radiating texture; **c** pore filled alumina-silicate (Cl) minerals where incipient goethite (Go) mineralization is occurring from the boundaries; **d** relict martite (Mt) grains showing mesh texture due to weathering, cavity filled clay mineralization (Cl) and ochreous goethite (Go)

Table 1 Chemical composition (assay %) of Fe ore of different size fractions

Size (mm)	Wt%	Total Fe	Fe_2O_3
– 1	100	58.54	83.01
– 1 + 0.5	39.92	59.88	85.67
– 0.5 + 0.15	20.39	58.68	83.50
– 0.15	39.70	57.63	82.88

2.1 Chemical Analysis

The results of the chemical analysis of different size fractions are given in Table 1. It has been observed that the assay % of total Fe and Fe_2O_3 is highest in the sample of – 1 + 0.5 mm size fraction. However, the variation in the assay % seems to be less with respect to each of the size fractions.

3 Experimental Results and Discussion

TBAFDS (Fig. 2) uses minimum fluidization velocity (Umf) difference between particles which is a function of the particle's density for the dry beneficiation process. In iron ore feed, particles with high Fe content have higher density and vice-versa and this density difference is the cause for iron ore feed particles to have a difference in Umf resulting in a dry beneficiation process. Chalavadi and Singh [8] discuss the principle of operation and separation on TBAFDS, and factors affecting the dry separation of iron ore in detail. Figure 3 shows the multi-stage dry separation of iron ore on TBAFDS. The first stage, which is operated at an airflow of 50 Hz, side tilt of 60 mm, and deck vibrating at 19 Hz produced a heavy 1 stream with Fe content enhanced from 58.28% to 63.62% with a yield of nearly 25%. With the objective to further increase the yield% of high content Fe% i.e. nearly 62% the middling 1 stream and light 1 stream of the first stage are processed for the second stage. The second stage was processed by increasing the deck vibration frequency to 19.8 Hz and slightly decreasing the air flow rate to 47 Hz but side tilt was not changed. Heavy 2 stream produced is mixed with heavy 1 and both middling 2 and light 2 streams were mixed and further processed as third stage. The third stage was focused on scavenging operation hence high air flow rate of 55 Hz was used. Deck vibration frequency was unchanged for the second and third stages, side tilt was put constant for all the 3 stages. Heavy streams of all the stages are mixed to produce an overall yield of 53.56% and Fe content of 62.25%. Further investigation of recovering high Fe content product can be done using dry magnetic/tribo separators by mixing the middling and light stream of the third stage.

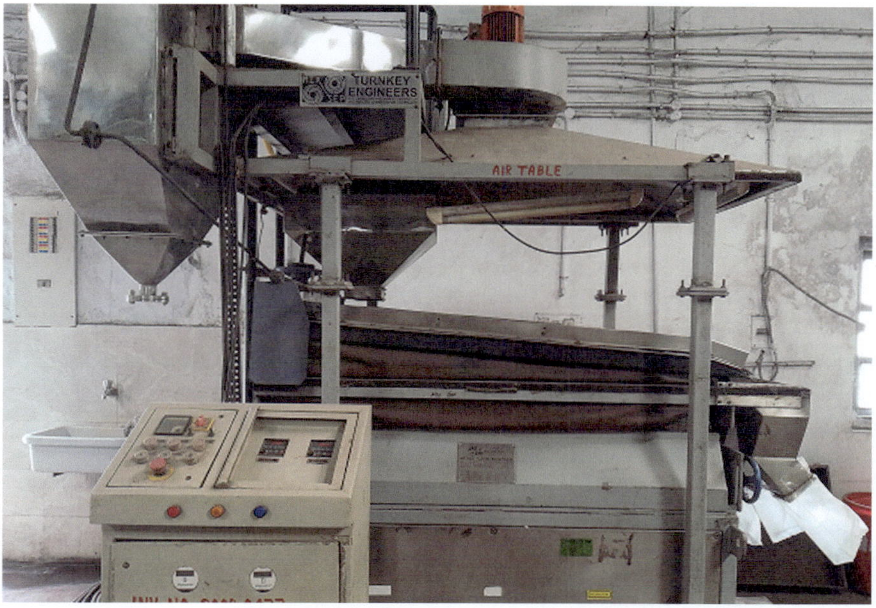

Fig. 2 Thin bed air fluidized dry separator

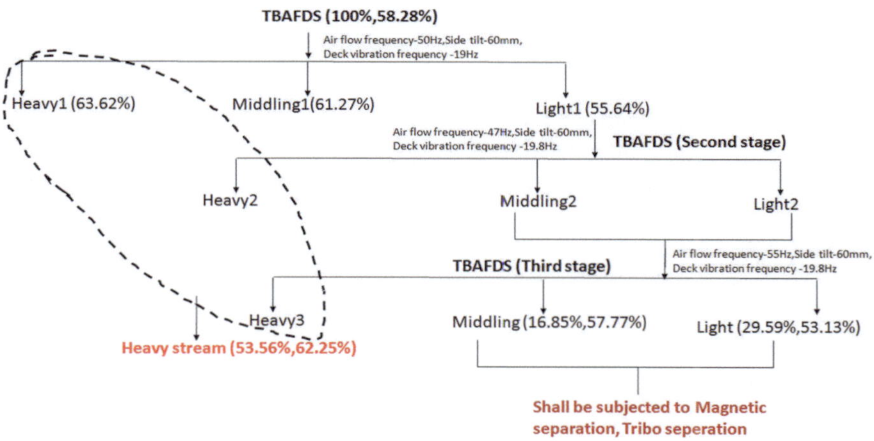

Fig. 3 Multi-stage processing of fine-size iron ore on TBAFDS

4 Conclusion

Operating TBAFDS in multi-stages i.e. 3 stages has increased the yield% of heavy stream having high Fe content. A nearly 4% Fe content increase was witnessed in the overall heavy stream with a yield of 53.56% compared to the yield of 25% in the first stage, almost doubling the yield. Owning to the limitations of magnetic separators employing TBAFDS will yield good advantages like high throughput and no issues of selectivity.

References

1. Zhang H, Chen L, Zeng J, Ding L, Liu J (2015) Processing of lean iron ores by dry high intensity magnetic separation, Separation Science and Technology 50(11), 1689-1694.
2. Chen L, Liao G, Qian Z, Chen J (2012) Vibrating high gradient magnetic separation for purification of iron impurities under dry condition, International Journal of Mineral Processing 102, 136-140.
3. He, J., Liu, C., Xie, J., Hong, P. and Yao, Y. (2017): Beneficiation of coarse particulate iron ore by using a dry density-based fluidized bed separator, Powder Technology 319, 346-355.
4. Oshitani J, Franks GV, Griffin M (2010) Dry dense medium separation of iron ore using a gas–solid fluidized bed, Advanced Powder Technology 21(5), 573-577.
5. Oshitani J, Kawahito T, Yoshida M, Gotoh K, Franks GV (2011) The influence of the density of a gas–solid fluidized bed on the dry dense medium separation of lump iron ore, Minerals engineering 24(1), 70-76.
6. Oshitani J, Ohnishi M, Yoshida M, Franks GV, Kubo Y Nakatsukasa S (2013) Dry separation of particulate iron ore using density-segregation in a gas–solid fluidized bed, Advanced Powder Technology 24(2), 554–559
7. Tripathy SK, Singh V, Murthy YR, Banerjee PK, Suresh N (2017) Influence of process parameters of dry high intensity magnetic separators on separation of hematite, International Journal of Mineral Processing 160, 16-31.
8. Tripathy SK, Banerjee PK, Suresh N (2014) Separation analysis of dry high intensity induced roll magnetic separator for concentration of hematite fines, Powder Technology 264, 527-535.
9. Chalavadi G, Singh RK (2021) Feasibility study of iron ore fines beneficiation by shallow bed air fluidized separator, Particulate science and technology 39(4), 391-400.

Role of NMDC in Mineral Conservation

G. Venkateswara Rao and S. K. Chaurasiya

Abstract Minerals and ores are site-specific, non-renewable natural resources with a limited supply. All natural materials on the earth are derived from the earth by forming or mining. Minerals are the raw materials for the manufacturing industry which were derived from the earth. Minerals play a vital role in the development of any nation. In India, the use of minerals has increased tremendously after its independence. Minerals in some areas are in danger of going extinct due to incorrect and excessive use. Therefore, mineral conservation is necessary. Mineral conservation can be accomplished by enhancing mining techniques, using low-grade ores, rejects, recovering related minerals, and requiring zero-waste technology. Minerals can also be conserved through the latest techniques of beneficiation, blending, sizing, concentration, Pelletisation, value addition, purification, and general customization of products. Converting waste to wealth is one of the main mineral conservation methods. Research and Development (R&D) plays a vital role in mineral conservation. The latest context of R&D in mineral conservation is recovering minerals from waste or rejects. This article deals with the role of R&D in the conservation of minerals, particularly low-cost bulk commodities like iron ore. Discussion includes mineral conservation through the development of a process flowsheet for the production of pellet grade concentrate from Banded Iron Formations (BIF) like Sub Grade Iron Ore (SGIO), Banded Hematite Jasper (BHJ), Banded Hematite Quartzite (BHQ) and iron ore slimes of NMDC mines. The beneficiation studies and research were done at the Research and Development Centre, NMDC Limited, Hyderabad.

Keywords R&D · Mineral conservation · Iron ore · BHQ · BHJ · Subgrade iron ore · Iron ore Slimes

G. Venkateswara Rao (✉)
Research and Development Centre, NMDC Limited, Hyderabad, India
e-mail: gvrao@nmdc.co.in

S. K. Chaurasiya
Research and Development Centre, NMDC Limited, Hyderabad, India
e-mail: skchaurasiya@nmdc.co.in

K. Randive et al. (eds.), *Current Trends in Mineral-Based Products and Utilization of Wastes: Recent Studies from India*, Springer Proceedings in Earth and Environmental Sciences, https://doi.org/10.1007/978-3-031-50262-0_10

1 Introduction

With 109.3 million tonnes of crude steel produced in 2018, India surpassed Japan to become the world's second-largest steel producer. Global crude steel production was 1885.0 million tonnes (Mt) in the year 2022, down by 3.92% compared to 2021. India's crude steel production for the year 2022 was 125.3 Mt, up by 6.01% in 2021. Japan produced 89.20 Mt in 2022, down by 7.37% in 2021 [1].

The extremely ambitious National Steel Policy 2017 (NSP 2017) seeks to establish a 300-mm crude steel capacity industry that is competitive worldwide by 2030–31. In India, steel has a 6.8X employment multiplier ratio and a nearly 1.4X output multiplier effect on GDP [2].

Typically, it takes 1.6 tonnes of iron ore and around 550 kg of coke to produce one tonne of pig iron, the raw iron that comes out of a blast furnace. Some of the coke can be replaced by injecting pulverized coal into the blast furnace. According to current estimates, the production of 1.885 billion tonnes of crude steel required the utilization of roughly 2.30 billion tonnes of iron ore, 1.21 billion tonnes of metallurgical coal, and 614 million tonnes of recycled steel. The raw materials requirement of India as per NSP 2017 is presented in Table 1.

According to NMI database, (which is based on UNFC system) as of 1.4.2015, the total reserves/resources of haematite have been estimated to be 22,487 million tonnes, of which 5422 million tonnes (24%) are under the "Reserves" category and the remaining 17,065 million tonnes (76%) are under the "Remaining Resources" category.

Another major iron ore resource that also occurs in the form of oxide is magnetite, which can be found in banded magnetite-silica formations that are either igneous or metamorphosed and may have sedimentary origins.

The estimated total reserves/resources of magnetite as of 1.4.2015 are 10,789 million tonnes, of which only 53 million tonnes are designated as "Reserves" and 10,736 million tonnes are designated as "Remaining Resources" [3].

Table 1 Forecast of major raw material requirement by 2030–31 (All values in Mt)

Sl. No.	Raw material requirement	Projections in (Mt)
1	Iron ore	447
2	Coking coal	180
3	Non-coking coal for PCI	35
4	Non-coking coal for DRI	82
5	Manganese ore	11
6	Chromite ore	5
7	Limestone and dolomite	91
8	Ferro-alloys	4
9	Refractories	3.2
10	Scrap	11

1.1 World Iron Ore Deposits Lifespan

The amount of steel consumed rises every year. The primary raw material used to produce steel is iron ore. Iron ore must weigh between 1.5 and 2.0 tonnes to produce one tonne of steel. How long does iron ore last if the output of steel keeps rising? The iron ore deposits appear to be pretty large at the moment, but some experts argue that the maths of continuous exponential demand growth could even make this resource appear quite finite. According to an exceedingly conservative projection of a 2% growth year, Lester Brown of the World Watch Institute has predicted that iron ore will run out within 64 years [4].

1.2 Why It Is Necessary to Conserve Minerals?

Minerals should be preserved by a concerted effort since they were generated in the earth's crust more than 2000 million years ago. Efforts should be made to ensure the secure conservation of low-grade minerals while seeking high-grade minerals. A new method to examine even low-grade minerals may emerge as a result of the increasing demand for minerals.

We need to conserve minerals because:

i. The geological process of mineral formations is so slow that the rates of replenishment are infinitely small in comparison to the present rate of consumption.
ii. Mineral resources are finite and non-renewable.
iii. Most of the minerals are unevenly distributed on the earth's surface.

1.2.1 Ideas for Mineral Resource Conservation

(i) Sustainable and planned use of mineral resources.
(ii) To enable the use of low-grade ores at low costs, improved technologies must be continually developed.
(iii) Mineral recycling.
(iv) Utilising waste and garbage as well as alternative replacements.

Beneficiation strategies will be devised and concentrate suitable for producing iron will be generated based on the Chemistry, Mineralogy, and Mineral Liberation studies. Depending on the relative physical characteristics of the desired mineral and the gangue present, there are various approaches to improve the concentration of the desired minerals. Increased concentration of the valuable mineral content is referred to as concentration. Specific gravity, magnetic susceptibility, electrical conductivity, and surface characteristics, such as hydrophobicity and hydrophilicity, are the primary physical characteristics employed in beneficiation. Gravity Concentration, Magnetic Separation, Froth Flotation, and Electrostatic Separation are a few different concentrating/beneficiation techniques.

1.3 Minerals from Waste

Mineral recovery from waste must be the focus of research. Low-grade ores, such as Banded Hematite Quartz (BHQ), Banded Hematite Jasper (BHJ), and Sub Grade Ores are discarded as waste in the mines during the excavation process. Iron ore slimes generated during iron ore washing and ultrafine produced/generated during mining operations are stored as trash. To create high-quality iron ore sources, these waste items can be employed as the starting raw material.

2 Role of Research and Development in Mineral Conservation

The full spectrum of activities including, geoscientific survey, exploration, mining, as well as beneficiation and co-production of low-concentration of strategic metals (especially as by-products from base metal processing), must be covered by research and development in the mineral sector. The development of new technologies will require focus in order to increase the viability of converting existing mineral resources into economically viable resources. Additionally, appropriate technologies must be created to enable domestic companies to make use of the country's substantial mineral resources as well as serve as replacements for minerals with insufficient reserves. Through research, proper appropriation of foreign technological advancements and adoption as indigenous technology must be enhanced. It is necessary to do research and development to increase process and operational efficiency.

Rock mechanics, ground control, mine design engineering, equipment deployment and maintenance, energy conservation, environmental protection, safety of operations, and human engineering require a special focus in terms of research and development. This essay discusses the role of research and development in maintaining minerals and ores, particularly iron ores.

3 Role of NMDC in Conservation of Minerals

Through a number of initiatives, including the recovery of iron values from slimes, Banded Hematite Jasper (BHJ), Banded Hematite Quartzite (BHQ), Sub Grade Iron Ore (SGIO), Lean Grade Tailings, High Silica Blue Dust (HSBD) and use of a greater quantity of fines in Sinter Fines, etc., NMDC is at the forefront of mineral conservation. The development of a process flow sheet for the beneficiation and use of goethite-rich iron ores and lean-grade tailings from the Donimalai Beneficiation facility is currently being discussed. This article deals with the above-mentioned conservation methods for the utilization of low-grade, lean-grade, and sub-grade iron ore.

3.1 Iron Ore Slimes from Kirandul—Beneficiation

The 'as received' sample has an elemental analysis of 61.40% Fe, 7.68% SiO_2, 1.52% Al_2O_3, and loss on ignition (LOI) of 2.63% and all other elements were confirmed to be within critical limits. Based on the mineralogical studies, it was found that Magnetic separation will yield good results for beneficiation and recovery of iron values from the Kirandul slimes by using a vertically Pulsated Wet High-Intensity Magnetic Separator (VP-WHIMS). The intensity of the magnetic field, the pulsing frequency, the ring rotation speed, the amplitude, and the slurry concentration are chosen as the operational parameters of the vertical pulsating wet high-intensity magnetic separator (VP-WHIMS). The current work focuses on recovering iron values using VP-WHIMS from piled slimes from Kirandul, India. The parameters viz. Magnetic Field Intensity, Ring Rotational Speed, Amplitude, and Feed Concentration were optimized. The grade and recoveries of the concentrate (%Fe and weight %) were observed and optimized. The test was repeated and run with optimum settings. The Product measured at optimum settings had a yield of 70.96% and with 66.20% Fe, 2.34% SiO_2, 0.95% Al_2O_3, and 1.92% LOI.

3.2 Iron Ore Slimes from Donimalai—Beneficiation

This study outlines a method for beneficiating slimes by reducing gangue content to a level suitable for producing hot metal using the blast furnace technique. Effective and efficient research has been done to make use of the iron ore slimes from the Donimalai iron ore mine's slime ponds. Utmost care was taken to collect a representative sample from the Donimalai Tailing Dam.

Representative samples were drawn for determination of physical properties, mineralogy, and chemistry and subjected to evaluation. The analysis of the "as received" slime sample revealed that all other trace elements were within the critical limits. The as-received sample assayed 49.40% Fe, 13.51% SiO_2, 8.44% Al_2O_3, and 5.16% loss on igniting. It was found that the received sample contains 7% moisture, a bulk density of 1.74 t/m^3, and a specific gravity of 3.7. Mineralogical studies reveal that the main ore mineral is Hematite and gangue minerals are quartz and Kaolinite. Screen Assay Analysis was performed using Tyler series standard sieves with the representative sample. Based on the chemistry and mineralogical studies, it was

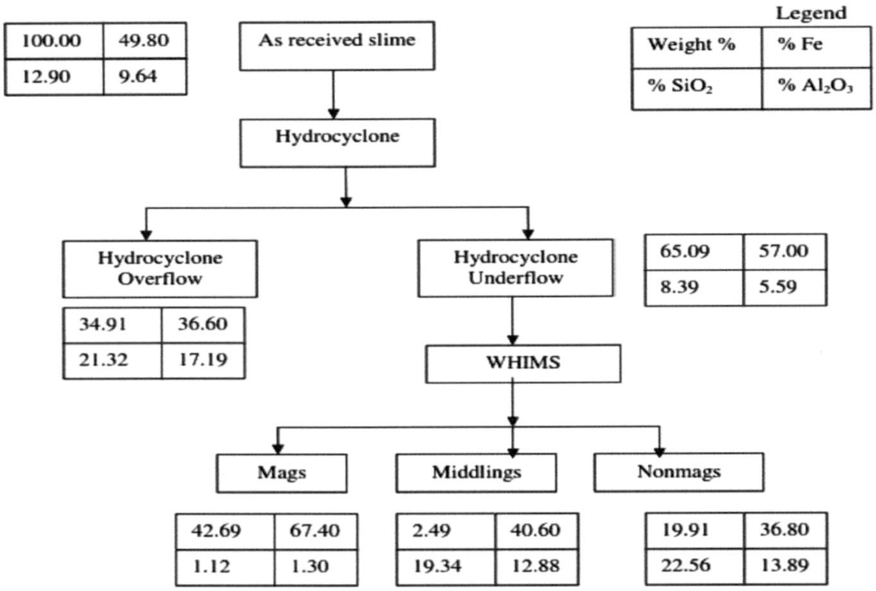

Fig. 1 Process flowsheet developed for recovery of iron values from Donimalai slimes

proposed to beneficiate the Donimalai iron ore slime sample by desliming by using a hydrocyclone and subjecting the hydrocyclone underflow to a wet high-intensity magnetic separator (WHIMS).

Hydro-cyclone optimization experiments were conducted for desliming. A sufficient quantity of hydrocyclone underflow was produced at optimized parameters for further studies. Hydrocyclone underflow was subjected to WHIMS and parameters were optimised.

A concentrate of 65.93% Fe, 1.45% SiO_2, 1.94% Al_2O_3, and 1.56% loss on ignition could be obtained with a yield of 45.18% and with a recovery of 60% iron values [5]. Figure 1 explains the process adopted for the recovery of iron values from Donimalai slimes with metallurgical balance.

3.3 Utilisation of Banded Hematite Jasper (BHJ)

The hematite and jasper layers alternately band BHJ in an irregular manner. The average thickness of the bands is seen to be between 1 and 2 mm, with individual band thicknesses ranging from a few millimeters to over a few centimeters. The bands also differ from white or grey to red, brown, or black.

BHJ sample was collected from Donimalai Iron Ore Mine, Karnataka. Representative samples were drawn from the received sample to study Mineralogy, physical properties, and chemistry. It was physically discovered that the sample was present in the hematite and jasper alternate layers. As per the mineralogical studies, hematite is the only ore mineral and jasper is the main gangue mineral. The sample's chemical analysis revealed that 41.20% Fe, 39.10% SiO_2, and 1.02% Al_2O_3. According to the chemical analysis, the alumina is already far below the necessary target specification of the concentrate that will be produced during beneficiation. According to the microscopic studies, the ore minerals are freed at a maximum size of 150 microns.

Indicative tests using gravity and magnetic separation tools, such as the Floatex Density Separator (FDS), gravity slime table, and wet high-intensity magnetic separator (WHIMS), were carried out based on the characterization results. The outcomes showed that the sample might be improved using both magnetic and gravity separation methods. On the other hand, the WHIMS data showed that jasper grains were reported at high magnetic intensity in the concentrate.

Thus, an attempt was undertaken to investigate the suitability of gravity spirals for beneficiating BHJ in the liberation size range. Based on the encouraging gravity separation results and the availability of innovative design spirals for treating tiny particles, gravity spirals were used for the beneficiation of BHJ.

The sample was ground to a top size of 150 microns and was subjected to Hydrocycloning and hydrocyclone underflow was pre-concentrated adopting wash water-less spirals. Tests were conducted at various operating parameters, viz feed rate, pulp density, and splitter position. The results indicated that it was possible to upgrade the sample from a feed grade of 41.20% Fe to 56% Fe with a yield ranging from 35 to 40%. The resulting pre-concentrate was then cleaned using wash water spirals. According to the test results, a final concentrate with an iron values recovery of 48.1% and a yield of 30.0% could be obtained, with an assay of 66.12% Fe and 4.50% SiO_2. Further attempts are being made to recover the iron units from the tails [6]. Figure 2 shows the process flowsheet adopted for the beneficiation of BHJ.

3.4 *Utilisation of BHQ*

The 'as received' Banded Hematite Quartzite (BHQ) sample assayed 35.70% Fe with 48.45% SiO_2 and 0.18% Al_2O_3. The sample was found to contain around 65% of + 50 mm size material and around 35% of −50 mm size out of which −10 mm size material is around 2%. In view of high silica, this material could not be subjected to conventional processing and hence requires beneficiation. The objective of this paper is to recover the maximum iron units adopting suitable beneficiation techniques, to assess the quantity and quality of the final concentrate, and to collect the relevant data for the process development.

The mineralogical study indicates that the 'as received' BHQ sample with alternate layers/bands/lenses of hematite and quartz consists of about 45% of ore minerals (by volume) and 55% of gangue minerals (by volume).

According to the mineralogical studies, liberation of hematite occurs at 100–200 micron size fractions. It was inferred from the screen-assay-analysis results that Fe

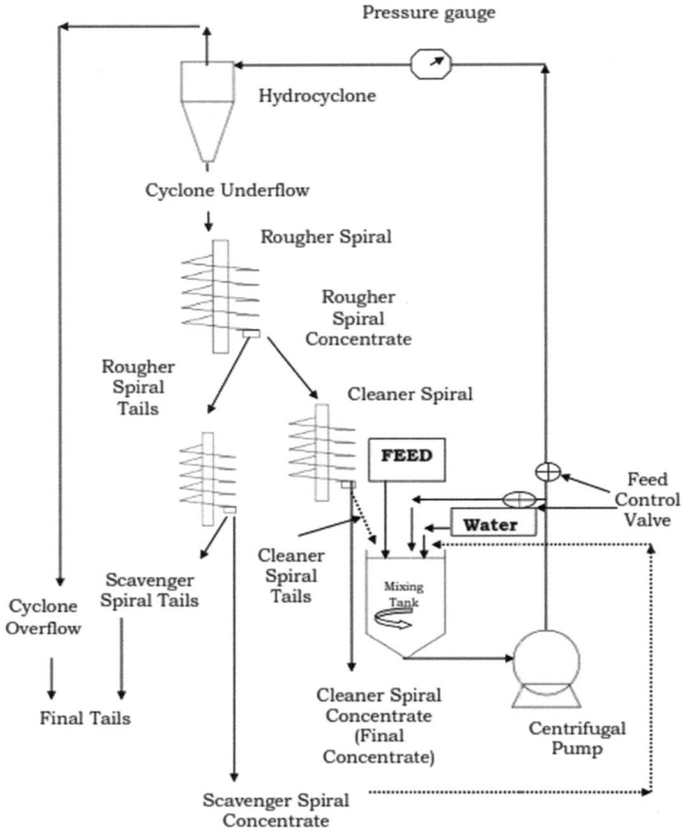

Fig. 2 Process flowsheet developed for beneficiation of BHJ

and SiO_2 were dispersed uniformly down to the finer size fractions when the sample was crushed to 30 mm, 1 mm, ground to 212, and 150 microns. This makes the processing and recovery of iron units more difficult and complicated [7]. Figure 3 shows the process adopted for beneficiation of BHQ.

3.5 Production of Pellet Grade Concentrate from Sub Grade Iron Ore (SGIO)

An attempt was made to use Sub Grade Iron Ore (SGIO) of Deposit 5, Bacheli Complex, Bailadila, Chhattisgarh, India. The run-off-mine (ROM) sample received was subjected to characterization and investigations.

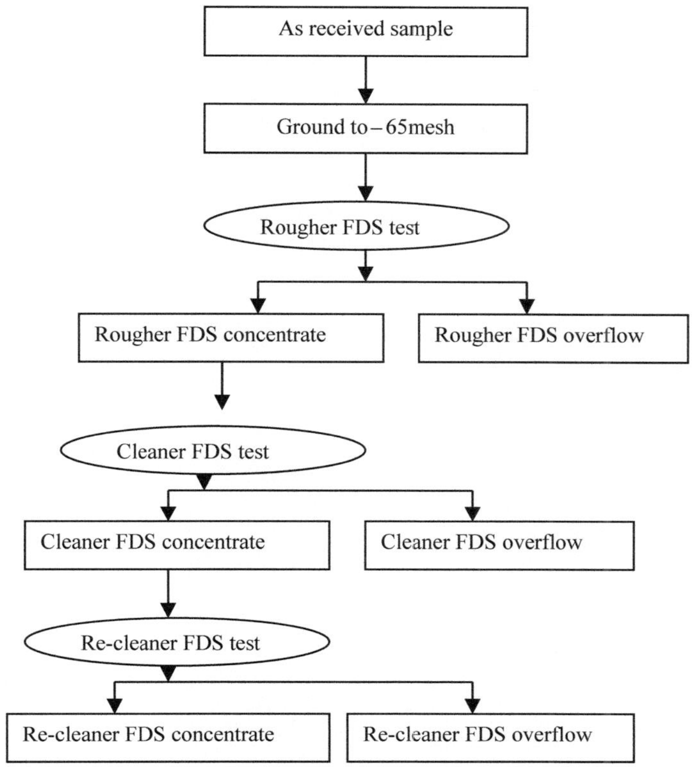

Fig. 3 A process adopted to beneficiate BHQ

The SGIO sample assayed 40.80% Fe, and 40.90% SiO_2, and all other critical elements were well within the limits. According to mineralogical research, quartz is the only gangue mineral and hematite is the primary ore mineral.

Hematite is getting liberated at a size range of 100–200 microns. For the "as received" ROM sample, Bond's Ball Mill Work index was determined to be 10.45 kWh/T.

A Multi Gravity Separator (MGS) was used to concentrate the stage-crushed and ground sample (less than 212 microns). Based on three operating parameters of the MGS the experiments were designed. 15 experiments were conducted by varying the parameters. The first cycle's fifteen experiments produced a lower concentration than was necessary. In order to obtain the highest possible iron values, parameters were optimized.

A concentrate of 55.8% Fe with a yield of 61.73% (yield) and an iron value recovery of 84.42% was achievable. The experiment was repeated at the optimized parameters and a sufficient quantity of pre-concentrate was produced for further studies.

Using the concentrate from the first cycle as feed material, 15 further trials were conducted in the second cycle to create a concentrate suitable for commercial-grade pellet concentrate. A concentrate with an assay of 66.67% Fe and 3.12% SiO_2 with a yield of 45.08% by weight and a concentrate recovery of 73% iron values was produced at optimized parameters. Figure 4 shows the process adopted for the production of pellet-grade concentrate along with metallurgical balance.

In-depth research was carried out along with process feasibility studies for utilization of BHQ, BHJ, SGIO, Iron ore slimes, and High Silica Blue Dust (HSBD). Process

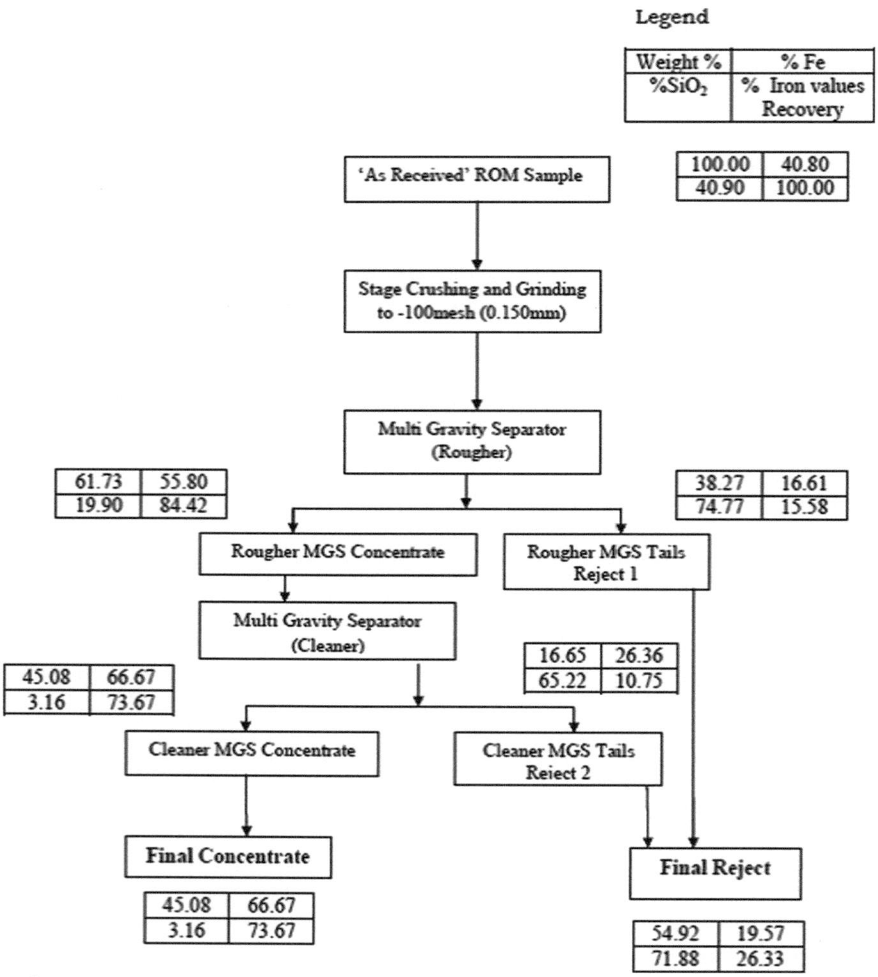

Fig. 4 Process flowsheet and Metallurgical balance of beneficiation of SGIO

flow sheets were developed to convert low-grade and lean-grade iron ores into pellet-grade concentrate. References [8–15] as a part of mineral conservation in general and iron ore in particular.

4 Conclusion

Approximately 500 MT of iron ore are needed for the domestic steel sector by 2030–31, according to the National Steel Policy (NSP) 2017. In order to meet these existing hematite resources in the reserve exploration followed by feasibility. One has to explore the beneficiation potential of hematite resources namely BHQ, and BHJ to conserve the mineral resources. Suitable technology has to be developed to utilize goethite-rich iron ore, iron ore slimes, and lean tailings in iron making. The Indian industry will greatly expand its beneficiation and agglomeration facilities in order to preserve the minerals, where R&D plays a crucial role. This is because the industry is aware of the iron ore reserves and resources, as well as the supply and demand of domestic steel plants. NMDC is developing various strategies to conserve iron ore and extend its lifespan.

Acknowledgements The authors appreciate the efforts of the management of NMDC Limited for allowing them to perform the research on the conservation of iron ore and allowing them to publish this paper.

References

1. Annon (2023) World steel in figures 2023 (infographic), World steel association, Brussels, Belgium, May 2023, www.worldsteel.org.
2. Annon (2017) The Gazette of India—Extraordinary, Part II, Section 3 (i), Ministry of Steel, National Steel policy 2017, Government of India, New Delhi, May 2017, 20–21.
3. Annon (2019) Indian Minerals Yearbook 2019, 58th edition, Part III: Mineral Reviews, Iron Ore, Advance release, November 2020, Indian Bureau of Mines, Government of India, Ministry of Mines, 16.1 to 16.33.
4. Lester R Brown (2006) Plan B 2.0, Rescuing a planet under stress and a civilization in trouble, updated and expanded edition, Newyork, W W Norton, 2006, 109.
5. G Venkateswara Rao, R Markandeya, SKSharma (2016) Recovery of Iron values from Iron ore Slimes of Donimalai Tailing Dam, Transactions of Indian Institute of Metals, January 2016, 69 (1), 143–150, https://doi.org/10.1007/s12666-015-0809-0, Springer International Publishing AG.
6. GE Sreedhar, DR Venkatesulu, NV Ramana Murthy (2008) Application of Gravity Spirals for Beneficiation of Banded Hematite Jasper (BHJ) Iron Ore of India, Proceedings of International Conference on Iron Ore 2008, Perth, Australia.
7. G Venkateswara Rao, T Anantha Raman, VKS Rathore (2010) Beneficiation of BHQ—A case study, Proceedings of An International Seminar on Mineral Processing Technology 2010, Jamshedpur, India.

8. G Venkateswara Rao, R Markandeya, Rajan Kumar (2017) Feasibility studies for production of Pellet grade concentrate from Sub Grade Iron Ore using Multi Gravity Separator, Journal of The Institution of Engineers (India): Series D, Metallurgical & Materials and Mining Engineering (In press). https://doi.org/10.1007/s40033-017-0147-Y, Springer Interational Publishing AG.
9. G Venkateswara Rao, R Markandeya, Rajan Kumar (2016) Optimisation of Process Variables for Recovery of Iron values from Sub Grade Iron Ore by using Enhanced Gravity Separation, International Journal of Engineering Research & Technology, 5 (1), 900-909.
10. G Venkateswara Rao, R Markandeya S,KSharma (2016) Feasibility Studies on Beneficiation of Sub Grade iron Ore from Deposit 5, Bacheli Complex, Bailadila, Chhattisgarh using Mozley Mineral Separator, Proceedings of XV International Seminar on Mineral Processing Technology, MPT 2016, TCS-TRDDC, Pune, 5–7, 488–499.
11. G Venkateswara Rao, R Markandeya, SK Sharma (2014) Modeling and optimization of Multi Gravity Separator for recovery of Iron values from Sub Grade iron Ore, Proceedings of XXVIII International Mineral Processing Congress, IMPC 2014, October 20–24, Santiago, Chile.
12. G Venkateswara Rao, R Markandeya, SK Sharma (2013) Application of Gravity Methods for Beneficiation of Sub Grade iron Ore, Proceedings of XIII International Seminar on Mineral Processing Technology (MPT–2013), IMMT Bhubaneswar, 24–28, September 2012, 397–405.
13. G Venkateswara Rao, R Markandeya, SK Sharma (2012) Process Amenability Studies Sub Grade iron Ore from Bacheli Complex, Bailadila Sector, India, Proceedings of XXVII International Mineral Processing Congress, IMPC 2012, New Delhi, India, 24–28, September 2012, 1842–1860.
14. Venkateswara Rao G, Markandeya R, Sharma SK (2017) Stratification Studies with Sub Grade Iron Ore from Deposit No. 10 and 11A, Bacheli Complex, Bailadila, Chhattisgarh, India. J. Inst. Eng. India Ser. D 98, 131–138. https://doi.org/10.1007/s40033-016-0117-9
15. G Venkateswara Rao, S K Chaurasia (2020) Beneficiation of Low grade Iron Ores, December 2020, Mines and Mineral Reporter. 13 (8), 32-38

Development of High Alumina Synthetic Aggregate Based on Impure Raw Bauxite

Gaurav Sinha, Satyaki Mandal, Parul Kumari, Santanu Basak, Saumen Sinha, and Indra Nath Chakraborty

Abstract The refractory industries in India are facing the challenges of unavailability of good quality raw materials, lack of continuous and assured supply as well as increased raw material price. It is difficult to produce high-quality refractory raw materials in the country due to the high impurity content in the available mineral resources such as bauxite, magnesite, etc. Hence, the manufacturing of premium refractory products is dependent on imports from China. On foot of its global commitment to reduce carbon emissions, the government of China has levied several restrictions on refractory raw material manufacturers which in turn affected production and deliveries. This in effect resulted in rising import costs and ultimately affected the cost of products. To circumvent this problem and to cater to the increasing domestic demand for quality products, an attempt has been made to explore ways and means to produce value-added products from the inferior quality raw bauxite available in India. The synthetic high alumina aggregates were manufactured by pyro-processing route from locally available bauxite, Bayer process alumina, and judiciously chosen additives. Extensive experimental work has been carried out to optimize the additive concentration and process to achieve the required properties in the aggregate. The produced synthetic aggregate is of superior quality with 93% Al_2O_3 & 5% porosity as compared to 86–89% Al_2O_3 obtained in Chinese bauxite. The refractory products

G. Sinha (✉) · S. Mandal · P. Kumari · S. Basak · S. Sinha · I. N. Chakraborty
Calderys India Refractories Ltd., Katni, India
e-mail: gaurav.sinha@calderys.com

S. Mandal
e-mail: satyaki.mandal@calderys.com

P. Kumari
e-mail: parul.kumari@calderys.com

S. Basak
e-mail: santanu.basak@calderys.com

S. Sinha
e-mail: saumen.sinha@calderys.com

I. N. Chakraborty
e-mail: indra.chakraborty@external.calderys.com

K. Randive et al. (eds.), *Current Trends in Mineral-Based Products and Utilization of Wastes: Recent Studies from India*, Springer Proceedings in Earth and Environmental Sciences, https://doi.org/10.1007/978-3-031-50262-0_11

manufactured using these aggregates have shown improved performance. Thus, the development of high alumina synthetic aggregates will open doors for the manufacturing of high-quality refractory products used for various customers like Iron and steel, Cement, Aluminium Industries, etc. As a result, the organization has reduced its dependency on the supply of Chinese bauxite.

Keywords Raw bauxite · Synthetic aggregate · Porosity · Chinese bauxite

1 Introduction

As per data published by IRMA in 2011, the steel industry is the biggest consumer of refractory products and accounts for about 75% [1]. Refractory aggregates like tabular alumina, magnesia aluminate spinels, calcium aluminate cement, etc., are used in both shaped and unshaped refractories. Key advantages of alumina in main refractory applications are high thermomechanical strength, chemical stability, high thermal shock resistance, and good abrasion resistance. Mullite-bonded corundum bricks consisting of 90–95% alumina are used in stacks. Corundum-based castables have replaced chrome-based castables for environmental reasons. High-purity alumina is a potential raw material for taphole clay due to high tapping temperatures like 1450–1550 °C and prolonged tapping time [2].

India has the fifth largest reserves of bauxite ores in the world. The major production of bauxite consists of 40–45% Al_2O_3 grade. The other grades of bauxite are 50–55% Al_2O_3 and less than 40% Al_2O_3 [2]. These types of bauxite grades are certainly not suitable for the refractory industry. India is dependent on China for high-grade bauxite and calcined alumina. However, the historical misuse of reserves has disabled the exporters to supply the required high-quality grades in the recent past. This led to rising extraction and production costs of high alumina bauxites and calcined alumina. Many export restrictions are put into place for environmental reasons or to preserve natural resources. As many of these raw materials are produced in a limited number of countries, export restrictions that are imposed in one country impact the availability of suitable grades of raw materials for other countries. Due to the rising threats, Indian refractory industries are bound to look for alternative mines or develop synthetic aggregates. The continuous advancement in steel and iron-making processes has increased demand for higher-quality refractories and ultimately synthetic alumina raw materials. The Calderys India Refractories Limited, Katni, Madhya Pradesh, India; where the present study has been carried out, imports Chinese Bauxite grade and brown fused alumina. The chemical properties of Chinese grade are shown in Table 1.

According to Table 1, calcined bauxite, from both the Indian sources, viz. Gujarat and Katni fall well short of the target. CaO and Fe_2O_3 contents of Gujarat bauxite are too high, whereas Fe_2O_3 content of Katni bauxite is too high to be accepted. Synthetic aggregate developed should have Al_2O_3 higher than Chinese bauxite i.e., around 90%. The iron and lime content of synthetic aggregate should be as low as

Table 1 Typical chemistry of calcined bauxite from various sources

Constituent (%)	Gujarat, India	Katni, India	China	Guyana	Required
Al_2O_3	82	75	87.2	90.1	
TiO_2	4.2	9.1	3.4	3.1	<4
SiO_2	8.2	8.3	6.7	4.6	<7
CaO	2.00	0.51	0.67	0.51	<0.5
Fe_2O_3	3.35	6.7	1.7	0.84	<1.8

1.8% and 0.7% respectively. Apparent porosity should be no greater than 7% and bulk density should be no lower than 3.10 gm/cc. To achieve better sintering with lower porosity at a lower temperature, sintering aids can be used [3, 4]. Recently Al_2O_3 doped with Mg, Sr, Ti, Lu, Zr or Pt have been used to improve densification [5]. Nagashima et al. [6] used MnO for densification of Al_2O_3 during sintering at a lower temperature [6]. The availability of high alumina aggregates with the desired properties mentioned above is a major concern for the company. Thus, the company took up the challenge to manufacture in-house high alumina synthetic aggregates at a reasonable cost from locally available bauxite. This paper explains the raw materials required for manufacturing and the pyro-processing route. The paper also presents a comparative study of conventional castable and low cement castable by incorporating Chinese bauxite and high alumina synthetic aggregates separately.

2 The Experiment

The project was carried out at the lab scale as well as at the plant scale. The raw materials required for high alumina synthetic aggregate are Calcined Alumina, Raw Bauxite, and Sintering Aids. The sintering aid is added to decrease the sintering temperature. These raw materials were mixed in a ratio so as to meet the chemical properties mentioned in the previous section. During lab trials, the raw mixture was heat treated at >1400 °C for 2 h in an electric furnace. After successfully meeting the desired chemical composition, a lab-scale comparative analysis between Chinese bauxite and synthetic high alumina aggregates by incorporating them into conventional castables and low cement castables was undertaken. The chemical and physical properties were tested as described in IS 10570. The properties like water demand to cast the castables, initial flow, and flow after 30 min were observed for all trials as per EN ISO 1927. Test specimens were prepared, treated, and tested for Bulk density (BD), cold crushing strength (CCS), and modulus of rupture (MOR) vas per IS 10570. The test specimens were heat treated at 1100 °C, 1400 °C, and 1550 °C for 3 h as per IS 10570. The physical properties like permanent linear change (PLC), CCS, and MOR were measured after heat treatment. Also, the abrasion resistance of specimens, heat treated at 800 °C for 3 h, was tested as per ASTM C 7014. Thermal spalling resistance at 1000 °C of test specimens heat treated at 800 °C and 1100 °C for 3 h were measured as per IS 1528 Part 3 by water quenching.

Once lab trials were successfully completed, a plant trial was taken up and all the above-listed properties were studied. The material was calcined at 1650–1700 °C in a rotary kiln at Calderys India Refractories Ltd. Katni plant.

3 Results and Discussions

3.1 Chemical Analysis

The chemical composition of Chinese bauxite and that achieved in lab and plant trials are mentioned in Table 2. According to Table 2, the in-house developed synthetic aggregate achieved approximately 93% alumina, lower silica, and titania content as compared to Chinese bauxite. The iron oxide and lime remain comparable. Also, the plant trial was successful since the operation process achieved comparable results with lab trials.

3.2 Physical Properties

The physical properties like apparent porosity (AP) and bulk density (BD) were measured as shown in Table 3. As inferred from the below table, the bulk density of synthetic aggregates is comparable. However, denser aggregates are manufactured. The porosity might have decreased due to the addition of sintering aid.

Table 2 Comparative study of the chemical composition of Chinese bauxite, lab trial, and plant trial

Parameters (%)	Chinese bauxite	Lab trial	Plant trial
Al_2O_3	86.5	93.56	93.31
SiO_2	7.5	1.55	1.68
Fe_2O_3	1.8	1.64	1.57
TiO_2	3.6	2.58	2.68
CaO	0.7	0.6	0.68

Table 3 Comparative study of AP, BD of Chinese bauxite, lab trial, and plant trial

Properties	Chinese bauxite	Lab trial	Plant trial
Apparent porosity, %	5.32	3.58	3.41
Bulk density (gm/cc)	3.25	3.28	3.33

3.3 Low Cement Castable (LCC)

The high alumina synthetic aggregates were substituted into the formulation of one of the company's low cement castable products using Chinese bauxite. The test results are compared with products with Chinese bauxite in the formulation. The results are tabulated below. Table 4 indicates that rheological properties are similar for the formulations.

The bulk density of castables cured at 110 °C was measured. From Table 5, it can be inferred that the density of synthetic aggregate-based castables is higher than Chinese bauxite-based castables. Synthetic-based castables have higher strength as compared to Chinese bauxite-based castables as shown in Fig. 1. MOR property for synthetic-based castables was slightly lower after oven drying at 110 °C. But this property increased gradually after firing at higher temperatures.

The PLC values were obtained at different firing temperatures 1100 °C, 1400 °C, and 1550 °C for 3 h as shown in Table 5. Both materials exhibit shrinkage after firing at 1100 °C and 1400 °C. However, Chinese-based castable exhibits expansion after firing at 1550 °C. Also, Chinese bauxite-based castables exhibited higher

Table 4 Comparative study of water demand, T_0 and T_{30} of synthetic aggregate-based low cement castables and Chinese bauxite-based low cement castables

Property (%0	Synthetic aggregate based	Chinese bauxite based
Water demand	4.2	4.3
Flow (T_0)	105	100
Flow (T_{30})	85	85

Table 5 Comparative study of physical properties of Chinese bauxite-based castables and synthetic aggregate-based castables (Low Cement Castables)

Property	Synthetic aggregate	Chinese bauxite
At 110 °C/24 h. BD (gm/cc)	3.055	2.905
At 110 °C/24 h. CCS (kg/cm^2)	996.75	740
At 110 °C/24 h. MOR (kg/cm^2)	219	255.5
At 1100 °C/3 h. PLC (%)	−0.34	−0.12
At 1100 °C/3 h. CCS (kg/cm^2)	1297.5	1071.5
At 1100 °C/3 h. MOR (kg/cm^2)	286	267.5
At 1400 °C/3 h. PLC (%)	−0.56	−0.965
At 1400 °C/3 h. CCS (kg/cm^2)	1329.25	987
At 1400 °C/3 h. MOR (kg/cm^2)	286	238
At 1550 °C/3 h. PLC (%)	−0.31	0.875
At 1550 °C/3 h. CCS (kg/cm^2)	1583.75	1150.75
At 1550 °C/3 h. MOR (kg/cm^2)	286	235.5

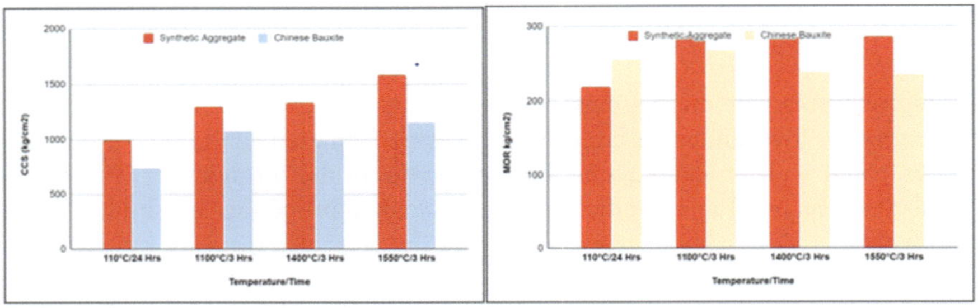

(a) (b)

Fig. 1 Comparative study of CCS **a** and MOR **b** values of Chinese bauxite-based castables and synthetic aggregate-based castables

shrinkage at 1400 °C than synthetic aggregate-based castables. The abrasion resistance of Synthetic Aggregate LCC was better than LCC made with Chinese bauxite. Spalling resistance was similar for both the LCC castables.

3.4 Conventional Castables (CC)

The high alumina synthetic aggregates were substituted into the formulation of one of the company's conventional castable products using Chinese Bauxite. The test results are compared with products with Chinese bauxite in the formulation and tabulated below. Table 6 indicates that water demand differs and flowability is similar for the CC products.

The bulk density of castables cured at 110°C was measured. From Table 5, it can be inferred that the density of synthetic aggregate-based castables is similar to Chinese bauxite-based castables. Both castable samples exhibit similar CCS strength at 110 °C, 1100 °C, and 1400 °C as shown in Fig. 2. However, synthetic-based castables exhibited higher CCS strength after firing at 1550 °C. MOR properties for synthetic-based castables are higher than for Chinese bauxite-based castables. The strength of both castables dropped after firing at 1100 °C due to the breaking of hydraulic bonds. The PLC values were obtained at different firing temperatures 1100 °C, 1400 °C, and 1550 °C for 3 h as shown in Table 7. Both materials exhibit expansion after firing at 1100 °C and shrinkage at 1550 °C. However, synthetic

Table 6 Comparative study of water demand and T_0 of synthetic aggregate-based low cement castables and Chinese bauxite-based conventional castables

Property (%)	Synthetic aggregate based	Chinese bauxite based
Water demand	9.3	9.5
Flow (T_0)	85	90

aggregate-based castables exhibited expansion at 1400 °C while Chinese bauxite-based castables shrank. However, Chinese-based castable exhibits expansion after firing at 1550 °C. Also, Chinese bauxite-based castables exhibited higher shrinkage at 1500 °C than synthetic aggregate-based castables. Test specimens made out of Synthetic Aggregates CC are more volume stable than that of CC made with Chinese bauxite. The abrasion resistance of Synthetic Aggregate CC was better than CC made with Chinese bauxite. Spalling resistance was similar for both the CC castables.

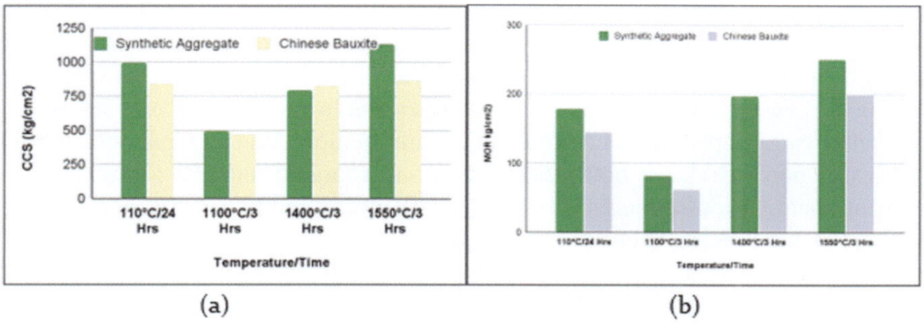

Fig. 2 Comparative study of water demand, T_0 and T_{30} of synthetic aggregate-based low cement castables and Chinese bauxite-based low cement castables

Table 7 Comparative study of physical properties of Chinese bauxite-based castables and synthetic aggregate-based castables (Conventional Castables)

Property	Synthetic aggregate	Chinese bauxite
At 110 °C/24 h. BD (gm/cc)	2.76	2.585
At 110 °C/24 h. CCS (kg/cm^2)	995	845.5
At 110 °C/24 h. MOR (kg/cm^2)	178	144
At 1100 °C/3 h. PLC (%)	0.28	0.09
At 1100 °C/3 h. CCS (kg/cm^2)	496.75	474.75
At 1100 °C/3 h. MOR (kg/cm^2)	81.5	62
At 1400 °C/3 h. PLC (%)	0.69	−1.06
At 1400 °C/3 h. CCS (kg/cm^2)	795.25	827.75
At 1400 °C/3 h. MOR (kg/cm^2)	196.5	135
At 1550 °C/3 h. PLC (%)	−1.09	−4.36
At 1550 °C/3 h. CCS (kg/cm^2)	1130.5	870.75
At 1550 °C/3 h. MOR (kg/cm^2)	250.5	200
Abrasion resistance, CC (prefired at 800 °C/3 h)	5.81	6.14
Abrasion resistance, CC (prefired at 800 °C/3 h)	18 cycles spalled	18 cycles spalled
Thermal Spalling, 1000 °C (prefired at 1100 °C/3 h)	18 cycles spalled	17 cycles spalled

4 Conclusion

The ever-increasing demand for high-grade alumina refractory materials from steel industries has increased the imports of raw materials from China. However, China imposed stringent restrictions on the mining and export activities of Chinese bauxite. This has resulted in motivating industries and researchers, especially in India, to develop in-house high alumina synthetic aggregate from calcined locally available bauxite grade. However, due to the inferior quality of locally available bauxite as compared to Chinese bauxite, it cannot be directly heat treated to produce 90–93% Al_2O_3. Instead, raw bauxite if mixed with other alumina-grade powders will produce high-grade alumina. Claderys developed a pyro-processing route initially at a lab scale and then a plant trial. The synthetic aggregate achieved approx. 93% Al_2O_3 and much lower impurities. The synthetic material achieved the required bulk density and lesser porosity. Later both grades were incorporated into the formulation of conventional and low-cement castables. The chemical and physical properties like AP, BD, CCS, MOR, PLC, abrasion resistance, and thermal spalling were determined. A comparative study shows that properties are either similar or improved for synthetic aggregate-based low-cement castable as well as conventional castable. Thus, the organization achieved the development of in-house high alumina synthetic aggregate. The properties of castable did not deteriorate. In this way, quality and cost-efficient substitution of Chinese bauxite by synthetic alumina aggregate is achieved. The model presented here is replicable and has the potential to revive the refractory manufacturing industry.

References

1. IRMA Journal (2005) Indian Refractory Makers Association, Kolkata, Vol XXXVIII, No. 3, September, 2005.
2. Horckmans Liesbeth, Nielsen Peter, Dierckx Philippe, Ducastel Antoine (2019) Recycling of refractory bricks used in basic steelmaking: A review. Resources, Conservation and Recycling, 140, 297–304. https://doi.org/10.1016/j.resconrec.2018.09.025
3. Culter IB, et al., (1957) J. Am. Ceram. Soc., 40, 134–139.
4. Koble R.L. (1958): J. Am. Ceram. Soc., 41, 55–62.
5. Yoshida Y. et al. Acta Mater, 53, 443–440.
6. Nagashima M et al (2008) Fabrication and Optical Characterisation of High-Density Alumina Doped with Slight MnO. J. Cer. Soc. Japan 116 (5) 645–648.

Characterization and Beneficiation of Pyrophyllite

P. G. Bhukte, G. T. Daware, M. J. Chaddha, T. P. Bhosale, and A. Agnihotri

Abstract Pyrophyllite $Al_2Si_4O_{10}(OH)_2$ stands as a common phyllosilicate mineral frequently co-existing with minerals like quartz, mica, kaolin, epidote, talc, feldspar, muscovite, siderite, and rutile. It is characterized by alumino-silicate minerals and contains moderate to high silica and alumina. The high quality, pyrophyllite is desirable in the industry such as refractory, ceramic, fiberglass, cosmetics, paper, plastic, paint, pesticide, fertilizer, rubber, and roofing because of its distinctive properties. During the mining of deposits, low-grade inferior pyrophyllites are discarded and remain unutilized due to their inferior characteristics. Keeping in view, the scarcity of high-quality pyrophyllite ores, inferior-grade ores necessitate treatments to render them suitable for industrial utilization. The enhancement of pyrophyllite quality through beneficiation techniques thus becomes imperative for its potential applications across industries. For the present studies, tests on beneficiation have been done on Pyrophyllite sample origin from a deposit located in Anantapur district, Andhra Pradesh. The pyrophyllite sample contains high silica (SiO_2–60%), low alumina (Al_2O_3–19%), and iron oxide (Fe_2O_3–3.5%). Mineralogically, it contains quartz, pyrophyllite, and muscovite minerals. The beneficiation tests have been done by using various techniques such as flotation, and hydrocyclone. The preliminary outcome of these tests is promising, demonstrating a noteworthy reduction in silica (SiO_2%) and a concurrent increase in the alumina (Al_2O_3%) content. Pyrophyllite holds potential as an economically viable substitute for various minerals such as kaolinite, talc, and feldspar. The quality and price of pyrophyllite depend mainly on the content of Al_2O_3 and impurities across a spectrum of industrial applications. Therefore, pyrophyllite with high alumina content and low impurities is desirable in the industry. To enhance its market desirability, pyrophyllite should ideally exhibit a moderate concentration of alumina (Al_2O_3%), remain devoid of impurities, and

P. G. Bhukte (✉) · G. T. Daware · M. J. Chaddha · T. P. Bhosale · A. Agnihotri
Jawaharlal Nehru Aluminium Research Development and Design Centre (JNARDDC), Amravati Road, Wadi, Nagpur, Maharashtra 440023, India
e-mail: pgbhukte@jnarddc.gov.in

A. Agnihotri
e-mail: director@jnarddc.gov.in

K. Randive et al. (eds.), *Current Trends in Mineral-Based Products and Utilization of Wastes: Recent Studies from India*, Springer Proceedings in Earth and Environmental Sciences, https://doi.org/10.1007/978-3-031-50262-0_12

possess maximal whiteness. The present paper endeavors to enhance the quality of low-grade pyrophyllite through the application of physical beneficiation methods.

Keywords Pyrophyllite · Characterization · Physical beneficiation · Upgradation

1 Introduction

Pyrophyllite, a hydrous silicate mineral boasting distinctive attributes, emerges as a viable substitute for various clay minerals such as kaolin, talc, and feldspar, demonstrating its versatility across industries spanning ceramics, refractories, fiberglass, cosmetics, paper, plastics, pesticides, fertilizers, and roofing applications [1, 2]. It may also substitute kaolinitic minerals in many industries such as ceramics and fillers because kaolin minerals are fast depleted as well as costly. A paramount attribute of pyrophyllite resides in its inherent hydrophobicity, a trait instrumental in the production of diverse roofing products, encompassing sealing agents, waterproof materials, and asphalt felts [2]. The pyrophyllite resources in India, quantified at approximately 59.61 million tons, have become integral to a multitude of industrial processes. Of this resource, approximately 42% (25 million tons) are categorized as reserves, with a geographical distribution primarily observed in Madhya Pradesh (48%), followed by Odisha (23%) and Uttar Pradesh (13%). The remaining resources (16%) are situated in Andhra Pradesh, Maharashtra, Jharkhand, and Rajasthan (Fig. 1). In India, the consumption of pyrophyllite is in refractory, insecticide, ceramic, etc. Among these, refractory grade, insecticide grade & ceramic mixed grade are the principal industries (Fig. 2) [3]. Pyrophyllite's industrial applications span refractory, insecticide, ceramic, and more. Among these, the refractory, insecticide, and mixed-grade ceramic sectors (Fig. 2) hold substantial prominence [3]. A cornerstone of pyrophyllite's utility lies in its Al_2O_3% content and impurity constituents. Consequently, the industries demand a precise specification, emphasizing the grade of Al_2O_3 and the impurity quotient, for their specific operational needs. The quality and pricing of pyrophyllite are inextricably linked to the proportion of Al_2O_3% and impurities present. Noteworthy impurities include silica, iron oxide, and titania, exerting pronounced influence on the final product attributes [1, 4]. The presence of iron oxide and titania notably induce color variations in the end product, underscoring the criticality of assessing their content in pyrophyllite ores in alignment with industry-specific requirements. The strategic upgradation of low-grade pyrophyllite ores hinges on the dual objectives of raising alumina content and diminishing impurities, especially iron oxide, silica, and titania. The methods employed for this purpose exhibit variability contingent on ore and gangue mineral characterization. These methods encompass physical, chemical separation, and hybrid methodologies. The ultimate goal is to engineer pyrophyllite ores that satiate the stringent industry specifications pertaining to Al_2O_3 grade and impurity levels. In this paper, the physical beneficiation studies for the removal of silica (SiO_2%), iron oxide (Fe_2O_3%), and enrichment of alumina ($Al_2O_3\%$) in raw pyrophyllite are discussed. These beneficiation strategies are pivotal in shaping the trajectory of pyrophyllite utilization across diverse industrial sectors. The study underscores the intricacies of impurity influence, quality determinants, and the possible beneficiation methodologies to realize pyrophyllite's maximum potential as a sought-after industrial resource.

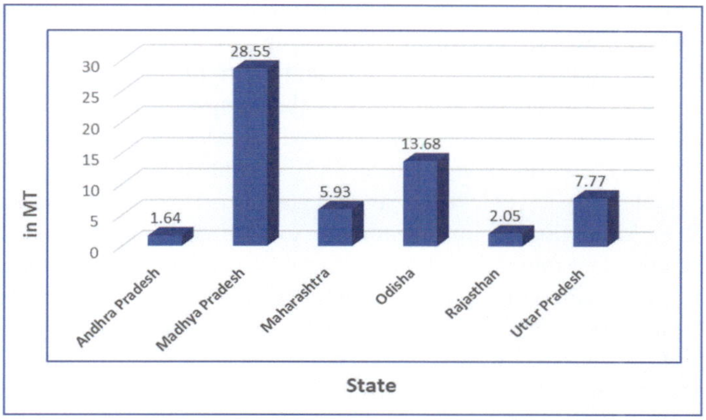

Fig. 1 Recourses of pyrophyllite (India)

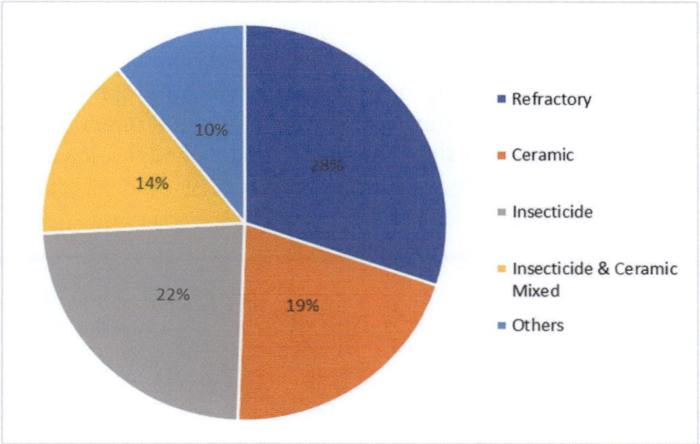

Fig. 2 Pyrophyllite consumption by different industries in India

2 Materials and Methods

The representative sample of Pyrophyllite obtained from the mining site underwent a crushing process, reducing it to a particle size of −15 mm using a jaw crusher, as illustrated in Fig. 3. To facilitate the subsequent characterization endeavors, a sample with a particle size of −74 microns was meticulously prepared through the employment of a universal mill or a Bond mill. This finely sized sample was then subjected to thorough homogenization using a homogenizer. The representative sample was acquired using the coning and quartering technique, ensuring its representativeness. The chemical analysis of the samples was conducted utilizing the wet chemical method. Notably, the Pyrophyllite exhibited distinctive characteristics, prominently featuring a high silica (SiO_2%) content ranging from 59 to 61%, coupled with relatively lower levels of alumina (Al_2O_3–19% to 20%) and iron oxide (Fe_2O_3–3.2%). Mineralogical investigations, executed via X-ray diffraction (XRD), unveiled the prevalent minerals within the Pyrophyllite. The analysis indicated that quartz and pyrophyllite constitute the dominant mineral phases, accompanied by other discernible constituents such as Chamosite, Muscovite, and Sudoite. A visual representation of the XRD diffractogram is furnished in Fig. 4. Upon a thorough mineralogical assessment, it became evident that the principal minerals within the pyrophyllite encompass pyrophyllite itself, quartz, muscovite, and tourmaline. Additionally, minor minerals were identified, including rutile, anatase, sphene, illite, and kaolinite, as observed in the research conducted by Das and Mohanty [5]. Our petrological investigations further corroborated the presence of pyrophyllite, quartz, and muscovite as the primary constituents, as depicted in Fig. 5.

Globally, high-quality pyrophyllite resources are very limited, however, there are abundant quantities of low-quality ores available. Since pyrophyllite is desirable in industry and may be an economical alternative to many minerals, it should be processed to cover the needs of the industry. Several techniques have been used to treat low-quality pyrophyllite ore and have proven their efficiency and economy.

Fig. 3 Pyrophyllite sample

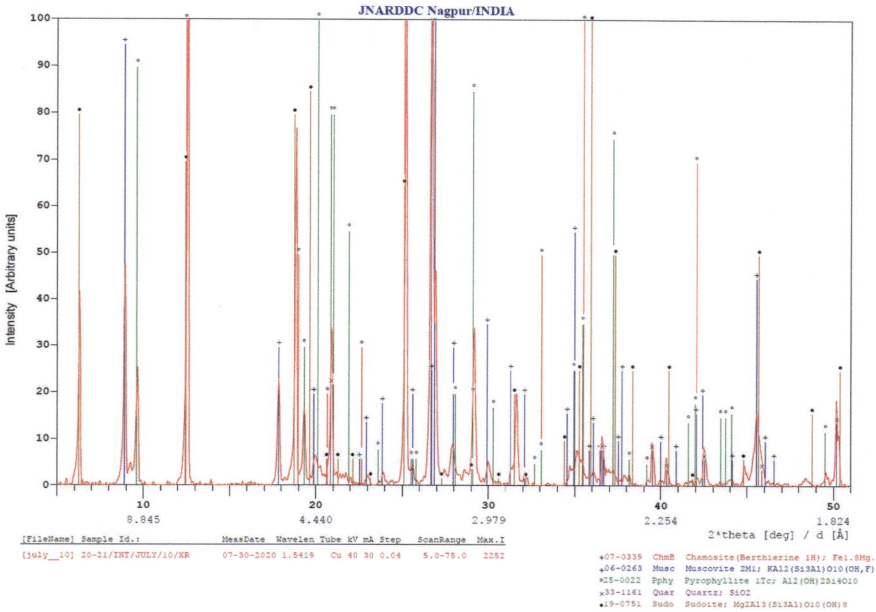

Fig. 4 XRD diffractogram of pyrophyllite sample

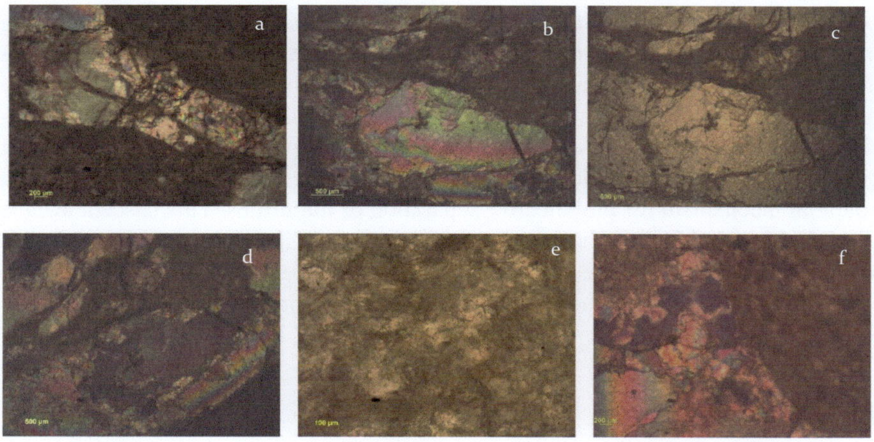

Fig. 5 Petrology of pyrophyllite sample. **a**, **b** The photomicrograph shows presence of pyrophyllite and muscovite associated with quartz, **c** morphology of pyrophyllite, **d** muscovite occur as elongated form intergrowth with quartz, **e** pyrophyllite mineral occur in radiating form, **f** presence of muscovite with quartz

These include physical separation techniques, flotation, and hydrocyclone. In this paper, these techniques will be discussed according to the contributions of researchers and the most important results obtained.

3 Beneficiation Studies

In order to align with the specific requirements of diverse industries, it becomes imperative to enhance the quality and luminance of pyrophyllite. This enhancement process entails the elimination of undesirable and pigmented minerals that might otherwise impede its optimal utility.

Physical Separation Techniques

Physical separation techniques play a pivotal role within the domain of mineral processing due to their remarkable efficiency and economic viability. These techniques are strategically designed to address the treatment of low-grade ores, with a primary focus on the eradication of impurities, most notably silica and iron, while concurrently fostering an enhancement in the alumina content inherent to the ore composition. Through an exhaustive exploration of pertinent scholarly literature, it becomes evident that notable strides have been undertaken in the pursuit of refining the quality of suboptimal-grade pyrophyllite. These endeavors have predominantly centered upon the utilization of physical separation methodologies encompassing magnetic separation, attrition-scrubbing, and flotation processes. Nevertheless, it is imperative to underscore that the realm of impurity separation through the agency of a hydrocyclone test rig remains an uncharted territory, devoid of comprehensive investigative endeavors hitherto. As such, an opportunity presents itself for pioneering investigations into the efficacy of employing a hydrocyclone test rig as a potential mechanism for the selective separation of impurities. This nascent avenue of exploration holds the promise of contributing novel insights and methodologies to the overarching landscape of pyrophyllite ore refinement, ultimately enriching the repertoire of viable techniques within the purview of mineral processing.

3.1 Flotation Studies

The primary aim of the beneficiation studies is to achieve the effective elimination of impurities, specifically silica, and iron oxide, while simultaneously investigating the performance of flotation processes for recovering pyrophyllite from its associated mineral mixtures. For this purpose, the chosen collector for the flotation investigations is Dodecylamine due to its proven efficacy in selectively attracting aluminosilicate minerals within specific pH ranges. Additionally, sodium silicate is employed as an auxiliary reagent in the process. The batch flotation experiments were conducted utilizing a Denver (Metso) D12 subaeration flotation machine equipped with a one-litre capacity. The core of these investigations involved the utilization of Dodecylamine, a cationic reagent, as the collector to facilitate the flotation of pyrophyllite. To initiate the process, the mineral samples underwent a series of preparation steps: initial comminution through jaw and roll crushers, followed by grinding in a laboratory ball mill. The flotation trials were performed subsequent to treating the mineral samples with the requisite quantities of reagents for predetermined durations. Throughout these experiments, parameters such as agitation intensity, pulp density, and pH levels were meticulously controlled and monitored. The pH of the slurry was adjusted before the introduction of the collector and was consistently observed throughout the experimentation period, with pH regulation achieved by the judicious addition of dilute hydrochloric acid (HCl). All the flotation trials were executed under a fixed pulp density condition. The outputs of the flotation process, namely concentrates and tailings, were meticulously collected, dried, weighed, and subjected to comprehensive analyses to evaluate the presence of distinct components, thereby enabling a thorough assessment of the final product's quality characteristics.

The flotation tests were carried out with the following parameters.

Size	:	−100, 200 mesh
Qty. of sample	:	500 gm +750 ml water
Solid slurry %	:	40, 20
PH	:	4.33, 4.44, 4.5, 4.53 (maintained by HCL 1:1)
Mixing of slurry	:	2, 3 min
Reagent	:	Dodecylamine (0.5, 1 ml, Condition Time 2 min, 3 min.) Sodium Silicate (2%) 3/8 ml (Condition Time 1, 3)
RPM	:	700–900
Flotation time	:	3 to 35 min

The findings of the study reveal a noteworthy alteration in the composition of the material. The application of dodecylamine reagent, coupled with an operational speed of 800 rpm, has led to a reduction in iron oxide content (Fe_2O_3–2–2.2%) and a concurrent enrichment of alumina (Al_2O_3–26–27%). However, the recovery rate of the processed material stands at 45%. Further investigation is imperative to establish precise parameters for the elimination of impurities. Regrettably, the outcomes pertaining to silica reduction are not promising, potentially attributed to the inherent characteristics of the original pyrophyllite and other contributing factors. The authors posit that an analysis of the crystal structure of pyrophyllite underscores its propensity for cleavage along specific planes. Owing to its relatively lower aluminum-to-silica ratio, pyrophyllite exhibits an enhanced responsiveness to cationic collectors. This could be attributed to the absorption of collector molecules on both internal and external surfaces, thereby facilitating improved pyrophyllite flotation in contrast to minerals like illite and kaolinite. It's pertinent to note that pyrophyllite flotation is significantly impacted by the extent of grinding. The observed decline in both the grade and recovery of Al_2O_3 can be rationalized by the influence of finely sized particles and the structural modifications that arise during the grinding process, a phenomenon previously documented by Erdemo-Glu and Sarıkaya [6]. Additional insights have been garnered from a study by Ali et al. [2], which underscores the impact of dry grinding on the flotation recovery of pyrophyllite utilizing an alcohol-based frother. The findings of this study suggest that prolonged dry grinding exerts an adverse effect on pyrophyllite flotation.

3.2 Separation Studies by Hydrocyclone Test Rig

The primary impurities present in pyrophyllite are notably elevated levels of silica and iron oxide. To effectuate the elimination of these impurities and to enhance the quality of the low-grade ore, an investigative initiative has been undertaken. This endeavor involves the utilization of a hydrocyclone test rig as a means of separation. A meticulously prepared pyrophyllite specimen, featuring a mesh size distribution of -100 $+150$ and -50 $+100$, has been meticulously crafted and procured for the express purpose of facilitating the ongoing research investigation.

The experiments were conducted on various parameters as below:

Size	:	-100, $+150$ mesh
Qty. of sample	:	3 kg
Solid slurry %	:	20
Pressure	:	5, 10, 15 PSI
Vortex finder	:	11.4 mm
Apex	:	3.2 mm

The findings indicate a noteworthy increase in the concentration of alumina (Al_2O_3%) to a range of 24–26%, up from the initial content of 19%. Simultaneously, there is a substantial increase in the content of silica silica (SiO_2%) to a range of 52–54%, down from the original 60%. The comprehensive retrieval of pyrophyllite falls within the scope of 50–60%. Further tests are required to ascertain the parameters.

4 Results and Discussion

Based on the outcomes of the conducted beneficiation studies, it has been observed that the application of physical beneficiation techniques is effective in diminishing the levels of silica and iron content. The utilization of beneficiation techniques offers the potential to enhance the number of ample deposits of pyrophyllite characterized as having low grades. The findings of these studies have demonstrated the substantial removal of impurities. Specifically targeting pyrophyllite designed for employment in non-metallurgical sectors, such as refractories industries, necessitates strict adherence to demanding physicochemical prerequisites and specifications, notably pertaining to constituents like silica, iron oxide, and alumina. The results obtained from laboratory-scale beneficiation tests provide evidence that the production of higher-quality pyrophyllite is feasible through this approach. However, it is imperative to conduct further comprehensive investigations to meticulously determine the techno-economic aspects of the entire process.

Acknowledgements The authors would like to express their sincere gratitude to Dr. Anupam Agnihotri, Director of JNARDDC for his unwavering encouragement and gracious permission to publish this research work. A heartfelt appreciation is extended to the authorities of M/s. Amoda Group of Industry for generously supplying the Pyrophyllite sample crucial for the investigations. Additionally, the authors would like to extend thanks to the dedicated staff of JNARDDC for their diligent efforts in conducting the laboratory work to this study.

References

1. Eng. Maaz A. Ali and Prof. Hussin A. M. Ahmed (2021); A Review of Low-Grade Pyrophyllite ore Upgrading Techniques; Arab Journal for Scientific Publishing (AJSP) ISSN: 2663-5798, 19–31
2. Maaz A. Ali , Hussin A. M. Ahmed, Haitham M. Ahmed and Mohammed Hefni (2021); Pyrophyllite: An Economic Mineral for Different Industrial Applications; Appl. Sci. 2021, 11, 11357, 1–18
3. Indian Bureau of Mines-IBM (2020); Indian Minerals Yearbook 2020 (Part–III: Mineral Reviews)
4. S.K. Jena, S. Singh, D.S. Rao and N. Dhawan (2015); Characterization and removal of iron from pyrophyllite ore for industrial applications (2015); Minerals & Metallurgical Processing, Vol. 32, No. 2, 102-110
5. B. Das and J.K. Mohanty (2009); Mineralogical Characterization and Beneficiation Studies of Pyrophyllite from Orissa, India; Journal of Minerals & Materials Characterization & Engineering, Vol. 8, No. 4, pp 329–338, 2009
6. M. Erdemo_glu and M. Sarıkaya (2002); The effect of grinding on pyrophyllite flotation; Minerals Engineering 15; 723–725

Limestone—A Review with Special Reference to the Iron and Steel Industry

Pradeep Kumar Jain and Gaurav Sharma

Abstract Limestone, Fluorite, and Calcium carbide are the materials used as flux in steel making. For steel making, a superior grade of limestone is required to control slag volume and improve productivity. As per Bureau of Indian Standards (BIS), the specifications of limestone for steel making are: CaO should be 53% minimum, MgO–1.5% maximum, SiO_2–1.5% maximum, total sulphur–2% maximum, and Alkalies–0.2% maximum. This grade is available in some quantities in Rajasthan, Himachal Pradesh, and Sikkim. Mostly, these resources are far away from the existing Iron and steel industries or plants, hence, movement from these areas involves high logistics costs. The steel industry imports this material from UAE, Oman, Malaysia, Vietnam and Iran. India has limestone reserves in Andhra Pradesh, Karnataka, Madhya Pradesh, Chhattisgarh, Bihar, Jharkhand, and Odisha but steel-grade limestone is limited to a few states only. The limestone requirement is about 100 kg per tonne of liquid steel produced. At 255 MTPA crude steel production, the requirement is 25.50 MT. An attempt has been made in the paper to discuss usage, applications, export & import, and demand & supply of steel-grade limestone in India.

Keywords Steelmaking · Superior grade limestone · BIS specifications · Imports

1 Introduction

The iron and steel industry are a cornerstone of modern infrastructure, providing the materials necessary for construction, manufacturing, and transportation. Within this industry, limestone plays a crucial role as a versatile and indispensable resource. Limestone, a sedimentary rock primarily composed of calcium carbonate, has long been recognized for its diverse uses across various industries. However, its importance in the iron and steel sector cannot be overstated. As a fluxing agent, limestone is added to the blast furnace during iron and steel production to remove impurities

P. K. Jain (✉) · G. Sharma
Indian Bureau of Mines, Nagpur, India
e-mail: pkjain3661@gmail.com

K. Randive et al. (eds.), *Current Trends in Mineral-Based Products and Utilization of Wastes: Recent Studies from India*, Springer Proceedings in Earth and Environmental Sciences, https://doi.org/10.1007/978-3-031-50262-0_13

and facilitate the formation of slag. This slag, a byproduct of the process, aids in the elimination of unwanted elements such as sulfur and phosphorus from the molten iron. Moreover, limestone serves as a vital raw material in the production of iron and steel. When combined with other materials like iron ore and coke, it undergoes a series of chemical reactions in the blast furnace, resulting in the production of molten iron. This molten iron is then further processed to obtain steel, a fundamental component in countless industries.

Beyond its role as a fluxing agent and raw material, limestone also contributes to the steelmaking process by adjusting the composition of the steel. By carefully adding limestone, manufacturers can achieve the desired chemical composition and properties of the final steel product, ensuring its suitability for specific applications [1].

In the wake of above, this chapter aims to delve into the multifaceted aspects of limestone's involvement in the iron and steel industry. By examining its applications, benefits, and impact on the production process, we can gain a deeper understanding of the significance of limestone in this critical sector. Through an analysis of existing literature, case studies, and industry practices, this study will provide valuable insights into the role of limestone in the iron and steel industry, shedding light on its importance and potential for further advancements.

Applications of Limestone in Iron and Steel Industry

Limestone has several applications in the iron and steel industry. Its versatile properties make it an essential component in various stages of the production process, contributing to the quality and efficiency of the final steel product [2]. Some of the main applications of limestone in the iron and steel industry as follows.

1. Fluxing Agent: Limestone is commonly used as a fluxing agent in the iron and steel industry. It is added to the blast furnace during the production process to remove impurities and facilitate the formation of slag. The slag helps in the removal of impurities like sulfur and phosphorus from the molten iron [3].
2. Raw Material: Limestone is used as a raw material in the production of iron and steel. It is combined with other materials such as iron ore and coke in a blast furnace to produce molten iron. This molten iron is then further processed to obtain steel.
3. Composition Adjuster: Limestone is also used to adjust the composition of steel. By adding limestone, the desired chemical composition and properties of the steel can be achieved. It helps in controlling the levels of sulfur, phosphorus, and other impurities in the final steel product.
4. Desulfurization: Limestone is utilized in the desulfurization process of iron and steel production. It reacts with sulfur compounds present in the molten iron to form calcium sulfide, which can be easily removed from the system. This helps in reducing the sulfur content in the final steel product, making it more suitable for various applications.
5. Slag Formation: Limestone aids in the formation of slag during the iron and steel production process. The slag acts as a protective layer on top of the molten metal,

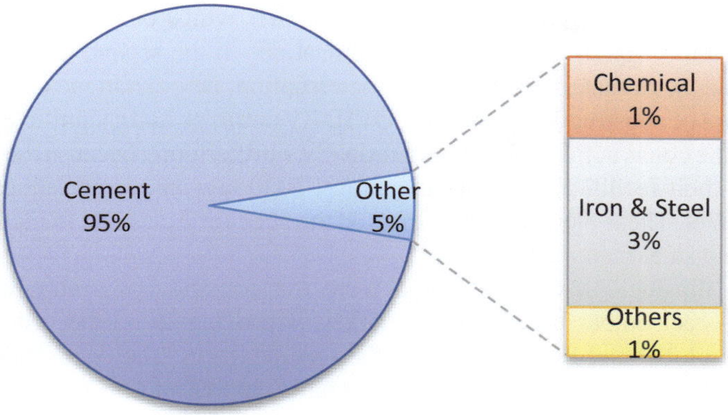

Fig. 1 Consumption pattern for limestone in India

preventing oxidation and heat loss. It also helps in the removal of impurities and enhances the overall efficiency of the production process.

About 95% of limestone is used by the cement industry, 3% by the Iron and steel industry, and 1% each by the chemical & other remaining industries. The consumption pattern of limestone in various industries in the country is depicted in Fig. 1:

2 Present Demand and Supply Scenario

Limestone is a widely used natural resource, and its demand is driven by various industries such as construction, agriculture, manufacturing, and the iron and steel industry, among others. The global demand for limestone has been steadily increasing over the years due to population growth, urbanization, and infrastructure development.

In terms of supply, limestone deposits are found in many countries around the world. Some of the major limestone-producing countries include China, the United States, India, Russia, Japan, Brazil, Germany, and Mexico. These countries have significant reserves and production capacities to meet both domestic and international demand. The supply of limestone is typically met through a combination of domestic production and imports. Countries with abundant limestone resources often have a surplus that can be exported to meet the demand in other regions. On the other hand, countries with limited limestone resources may rely on imports to fulfill their requirements.

The Global Limestone Market was valued at 780,426.91 kt in 2020, and it is expected to grow at a CAGR of more than 3% during the forecast period (2021–2026). The COVID-19 epidemic hindered economic growth in several nations, viz.

China saw repercussions in the first quarter of 2020, whilst other nations where the impact began in March had significant implications in the second quarter of 2020. Due to issues with material supply chain interruption, labor shortages, and the city-wide closure of construction sites, the COVID-19 outbreak had a significant negative impact on the construction sector internationally. Furthermore, because the producers shut down their facilities in European nations, 2020 saw an overall reduction in the output of steel. These elements have affected the need for limestone in industries like building and steel manufacturing.

On the flipside, increased demand from the construction sector, along with expanding steel output in the global market, is proving to be the primary driver in both the global and local markets in the medium term. The Asia–Pacific region is predicted to dominate the market and to have the fastest CAGR throughout the forecast period.

When using the blast furnace (BF) and basic oxygen furnace (BOF), about 270 kg of limestone is utilized to manufacture 1000 kg of crude steel. When an electric arc furnace is utilized, however, around 88 kg of limestone is needed to manufacture 1000 kg of basic steel. The graph for the production of crude steel versus apparent consumption of limestone is depicted in Fig. 2. Also, the consumption factor for limestone in the Iron and steel industry is given in Fig. 3.

India's economy is expanding, and there is a great prospective for growth in the future. India's growth rate decreased to 4.2% in 2019. As a result of the COVID-19 pandemic, the country's GDP growth is predicted by the IMF to be −10.3% in 2020 and 8.8% in 2021, subject to the post-pandemic global economic recovery. The country's growth rate has decreased as a result of the standstill brought on by flattening demand, greenfield investment, and industrial production. The majority

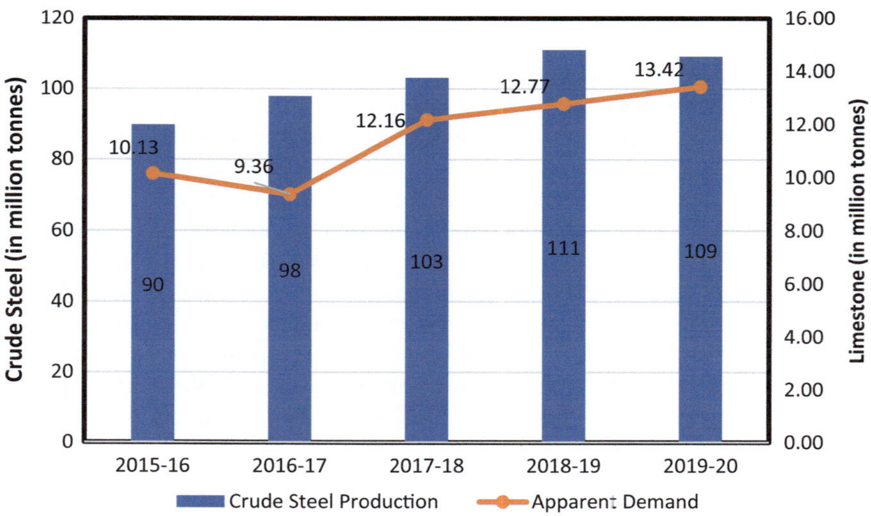

Fig. 2 Crude steel production versus demand of steel grade limestone

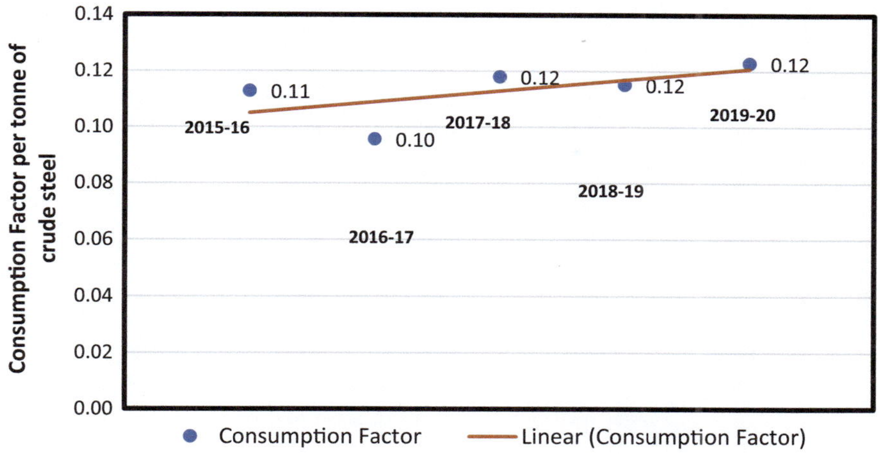

Fig. 3 Consumption factor for limestone per tonne of crude steel

of the commercial projects are likely to drive construction activities after the lock-down ends because they are either public projects with pre-budgetary allocations or public–private partnerships, such as smart city projects, even though the construction industry is predicted to slow down in the 2020–2021 fiscal year. Long-term expansion in the construction industry is thus anticipated to be rapid, which is anticipated to have a beneficial impact on the demand for steel and, in turn, for limestone. Eventually, these patterns should considerably increase demand for limestone. India is the second-biggest steel manufacturer in the world. The nation's steel output has been rising quickly due to the growing demand for the material from the infrastructure and automotive industries as well as government measures (such the National Steel Policy) that encourage the expansion of the industry. The country's steel output is projected to reach 255 million tonnes by 2031; as a result, the steel industry's requirements for limestone is anticipated to rise in the upcoming years. About 1.2% of the total production of limestone during 2020–21 was of BF & LD grade (both including) and the remaining 98.8% is of other grades. The grade-wise production during 2015–16 to 2020–21 is depicted in the Fig. 4.

The production, net import, and estimated consumption/apparent demand of steel-grade limestone during 2015–16 to 2017–18 in the country are depicted in Fig. 5. The apparent consumption of steel-grade limestone has been estimated on the date of production, export, and import of flux grade limestone (Apparent Consumption = Prod. + Import − export).

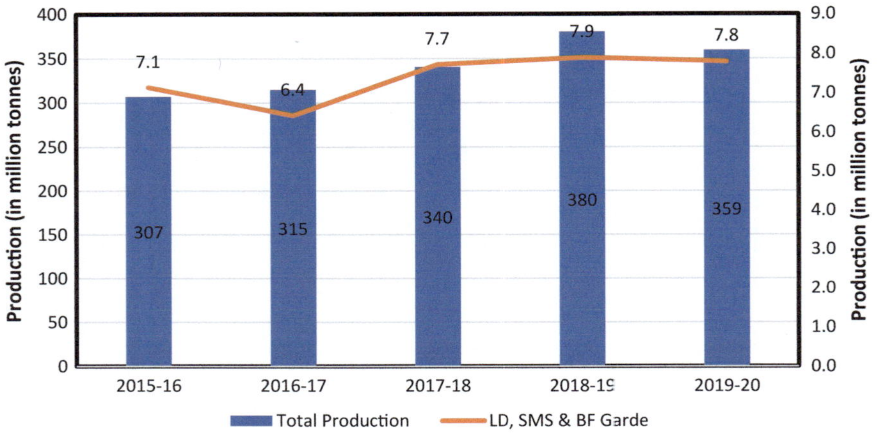

Fig. 4 Production of limestone

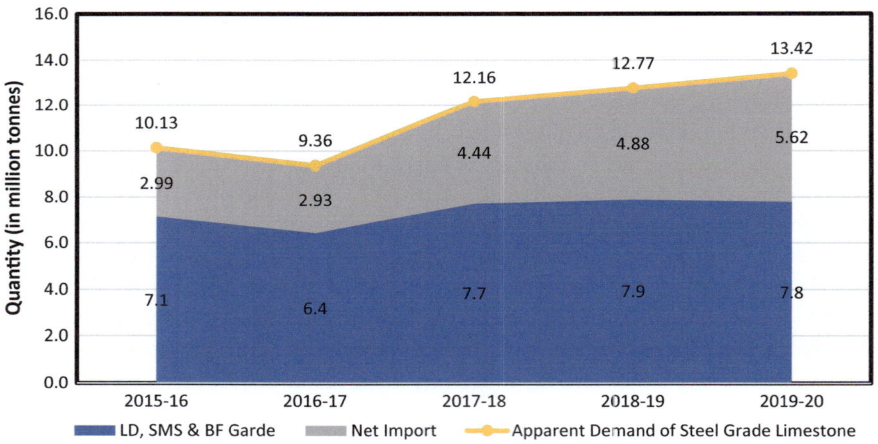

Fig. 5 Production, net import, and demand of steel grade limestone

3 Future Demand and Supply Scenario

The apparent demand for cement-grade limestone is growing at a CAGR of 7.3% during the last 5 years. If the growth continues to rise @7.3% in the future, then the demand for steel-grade limestone will be 19.06 million tonnes by 2025 [4]. The total requirement of limestone will be met through imported (40% of total requirement) as well as indigenous supplies (60% of total requirement). Here, it is assumed that limestone production will grow @ 7% on a YoY basis. The future demand projection of steel-grade limestone is depicted in Fig. 6.

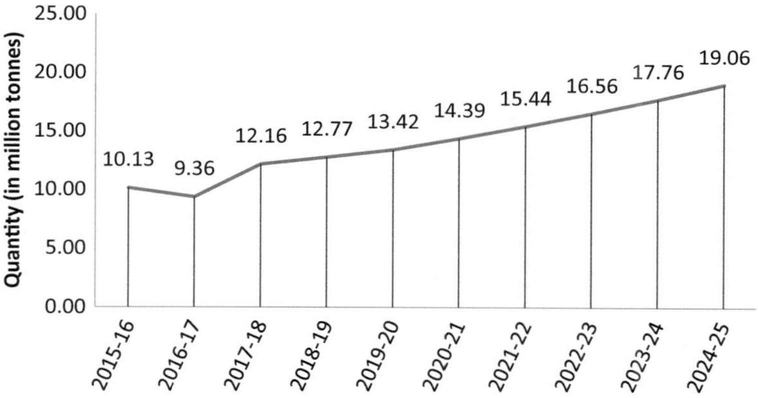

Fig. 6 Demand for steel grade limestone up to 2025

4 Future Challenges

- **Upscaling of exploration activities** in Greenfield exploration targets, as well as brownfield exploration/mining projects after ascertaining future demand by domestic Iron and steel industries.
- **New technologies** must be developed in order to transform the world's mineral resources into sustainable economic assets.
- **Exploring the avenues for substitution** of steel grade limestone in the Iron and steel industry.
- **Import substitution**, particularly for steel grade limestone which is not adequately available in the country.
- **Ensuring easy and abundant availability** of steel-grade resources from domestic sources.
- **Auctioning of more and more blocks for mining** with pre-embedded statutory clearances.
- **Demand for steel-grade limestone** could be significantly reduced by a range of 'circular economy' approaches.

5 Conclusion

Limestone is an essential component of the Iron and steel industry. It is used as a flux in steel making. India produces around 2% of the steel-grade limestone annually. The production of steel-grade limestone is around 7.5 million tonnes. However, the apparent demand in the country is around 11.5 million tonnes. There is a gap of around 4 million tonnes, which is met through import. Limestone is not a critical mineral in the short run but it may be critical in the long run due to the unavailability of steel-grade limestone, which is the backbone of the Iron & Steel industry.

References

1. IBM Report (2020) Limestone and other Calcareous Minerals, Indian Minerals Yearbook 2020, Part III: Mineral Reviews, 59th ed. Indian Bureau of Mines, Nagpur, India.
2. Satyendra (2017) Limestone-Its Processing and Application in Iron and Steel Industry, https://www.ispatguru.com/limestone-its-processing-and-application-in-iron-and-steel-industry/.
3. IS 10345 (2004) Flux grade limestone for use in Iron and Steel Making.
4. PWC Indian Steel Association (2019) Indian steel industry: Growth, challenges and digital disruption.

A Pragmatic Approach Towards the Exploration of Industrial Minerals

Alok Sood

Abstract The significant hidden potential of additional resource identification, characterization, and categorization of various grades of industrial minerals in India is well recognized. This is a major contributing factor hindering innovation and the discovery of new industrial minerals. The country, as a result, ends up importing high-end value-added mineral-based products. China clay, like many other industrial minerals finds application across various sectors of the industry like textile, rubber, paper, ceramics, and refractory, all of which require distinct properties. Various state and central government agencies have carried out exploration of China clays, however, that data and presentation need to be pragmatic keeping in mind the user's requirements. In most of the cases, only chemical composition is reported in exploration reports, which has very limited application and relevance. However, in practice, the data on industrial minerals must include their physical properties, presence of oxide phases, purity of content, etc. which could indicate their grade, specific applications, and therefore, their economic value. This is unlike the base metals, iron ore, limestone, etc., wherein the economic viability and the ultimate usage of the minerals are largely determined on the basis of chemical composition alone. This chapter suggests some measures for the exploration agencies in this regard.

Keywords Industrial minerals · China clay · Exploration

1 Introduction

The significant potential of additional resources of industrial minerals in the country is well-known to the professionals of the mineral industry. This potential is not only confined in the deposits yet unexplored, but also in the ones which have been partially explored. This is a result of gaps having inadvertently been left because of a lack of appropriate quality tests conducted, and a rather conservative approach adopted in

A. Sood (✉)
Mining Business Consulting, Jaipur, India
e-mail: sood.alok@ymail.com

K. Randive et al. (eds.), *Current Trends in Mineral-Based Products and Utilization of Wastes: Recent Studies from India*, Springer Proceedings in Earth and Environmental Sciences, https://doi.org/10.1007/978-3-031-50262-0_14

reporting potential resources in the early stages of exploration. The objective of this paper is to highlight these in order to stimulate innovative thought for the exploration agencies to recognize and fill these gaps in future exploration work. That will enable the industry to make quicker decisions in acquisition and development, rather than these deposits remaining idle for a long time or even getting discarded.

The two key elements mentioned above have largely hindered the development of some of the already explored mineral resources. Many industrial mineral deposits, despite having been explored in the past, still have gaps in their characterization, and categorization into various grades—a must for industrial minerals for identification of their potential use. This is one of the critical aspects largely ignored, in the traditional early-stage exploration of industrial minerals including China clay. Another key element is the potential quantity of the available resource which the exploration reports elucidate inadequately. This paper is an attempt to shed some light on how these two important aspects contribute to mineral deposits remaining idle or unexploited for long periods after exploration, and how these challenges can be mitigated.

2 The Impact of Issues

Most of the exploration for industrial minerals has been carried out by agencies like GSI, IBM, state mining departments, etc. The grade reporting is mostly limited to chemical analysis alone, without any or inadequate testing of physical properties which play a vital role not just in identifying whether the clay is china clay or not, but also in defining its grade and therefore the end use. The China clay finds use in multiple industries, e.g., textile, ceramics, rubber, paper, chemicals, electricals, paint, etc. However, the initial exploration hardly identifies properties that define the basic characteristics of these clays even for their broad categorization [1, 2].

How resource quantity predictions are made especially after the early-stage exploration which defines whether or not any further exploration/exploitation will be justified, is the other major issue at hand. Many China clay deposits of economic significance lie idle due to the fact that the initial estimates are unduly conservative. This situation is grim in India because there is no mechanism available for second-stage exploration of the already explored deposits of most industrial minerals because these projects do not fall in NMET funded category. Secondly, and more importantly, due to the under-reporting of mineral deposits and lack of information on their value-addition capability, the entrepreneurs find these ventures less lucrative.

This largely hinders the development of smaller deposits and also fails to attract innovations in new industrial mineral-based raw materials. The home country, therefore, ends up importing high-end value-added mineral-based products from other countries. It is just a matter of coordination within exploration and such research organizations that can provide support in the testing and characterization of China clay to properly unravel the true potential in early-stage exploration itself [3].

Therefore, there is a need to report appropriately during the early stage, shrugging off the extremely conservative approach of predicting the resources of these minerals. Reporting can be done in a forward-looking manner, rather than mere recording the data. Once this requisite is fulfilled, even the quantum of work done can be modified to suit the need for such reporting. The reporting by a geologist/an exploration agency has to be aligned to the business needs in order to contribute proactively to the growth of mineral-based industry, keeping in mind that just a standard approach in exploration for all minerals does not meet the end-user requirement, thereby defeating the purpose of exploration.

The objectives of the early-stage exploration for clay are:

(i) To establish what kind of clay exists in a particular area, as an example. Whether it is ball clay, china clay, fire clay, or bentonite? For that, apart from the chemical analysis, the physical properties such as plasticity, shrinkage, mineral composition, L, a, and b values, firing characters, etc. should also be determined for defining the type of clay, its quality, and possible end-use.

(ii) To report not just what has been established as a resource/reserve, but also report the possibility of addition to the reported resources with a hint of the anticipated order of magnitude of the blue-sky potential. This will have to be based on geological information, knowledge, and experience of the explorer, who must provide what might be termed as "resource potential" (not as reserve or resource, which has to be essentially reported by following strict professional reporting standards).

It is important because any business or an entrepreneur would look for the potential in a business and possible value addition without which many, if not most, mineral deposits look completely unattractive for investment. If the resource is sizeable and has a potential for value-addition, risks can be taken in business. A very conservative approach kills such interest entirely.

> Everyone can tell you the risk. An entrepreneur can see the reward—Robert Kiyosaki
>
> (He is an American investor, businessman, author, motivational speaker, and financial commentator who has a net worth of $100 million. Best known for a series of books called Rich Dad Poor Dad).

That is the mindset of an entrepreneur, and that is what every exploration geologist should understand, and learn to work with, in the best interest of the development of mining ventures, and also in doing justice to his role and career.

3 Understanding with Specifics

3.1 The Need and Justification for Comprehensive Testing

Why chemical analysis is not enough and physical properties also must be defined in order to define the type of clay being reported is exemplified by the clay deposits in Rajasthan, where the confusion regarding ball clay and/or china clay, and the issues with resource potential identified are clearly brought out [1, 3]. First of all, in many deposits, there is no specific mention of which type of clay is available in the area. Where such differentiation has been done, there is a lack of understanding of the characteristics of these clays. For example, china clay of residual origin has been mentioned as occurring away from the source and is identified as being "different from primary clays"! Individual clay deposits have not been specified whether they are China clay or ball clay, but only regional classification of clays has been provided like those of Ajmer and Udaipur divisions are China clay, for example [2].

Whereas ball clays are clearly of secondary origin, many high-grade China clay deposits are of primary origin (residual) e.g., China clay deposits of Jaipur and Sikar districts near Kotputli, where very high-grade and clean China clay deposits are found [3–6]. There are examples of secondary China clay deposits that are formed by the sedimentation process and are relatively not so clean, but they have good concentrations of Al_2O_3 and kaolinite. Some of these China clay deposits can be finer-grained and show good plasticity as well [2].

Tables 1 and 2 given below have been compiled for a very broad comparison between the physical and chemical properties of China clay and Ball clay, which are the two major varieties of clays most widely used for various industrial applications.

The illustration below shows the comparison of the grain size distribution in typical **China clay** and ball clay. One can clearly notice just one peak in **China clay** while there are two in ball clay, and the grain size distribution is distinctly different in both (Fig. 1).

Clays are generally recognized as very complex minerals not only in terms of their chemical and mineral composition but also their physical properties. Even in engineering applications clays are considered to be very complex and pose a lot of challenges to civil engineers when present on the project construction sites, especially dams and tunnels.

3.2 The Need and Justification for Optimism in Quantity Assessments

Assessment of quantities of deposits like clays can be pretty tricky considering the fact that exploration activities only occasionally involve any drilling. The state of Gujarat is an exception where the Commissionerate of Geology and Mining has conducted core drilling in many villages to constrain the dimensions on clay deposits. However,

Table 1 Key differences between China clay and Ball clay—physical properties [3–9]

S. No.	Characteristics	China clay	Ball clay
1	Purity	Unique for its purity of kaolinite content which is provided usually by high degree of leaching	It contains high amounts of quartz/ silica and mica along with kaolinite, often with higher TiO_2, CaO, MgO and Fe_2O_3
2	Grain size	Coarser	Finer
3	Plasticity	Non-plastic and lower shrinkage	Highly plastic and excessive shrinkage
4	Chemically	Chemically inert	Not so inert
5	Defining character	Defined on whiteness and brightness	Defined on plasticity and malleability
6	Application	Used as coating and filling applications in pigments, paper, paint etc.	Used in binding together in firing, and in ceramic applications as it provides shape and does not shake prior to firing
7	Composition	Kaolinite 20–80%, Quartz 12–55%, Mica 5–15%	Varied mineralogy include clay minerals
8	Plasticity	Low	High
9	L	92.58	73.05
	a	1.02	3.01
	b	3.33	14.28

Table 2 Key differences between China clay and Ball clay—chemical analysis [3–9]

S. No.	Parameter (%)	China clay	Ball clay
1	SiO_2	45.8	52.84
2	AlO_3	37.5	29.14
3	Fe_2O_3	0.4	2.7
4	TiO_2	0.55	3.43
5	CaO	0.24	0.64
6	MgO	0.29	0.48
7	Na_2O	0.25	0
8	K_2O	0.45	1.46
9	LOI	13.6	10.2

Tables 1 and 2 have been compiled from various sources just to provide a broad comparative idea only

resource estimates are to be made on the basis of surface observations and limited drilling only. Most estimates made in such a scenario are quite conservative. There are instances of several deposits where China-clay resources have been estimated to be a few thousand to tens of thousands of tonnes or at the most hundreds of thousands of tonnes. These estimates do not even justify further exploration with the ultimate

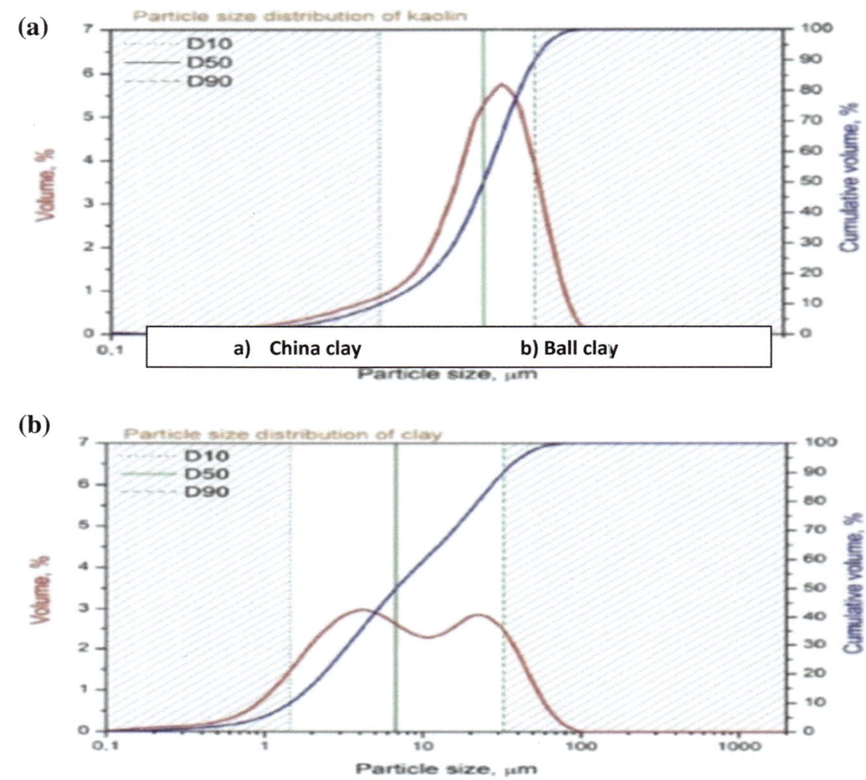

Fig. 1 Particle size distribution comparison of China clay and ball clay source of particle size curves [10]

objective of setting up of industry, let alone investment in R&D, which is so critical for high-end industrial mineral applications.

While the limitation for an exploration geologist is clearly understood, this also needs to be seen in light of the requirement of a decision on whether such extreme conservatism in reporting only what is safe to predict is serving the purpose. Further exploration and/or development of a mineral-based project is entirely dependent on how well the potential is understood, and how well it is presented in a comprehensive manner after the first-stage exploration itself. This is the challenge for a geologist to overcome.

It is true that the estimate one makes for the resources/reserves has to be made in line with prevailing guidelines that restrict forward-looking statements. But, on the other hand, there is also a need to keep an opening for the yet unknown/inadequately explored part of the potential to be understood and reported in a way to keep the interest alive, where there is such potential. At many places, in Rajasthan and Gujarat, the reported estimates of China clay resources/reserves are within thousands of tonnes. Such estimates not only present an inaccurate picture of the potential but also end up killing the interest of the business entities, and as such deposits become practically insignificant.

This is where a geologist needs to adopt a pragmatic approach by reporting resources/reserves only for the measured portion of the deposit, but at the same time, must report the blue-sky potential, largely based on the professional expertise and experience indicating the upside potential in terms of "resource potential". A suitable disclaimer can be put in explaining the background of the estimate made. The difference it makes is that it opens up the possibility of further exploration, where justified because a forward-looking estimate has been made based on what can be seen on the ground and what is possible based on the fundamental principles of geology. It is possible, especially in the case of industrial minerals because they are, in many cases, "rocks" actually, and not vein-type deposits of an irregular nature.

Once these fundamental principles behind "resource potential" are highlighted, and a statement is made clarifying that the "resource potential" is purely based on these fundamentals the forward-looking estimate is well justified. These projections will be found true in cases where no geological accidents (rare, but possible) have affected the mineral deposit.

For example, if a clay deposit is seen in a road, Nalla, other cuttings, or nearby existing pits/mines and similar geology is seen say for hundreds of meters with gaps of no exposure in between, it is safe to presume that the deposit is continuous and there is nothing that stops a geologist to consider that portion also in "potential resource". Such bedded deposits sometimes continue even for kilometers!

Even in the case of primary clay deposits which are a result of alteration of igneous/metamorphic rocks, sometimes the continuity of the source rock can be seen for hundreds of meters or even a couple of kilometers. Even if kaolinization has taken place in pockets the same set of parameters can be extended in the estimation of "potential resources".

The point of critical importance being made here is that if there is a significant resource potential number seen in the estimation, risks can be taken in business to invest in the exploration of such deposits. The risk vs. reward can be worked out provided there is a potential in the deposit. But if the geologist adopts a conservative approach out of lack of confidence/fear, he ends up defeating the very purpose of the exploration.

4 Case Studies

There are relevant examples where the adoption of a proactive approach has helped the concerned organizations in backing the judgment on potential resources and approving the exploration programs which led to major resource discoveries and large-scale expansions in operations. Names and details of some projects have not been disclosed due to confidentiality issues.

(a) While reviewing the stage of a gemstone mining project in Zambia, in 1997, it was advised by the company geology head that the deposit was hosted in metamorphosed ash beds with limited depth extension. The author having worked on the regional geology of the area differed with that opinion. The so-called "ash beds", were traceable for several km in strike length even outside the company's lease area, and going by the thumb rule, such rock formation could not have been restricted to only an anticipated 50 m depth, but should continue for several 100 m in depth. When this model was presented, the management could see the upside vs. the risk of a very small investment in exploration. The exploration program was approved and it ended up proving the model. That mine today is the world's largest precious gemstone mine of the size of a large metal open pit mining operation.
(b) There is a case from the industrial mineral sector that can be highlighted here. An industrial unit was undecided for several years on the question of expansion of its operations because of limited resources defined and envisaged. The potential of the deposit was unraveled after a careful study of the regional geology along with the mine information and a proposal was made for a large exploration program based on the prognostic model of the deposit, which was proven right by the drilling conducted. It doubled the reserves, and the resources swelled 5 times with less than half of the program completed within 1 year. All this was achieved despite tremendous opposition to the proposal in the management committee due to pre-existing notions based on a very conservative approach. The program was approved entirely based on the professional judgment of the possibility of a huge upside.
(c) In a recent project on another industrial mineral, an area believed to be hosting pocket-type deposits with small scattered occurrences has been proven to be a consistent one and of a large scale. This has opened up new possibilities for the business. The follow-up work is in progress as this information is discussed here.

5 Conclusion

Most industrial minerals are comparatively low-value minerals in their raw form. Mineral processing adds value to an extent and further R&D is required for real value addition which increases the value of these minerals multi-fold through the development of high-end application products. But if the potential deposits are available in limited quantities, like thousands or lakhs of tonnes only, it is nearly impossible to attract entrepreneurial interest. Such interest is generated only when there is long-term assurance of raw material supply from a mine. Coupled with this, the quality parameters of the industrial mineral in question should also be known. At least broadly defined in an exploration report, which hints at the potential availability for meeting the quality parameters for specific applications.

It is with this view, that it is recommended that inclusion of quality/relevant properties testing of industrial minerals must be included in all early-stage exploration reports, along with not just the reserves/resources but also the "potential resource" or the blue-sky potential, which is critical for entrepreneurial interest in these partially explored deposits. Inclusion of these requirements in the reports will go a long way in mitigating the challenge of many explored deposits remaining unexploited for a long term, and worse still, getting discarded simply because the real potential of the deposit could not be brought out.

Later stages of exploration can be more conservative and precise because by that time there is enough information to prove the quality and quantity more precisely, and also less likelihood of potentially hidden better quality and larger quantities.

References

1. Geological Survey of India (2001) Miscellaneous, Geology & mineral resources of Rajasthan, publication no. 30, Part 12, 2nd Revised edition.
2. IBM (2002) Monograph on China clay (Kaolin), Publication Cell, Mineral Economics Division, Indian Bureau of Mines, October 2002.
3. Geological Survey of India (1968) Bulletins of the Geological Survey of India, Series—A, Economic Geology, No. 29: Clays.
4. Department of Mines & Geology, Rajasthan (April 1960) Clays of Rajasthan.
5. Dashora Vijay, Shekhawat MS (2020) Particle Size distribution study of China clay of Bhilwara area, Rajasthan using Laser diffraction technique.
6. Geological Survey of India (2012) Geology & mineral resources of Gujarat, Daman and Diu, Miscellaneous publication no. 30, Part 14, 2nd Edition.
7. Geological Survey of India (2012) Geological & mineral map of Gujarat, Daman and Diu.
8. IBM (2021) Indian Minerals Year Book, 2020 (Part III: Mineral Reviews), Indian Bureau of Mines, Nagpur (July 2021) 59th Edition, Advance Release.
9. Industrial Extension Bureau—A Govt of Gujarat Organization (June 1996) Opportunities for value-added China Clay projects in Gujarat.
10. Dowuona Andrew, Yaya Abu, Nyankson E, Efavi Johnson, Damoah Lucas, Dodoo Arhin, David, Apalangya Vitus, Annan Ebenezer, Tiburu Elvis, Onwona-Agyeman Boateng, Tomiczek Błażej (2018) Investigating triaxial electrical properties of ceramic composites for electroporcelain insulators. Journal of Ceramic Processing Research. 19.

Import Substitution of Raw Material in the Refractory Industry

Suparna Basu, Sushavon Sarkar, Sumit Samanta, Santanu Basak, Saumen Sinha, and Indra Nath Chakraborty

Abstract The world has almost entirely relied upon China for the supply of refractory raw materials like dead burned and fused magnesia, tabular alumina, refractory bauxite, chromite, and graphite over the past decades. However, China's environmental crisis and shortage of high-quality mineral resources in India, have wreaked havoc on the raw material supply chain taking prices to newer heights. A major concern was the availability of low-iron refractory bauxite and the need for research on the commercially viable fusion of Indian magnesites. In order to cope up with this need, more number of composite synthetic materials like zirconia-mullite, silicon carbide, Al_2O_3–Cr_2O_3, MgO–CaO, MgO–Cr_2O_3, spinel, etc. were gradually getting used. Therefore, in order to remain competitive in the current scenario, there is an urgent need to enhance the domestic raw material production, development, and use of substitutes by way of continuous R&D and extensive brainstorming for recycling of refractory wastes in product formulations. In India, continuous efforts are being made to study the efficacy of indigenously developed new materials over traditional raw materials. In this chapter, we present one such study in which, descriptive laboratory tests were conducted before we had decided to explore this idea, and so far, there have been positive results pertaining to the product performance. In this chapter, we present an investigation carried out at Calderys India Ltd., in which the substitution of imported aggregates, by the indigenously developed aggregates, has been evaluated.

Keywords Recycling · Refractory · Substitution · Imported aggregates · Indigenous equivalents

S. Basu (✉) · S. Sarkar · S. Samanta · S. Basak · S. Sinha · I. N. Chakraborty
Calderys India Refractories Ltd., Nagpur, India
e-mail: suparna.basu@calderys.com

K. Randive et al. (eds.), *Current Trends in Mineral-Based Products and Utilization of Wastes: Recent Studies from India*, Springer Proceedings in Earth and Environmental Sciences, https://doi.org/10.1007/978-3-031-50262-0_15

1 Introduction

The global refractory industry including India, depends greatly on the import of key raw materials from China. India depends on China primarily for bauxite, magnesite, silicon carbide, etc. Dependence on Chinese raw materials have increased since the beginning of this millennium since high-quality raw material was made available at a lower cost. In the recent past, however, the situation has changed due to the policy changes imposed by the Chinese government, e.g., the imposition of high taxes on the export of refractory raw materials. In addition to this, upward price revision by the Chinese raw materials supplier has been too frequent. The changed logistic scenario makes the delivery time of the raw materials uncertain. Against this backdrop, it has become difficult for the Indian refractory makers to commit the price as well as delivery time to refractory users. Indian refractory makers have been forced to look into viable as well as long-term solutions to the raw material problem. In this paper, we highlight the actions taken by Calderys India to circumvent this issue. Since Calderys India primarily is an aluminosilicate refractories company, our focus in this paper would be related to Calcined Bauxite, Brown Fused Alumina, and Silicon Carbide.

2 Chinese Raw Material Characteristics

2.1 *Bauxite*

First of all, we discuss the characteristics of the Chinese bauxite. Table 1 reports the chemical analyses of the Chinese Calcined Bauxite, Indian bauxites that are closest to the Chinese bauxite quality [1]. Apart from the chemical features of these raw materials, the table also reports their Apparent Porosity (AP) and Bulk Density (BD) since they are important features and have a significant bearing on the properties of the refractories based on them as well as their performance during application in the field.

It is evident that the major shortfalls in the bauxite from Indian sources are their high Fe_2O_3 and CaO contents. Both of these constituent acts as fluxes in an aluminosilicate system owing to the presence of eutectic at as low as 1382 °C and 1265 °C

Table 1 Typical Properties of Calcined Bauxite from different sources

Bauxite source	Chemical analysis, %					AP %	BD gm/cc
	Al_2O_3	Fe_2O_3	TiO_2	CaO	SiO_2		
China	86.94	1.75	2.96	0.64	6.9	5.82	3.26
Highly pure Calcined Bauxite, domestic	89.43	0.95	3.1	0.51	4.6	8.2	3.27
Medium purity Calcined Bauxite, domestic	86.2	2.17	3.2	1.1	7.18	7.7	3.24

in Fe_2O_3 and CaO-containing aluminosilicate systems, respectively. The concentrations of these constituents, hence, should be kept at the lowest possible level. The presence of silica and titania in aluminosilicate systems, especially when CaO and Fe_2O_3 concentrations are low, is not an issue since they form desirable compounds, viz. Mullite and Tielite, respectively. These compounds improve the thermal spalling resistance of the products. Against this backdrop, it can be stated that the imported calcined bauxite from non-Chinese sources is competent as a substitute for Chinese bauxite in aluminosilicate formulations. Nevertheless, the dependence on Chinese raw materials will not come to an end completely. Therefore, the preferred choice would be to devise a solution through calcined bauxite from Indian sources.

2.2 *Silicon Carbide*

The other refractory raw material, that has been in focus, is Silicon Carbide. Silicon carbide is widely used in refractory formulations owing to its high thermal conductivity and low thermal expansion. These properties of silicon carbide make the refractories based on it or containing it resistant to thermal spalling. Apart from this feature of silicon carbide, it also improves the refractory performance in certain industry segments by effectively resisting the operating environment. The raw materials used for Silicon carbide production are cheap but its production is a highly energy-consuming process. Though India has high-quality raw materials for SiC production, energy cost makes its production expensive. Because of the high quality of Indian SiC, it is a preferred option for refractory formulations. However, the high cost of Indian SiC sometimes becomes a deterrent for its usage in certain refractory formulations. This has been the main reason for the Indian refractory industry to import silicon carbide, primarily from CIS countries and China. With the changed logistic scenario, the aim is to reduce the dependence on Silicon Carbide import. To tide over the situation, the Indian refractory industry has been working on incorporating recycled Silicon Carbide in their formulations [2]. Table 2 reports the typical characteristics of Indian as well as recycled SiC.

Table 2 Silicon carbide characteristics

Silicon carbide source	Composition, %					AP %	BD gm/cc
	SiC	Fe_2O_3	TiO_2	Al_2O_3	SiO_2		
Indian Silicon carbide	97.53	0.2	Trace	0.6	1	4.87	3.03
Recycled Silicon carbide	85.1	0.5	0.1	0.9	5	11.2	2.68

3 Actions by Calderys India

It is evident from Table 1 that one-on-one substitution of Chinese by Indian bauxite in any product formulation would alter the chemistry of the product, which unfortunately is the primary focus of the Indian refractory users. As explained earlier, the refractory properties of Indian bauxite-based formulation also would be inferior compared to those based on Chinese bauxite. Similar would be the case if Indian SiC is substituted by recycled SiC in refractory formulations.

During the present study, a 2-prong approach was adopted at the Calderys India. The R&D focus was only on relevant properties for specific applications, instead of the traditional ones, such as product chemistry and strength. Also, the product formulation concept was changed so that the relevant properties are not adversely affected. Products with different chemistry, thus, yielded similar performance even when they were based on different sets of raw materials. The other approach had been on the marketing front, where the Calderys India marketing team could convince the customer to focus only on the refractory performance instead of refractory properties traditionally monitored. This change had yielded a dividend. By incorporating raw materials from different sources in the product formulations, we could maintain the *on-field* refractory performance as well as contain the cost at the existing level. More so, in some cases, the product cost could even be reduced. In the next section, we discuss a few cases where the bauxite and SiC from more reliable sources have been used. The results obtained on the field reveal that products with differing "traditional properties" would yield equivalent performance. The section discusses a few cases highlighting the on-field refractory performances.

3.1 Blast Furnace (BF) Trough Castable

BF trough is Al2O3–SiC—C based castable. In this formulation, Chinese bauxite was replaced by Indian bauxite and also virgin SiC was replaced by the recycled one. Along with these changes, the concept formulation also was changed and the same had yielded dividends. Table 3 reports the characteristics of the BF, where the trial with these materials was conducted, and also the onsite performance of these materials. In the year 2019, the product was based on Chinese aluminous aggregate and Indian SiC. Part substitution of these constituents was made in 2020 and it is evident that the performance virtually remained unaltered. In the year 2021, high-grade materials were completely substituted by the "lower" grade raw materials, along with the product concept. The performance on the field virtually was comparable to the one in 2019. So, when the raw material is changed, along with the product formulation concept, the refractory performance can be maintained. While making these changes, the product supply to the customer not only remained uninterrupted owing to assured raw material supply from the indigenous source, the product cost

Table 3 BF Characteristics and the on-field Performance

Year	2019	2020	2021
BF capacity, m^3	2250	2250	2250
Slag volume, kg/ MTHM	400	410	470
Operating temperature, °C	1480–1490	1490–1500	1500–1510
Slag chemistry, %	CaO: 32–34, SiO_2: 32–34, MgO: 9–10, Al_2O_3: 18–19		
BFT castable lining thickness	Impact zone: 400 mm, non-impact zone: 300 mm	Impact zone: 400 mm, non-impact zone: 300 mm	Impact zone: 400 mm, non-impact zone: 300 mm
Major repair duration	96 h	96 h	96 h
Trough life, MT of hot metal	138 K with one minor repair	135 K with one minor repair	132 K with one minor repair

was also reduced by ~ 2% in 2021, compared to the cost of the product based on the Chinese bauxite.

3.2 *Aluminium Industry*

The aluminium industry is unique in the sense that its melting point is in the range of 900 °C. The exact melting point depends on the alloying element present in it. Still, it probably is the most difficult industry for the refractory owing to the high oxygen affinity of aluminium metal, the molten metal itself reacts with aluminosilicate refractories. Fall out of this reaction is "Corundum Build Up", primarily in the Belly Band area. The major requirement of the refractory is the prevention of corundum buildup. Both the raw material selection process as well as the product formulation concepts of the refractory is crucial for its acceptable performance in molten aluminium contact. Typically, calcined bauxite-based refractories are the preferred choice for molten aluminium contact.

In a product formulation of ours, for molten aluminium contact, Chinese bauxite was substituted by calcined Indian bauxite. Along with this change, the product formulation concept also was worked with. The altered product was used in molten aluminium contact in melting/holding furnaces and also in a ladle. Table 4 reports the characteristics of these units where the refractory was tried and also its performance.

The lives of Chinese bauxite-based refractories of ladle bottom and furnace hearth are 12 months and 5 years, respectively. Similar life has been obtained in the ladle and the appearance of the refractory in the melting/holding furnace shows that the targeted life would be achieved while using the refractory based on changed raw material. Above all, the corundum buildup, which is crucial for molten aluminium

Table 4 Characteristic and performance of the refractory with alternate raw materials in melting/holding furnace and ladle

Characteristics	Melting/holding furnace	Ladle
Capacity, MT	20	6
Operating temperature, °C	850	850
Other critical operating condition	None	Cryolite transfer into ladle from Pot Cell
Location where the refractory was tried out	Furnace hearth	Working lining bottom
Lining thickness, mm	300	200

contact refractory, has not been observed in the trial material. This implies that the refractory could prevent the attack of molten aluminum.

Similar to the previous case, the change of the raw material base not only ensured uninterrupted refractory supply to the refractory impact, but it also had a favorable impact on the cost. Depending on the exact quantum of Chinese bauxite substitution, the cost is reduced in the range of 6–7.1%, compared to the price of the product based on Chinese bauxite.

3.3 *Cement Industry*

Refractories of Preheater cyclones contain a fairly high concentration of SiC. Silicon Carbide is incorporated primarily to resist coating build-up on the refractory, which hinders the raw meal flow into the kiln. In case of severe build-up, the kiln operation needs to be stopped. Silicon Carbide quality as well as its concentration in the refractories for preheater cyclones is crucial for cement kiln productivity. In the recent past, Indian Silicon Carbide of desired quality became a major hurdle for the production of preheater cyclone refractories. The option, hence, was to explore the possibility of incorporation of recycled SiC in the product formulation. Since there are significant differences in the qualities of virgin and recycled SiC (Table 2), extensive work was undertaken by R & D for this change. The outcome has been a significant change in the concept of the formulation design itself.

The product was used in smoke chambers, riser ducts, lower cyclones, feed pipes, gas ducts, and calciner. During the period, when the product was based on purer Indian SiC, the fuel used in the kiln was imported and indigenous coal. While using the changed product, Pet Coke had been the major fuel. It is well known that Pet Coke contains significantly higher levels of Sulfur compared to coal and thus, the kiln environment gets affected. Such change in the working environment enhances the buildup tendency on the refractory surface.

It has, however, been observed that the recycled SiC-based refractory, of course with a different formulation concept, has worked equally well and has resisted the buildup formation. The life of virgin SiC-based refractory in these locations is typically 2 years. Similar life has also been obtained with the product based on recycled SiC. This change in the raw material base has reduced the product cost by over 13% when compared with the product cost based on the prevailing price of Indian SiC. In short, the usage of raw material from alternative sources retained the performance of refractory at the same level at a lower cost.

4 Summary

It is evident from the results of the present work that the substitution of imported refractory raw materials is a distinct possibility without sacrificing performance. It also needs to be emphasized that simple one-on-one substitution does not yield the desired performance. Extensive work is required by the R&D team and in some cases, the product concept itself needs to be changed so that the refractory performance remains unaffected after incorporation of raw materials from different sources. Thus, by eliminating the import of raw materials, the need for logistics will contract, lead times will come down, and environmental impacts will be minimized while at the same time, it will ensure that the Government's mission is fulfilled.

References

1. Ghosh A (2018) TIFAC Report on Bauxite Status, Challenges and Opportunities and Road Ahead, CGCRI, Kolkata, and JNARDDC, Nagpur.
2. Swain R, Padhy LN, Bhima Rao R (2011) Beneficiation Studies on Bauxite Mining Waste: A Value Addition for Refractory Industries, Iranian Journal of Materials Science & Engineering, 8(3).

R4 is the Key!

Gaurav Sinha, Satyaki Mandal, Bhavna Kumari, Santanu Basak, Saumen Sinha, and Indra Nath Chakraborty

Abstract Deposits of premium quality raw materials are depleting at a faster rate due to over exploitation and measures must be taken to preserve them for future generations. Calderys India Refractories Limited makes all the efforts to develop environmentally friendly technologies to save minerals that are on the verge of extinction. In the context of refractories R4 project targets to reduce virgin raw material utilization by reusing the dismantled lining material, developing a process of making the dismantled material suitable for refractory formulations, i.e. repair and finally repeat the process, viz. Recycle. Prior to the development of the R4 concept, dismantled refractories were used for land filling or road construction. The project aims to develop techno-economic solutions by producing high-performance greener products using materials salvaged from dismantled refractory lining. The project work involves identification of the alternative raw materials and their right source, performing extensive testing, and product reformulation. It also includes responsible supply chain management and upskilling the sub-vendors to increase their technical know-how and ability to supply the right grades of raw materials. This process conserves precious resources, requires less mining, reduces quantities ending up in landfill, and uses less energy. Discarded used refractories are basically of two types, bricks and monolithics. Used monolithic lining often has metallic elements embedded in it, which makes it difficult to reprocess compared to bricks. At the same time, monolithic products offer immense potential for the use of recycled refractories. The recovered refractory materials originate from diverse industries such as iron and steel, glass, electrical insulators, aluminium, etc. and hence it is critical to analyze and evaluate them for the presence of any adverse constituents to decide the right usage. In this way, R4 aims to find a balance between sustainability and profitability. Our consistent efforts have given us a CAGR for recycled material consumption of 10.8% for the period of 2015–2020.

Keywords Reduce · Recycle · Reuse · Sustainability · Raw materials · Monolithic refractory

---

G. Sinha (✉) · S. Mandal · B. Kumari · S. Basak · S. Sinha · I. N. Chakraborty
Calderys India Refractories Ltd., Katni, India
e-mail: gaurav.sinha@calderys.com

K. Randive et al. (eds.), *Current Trends in Mineral-Based Products and Utilization of Wastes: Recent Studies from India*, Springer Proceedings in Earth and Environmental Sciences, https://doi.org/10.1007/978-3-031-50262-0_16

1 Introduction

The literal meaning of "Refractory" is stubborn/headstrong. In the context of technology, however, refractories are construction materials used for the construction of kilns in furnaces. Conceptually the refractories bear similarities with the materials used for civil construction. Refractories, however, differ from the civil construction materials in terms of their resistance to chemical attack, mechanical stress, thermal shock, abrasion, etc. at elevated temperatures.

Refractories are required for the production of any synthetic material. In case a material is not available in nature, refractories are required for their production. Refractories, hence, are required for the production of Iron, steel, non-ferrous metal, petrochemicals, fertilizer, cement, glass, etc. Among them, steel is by far the biggest consumer of refractories. The refractory market growth, hence, is linked directly to the growth of the aforementioned industries. The global refractory market is expected to grow at a CAGR of 4.1% from 2018 to 2026 [1]. Indian refractory market growth is expected to beat the global trend and grow at a CAGR of 5.4% during the same period. Figure 1 shows the distribution of refractory products consumed in steel, glass, ceramics, non-ferrous, and other types of industries. As per data published by the Indian Refractory Makers Association (IRMA) in 2011, the steel industry is the biggest consumer of refractory products and accounts for about 75% [2, 3].

Refractories are usually classified based on their chemical nature, i. e. acidic, basic, and neutral. The chemical nature of the refractories decides the raw material base. For example, the refractories based on magnesia (MgO) and calcia (CaO) are categorized as basic, whereas silica is the only acidic refractory material. Chromite and carbon are truly neutral refractory but the products based on aluminosilicate, i.e. those containing Al_2O_3 and SiO_2 as constituents, also are categorized as neutral. Aluminosilicate refractories constitute the largest volume of the refractory market. Materials used for their production are magnesite, dolomite, clay, bauxite, sillimanite, chromite,

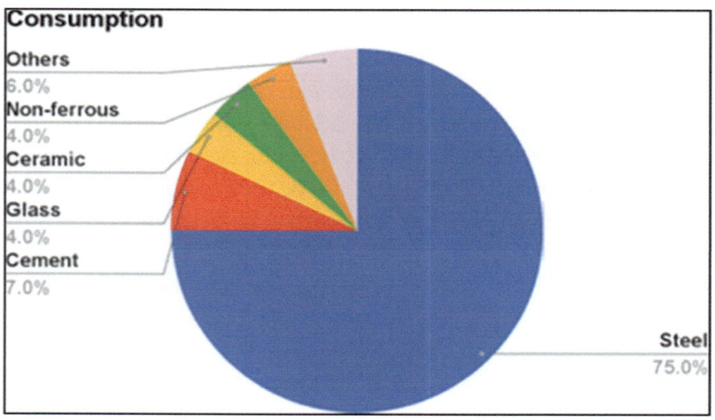

Fig. 1 Refractory usage pattern by different industries [2, 3]

graphite etc. Apart from these natural raw materials a wide range of synthetic materials, like, SiC, Si_3N_4 are also frequently used in refractory formulations.

India is blessed with large natural deposits of the majority of the aforementioned raw materials. Unfortunately their chemistry, however, does not conform to the requirement for refractory production. As a result, Indian refractory producers depend massively on imported raw materials for refractory production. Against this backdrop implementation of R4 concept, viz. Reduce—Reuse—Recycle—Repair, virtually becomes obligatory for reducing the dependence on import as well as preservation of the domestic natural mineral deposits. Apart from these aspects, the implementation of the R4 concept would eliminate the waste disposal problem to a great extent, and limit greenhouse gas generation attributable to the reduction of the calcination process, which is imperative for natural raw material to be useful for refractory formulations.

The aim of a circular economy is to ensure that the natural resources used for manufacturing of every product are fully recovered and re-injected into the economy without having to be extracted again. Such 'eco-design' currently generates a great deal of innovation and allows substantial savings. While replacing the natural raw materials with the recycled, it is important to remember the special properties they have and product performance is not affected. Recycling and reuse of the raw materials, thus, is the solution.

The project aims to develop techno-economic solutions by producing high-performance greener products using materials salvaged from dismantled refractory lining. The proposed methodology, thus, aims to take forward a balance between an environment and development.

2 Implementation Process of Used Refractories

R4, the sustainability project, aims to produce greener products by identifying the right source of material, designing the supply chain, and R & D to produce greener products. The R4 cycle (as shown in Fig. 2a) starts from raw material that has 2 sources: Virgin material from mines, which is associated with supply chain risk and environmental risk, and Used refractory. The consumed refractory, which is waste for customers, is processed by sorting and separated—classified into different grades. Low-grade material is used for landfilling and high-grade material is used for refractory manufacturing as a partial substitute of virgin raw materials. The material is characterized in the lab and intensive work for product reformulation is carried out to utilize them as raw materials. The eventual product can now be categorized as a high-performance green product.

High-supply risk raw materials are identified and targeted for replacement in the product formulations. According to the chemical and physical properties of the materials, the potential sources of recycled materials are identified. The representative samples are evaluated in Calderys India's laboratory. Once the material meets all the critical targeted parameters, a decision is taken regarding their procurement. More

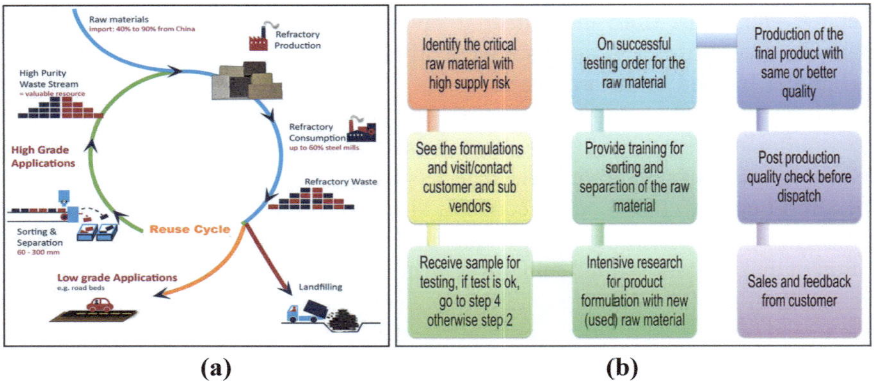

Fig. 2 **a** R4 cycle for refractory, **b** step-by-step implementation process

frequently than not extensive R & D activity is required for the incorporation of even acceptable recycled materials in the product formulations.

In case it is realized that pre-sorting is a requisite step, training sessions on sorting and separation from impurities are conducted at the vendor's premises. The usual steps for pre-sorted materials are crushing, screening, magnetic separation, color, and removal of impurities such as iron particles or fibres, slag coatings, etc. A Quality Assurance Plan is shared with the vendor to ensure product consistency. Once the recycled material is incorporated in the product formulations, initially its performance is monitored so that course correction can be made, if required. The pictorial representation of the above-described process is shown in Fig. 2b.

The supply chain of reclaimed materials, in India, is an unorganized sector and thus, faces challenges like a lack of continuous and assured supply by vendors. The solution for the first could be establishing direct contact with customers to obtain used refractory materials, and developing the vendor by exchanging the technical know-how. The other challenge is the inconsistent quality and cost competitiveness of reclaimed products. Segregating the impurities mixed refractories is a labour-intensive task. Also, the workforce may not have sufficient knowledge and technical skills for sorting. The major hurdle on this front is "Small-scale vendors" lack technical know-how as well as pre-processing facilities. In such a situation "Hand Holding" becomes mandatory and in such a situation Calderys India organizes frequent technical training of vendors.

With the manual sorting of small batches and implementation of QAP of each small lot, economically viable recycled material of the highest quality can be ensured.

3 Classification of Recycled Refractories

Recycling can broadly be classified into 2 categories, closed and open loop. Closed-loop recycling is an ideal situation since 100% of the used product goes back for production of a new version of the same product. Glass and aluminum products are good examples of closed-loop recycling. The yield of open-loop recycling, on the contrary, is as raw material for the production of inferior products.

At the first glance, the utilization of recycled refractories appears to be in the Open Loop Recycling category since the refractories are usually contaminated by slag and gaseous constituents of the kiln/furnace operating environments. Incorporation of such material directly in the same product formulation naturally would degrade its quality. This is where the process knowledge of the refractories user industry comes in useful. Based on the refractory user's process knowledge, the contaminated part is removed and the salvaged materials can be used in the same product formulation without sacrificing the quality. The material that has been removed, can be used as slag conditioner in steel plants or landfill. In this process, though 100% of the material is not reused but economically this appears to be a better option. The other option for used refractory is to adopt open loop recycling and use the same for inferior quality products.

Recycling refractories is one of the major options to counter the unpredictable raw material quality as well as supply from China. To cope with the emerging raw materials market trends, Calderys India put consistent efforts to increase the consumption of recycled raw materials as shown in Fig. 3. Though the recycled material consumption volume in 2020 was lower, compared to 2019 due to the pandemic, in terms % of the end product usage, reclaimed consumption in 2020 increased by 10% compared to that for 2019.

This is a long-term as well as continuous project. The project has been ongoing since inception and results are ever-growing. The generated revenue and customer satisfaction are the key success factors. The Supply Chain, Quality, Operations and sales, and R & D teams are the pillars of this project.

Figure 4a, b illustrate the consumption of various grades of used refractories from various refractory user industries. In this context, it is prudent to mention that bricks

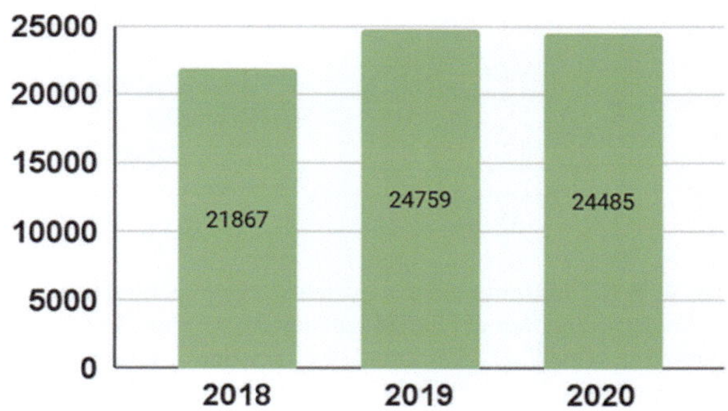

Fig. 3 Consistent increasing trend of consumption of reclaimed refractory materials from 2018–2020 by Calderys

are easy to sort and separate from impurities. Bricks are recycled from major contributors like Iron & Steel, Cement, etc. High alumina bricks find application in transport ladles and electric arc furnace roofs. Types of bricks, that are available in India and can be recycled conveniently, are Chamotte, sillimanite grog, insulating bricks, Alu-Cr grog, Mag-Chrome grog, magnesite bricks, Zirmul grog, 70% Alumina Mullite brick grog and other 50–70% Alumina brick scraps. Castables can also be recycled. The Iron and steel industry plays a major role in providing waste castables. They are recovered from Ferrous as well as Non-Ferrous industries where alumina ranges from 40 to 90%. Castables are primarily recovered from non-ferrous and DRI industries but Fe_2O_3 content is expected to be higher due to process effect. Other types of materials used in refractory manufacturing are spent alumina obtained from the Petrochemical and fertilizer industries, fly ash obtained from cement plants, roller scrap obtained from the tiles industry, slide gate grog, high alumina balls and tablets from non-ferrous & petrochemical industries, ZAC from the glass industry. An intensive search is being carried out to obtain suitable grade silicon carbide material. Tiles like Ceralin or SiC tiles are recovered from various industries. While adopting open-loop recycling, it is observed that < 1 mm size fraction of the recycled material is adulterated by slag and other contaminants, and hence, they need to be rejected, provided the targeted use of the recycled material is the same product. The < 1 mm yield, hence, needs to be utilized for other purpose [4].

It can be observed from Fig. 4c that the consumption of insulators and tiles has increased since 2018. This can be attributed to changes in the product mix. Whereas consumption of recycled castables has remained almost the same over the same period. Increased focus on refractory recycling has enabled to reduction of waste as well as greenhouse gas generation, reduced production cost, better utilization of limited domestic natural mineral deposits, and above all reduced dependence on unreliable imports from China.

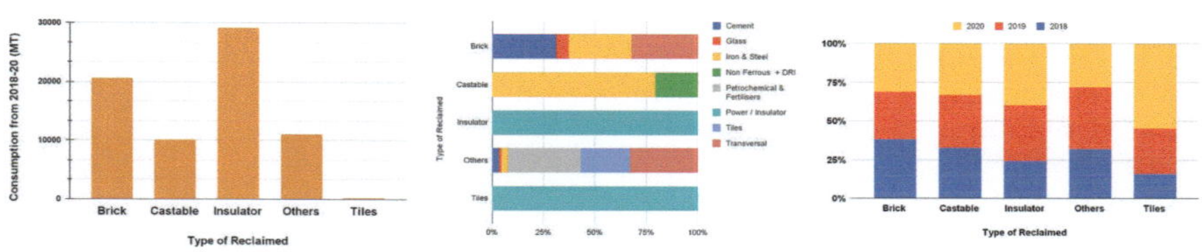

(a) (b) (c)

Fig. 4 **a** The consumption of recycled bricks, castables, tiles, insulators and others from 2018–2020 by Calderys, **b** different types of reclaimed materials like bricks, castables, insulators, tiles collected from from cement, glass, iron and steel, non-ferrous and DRI, Petrochemical and Fertilizers, Power/ Insulators, iles, Transversal industries by Calderys, **c** distribution of different types of refractory materials consumed in 2018, 2019, 2020 by Calderys

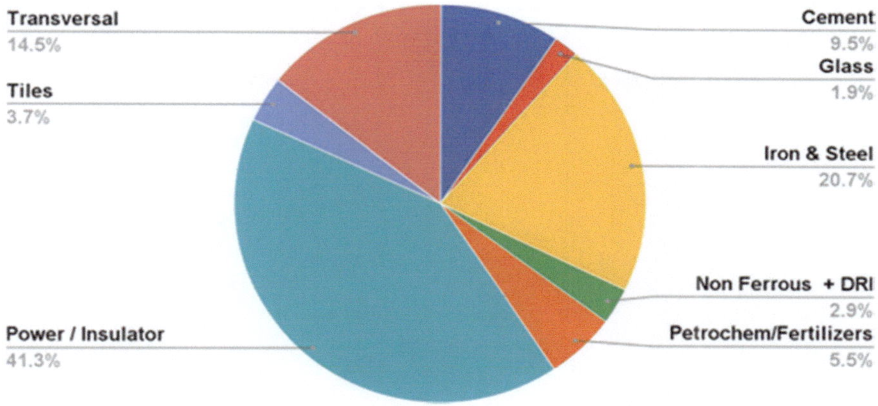

Fig. 5 Distribution of different industries contributing to the supply of reclaimed materials

As inferred from Fig. 5, power/insulator industries have the highest contribution to the supply of waste refractory materials. Penetration by alkali and sulfate compounds is commonly observed in the used refractories from glass, cement kiln, and lime industries. It is well known that these penetrated contaminants act as flux in aluminosilicate systems. Separation of the aforementioned penetrated contaminants from the used refractories is not possible by simple mechanism and thus, the utility of used refractories from glass, cement, petrochemicals, and fertilizers is not very significant.

As shown in Fig. 6, the collection of recycled refractories from Western India is the highest and it contributes 51.3% to Calderys India's basket. Whereas central, northern, and eastern India contribute 27.2%, 5.5%, and 3.1%, respectively. In the present scenario, hardly any material is procured from the southern zone.

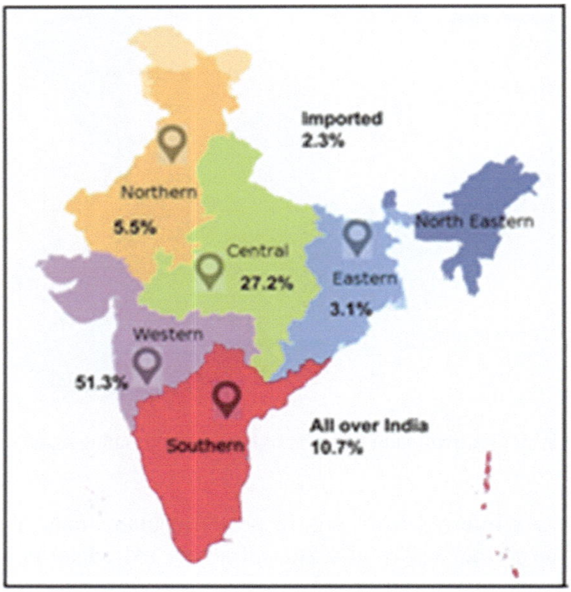

Fig. 6 Distribution of reclaimed refractory materials generated from different parts of India and imported by Calderys

4 Conclusion

The ever-increasing demand for refractories has in turn increased the demand for raw materials. On the contrary, the availability of raw materials has become scarce owing to the depletion of local natural sources. As a result, refractory makers have relied on synthetic as well as imported raw materials. Either of these actions makes the product economically non-viable as well as brings uncertainty due to changes in the regulations of exporting countries. Against this backdrop, R4—Reduce, Reuse, Recycle is seen as one of the viable solutions and the correct implementation of the process has refractory producers.

R4 holds the very concrete promise of a significant decoupling of production from the exploitation of natural resources. Refractory consumers, against the stricter environmental regulations, have found a sustainable way to dispose of the waste refractories. The primary advantages have been less abundant materials are conserved and mining activities have been reduced considerably. In this way, we protect forests and prevent soil erosion, lower energy consumption, reduce greenhouse gas emissions, and prevent dust and land pollution.

R4 is implemented by identifying the "at risk" raw materials. The procured recycled materials, which are considered "waste" by the refractory users, are evaluated for their suitability in various product formulations. Needless to mention, the "as available" used refractories cannot be used in refractory formulations. A well-planned QAP, which involves pre-cleaning work like sorting, separation crushing, etc., needs to be implemented to increase their utility.

The training sessions are conducted for vendors as well as their workmen to assist them in the above activities. In this way, the project aims to develop techno-economic solutions by producing high-performance greener products using materials salvaged from dismantled refractory lining. Owing to the implementation of the time-tested process, the internal consumption of reclaimed materials, in Calderys India, has increased significantly over the years. This has helped Calderys India to contain the product cost. R4, thus, has numerous benefits, including sustainable development, and a win–win situation for all.

References

1. Persistence Market Research (2018) Global Market Study on Refractories: Monolithics & Other Unshaped to Remain Most Attractive Product Form Through 2026, Persistence Market Research, June 2018, PMRREP3510.
2. Gochhait, Dr. Saikat. (2013) Role of ICTs in the marketing of Refractory Products as Innovative Success Factor, published in a refereed journal, Herald Journal Marketing and Business Management (ISSN Online: 2350–2207), 2 (3), 081–088 September, 2013. Herald Journal Marketing and Business Management 2350–2207. 2. 081–088.
3. IRMA (2005) IRMA Journal. Indian Refractory Makers Association, Kolkata, XXXVIII, No.3, Sep 2005.
4. Horckmans Liesbeth, Nielsen Peter, Dierckx Philippe, Ducastel Antoine (2019) Recycling of refractory bricks used in basic steelmaking: A review. Resources, Conservation and Recycling, 140, 297–304. https://doi.org/10.1016/j.resconrec.2018.09.025

Recovery of Placer Industrial Minerals from Red Sediments of Badlands Topography and Its Value Addition

Raghupatruni Bhima Rao

Abstract Red sediments of Badlands topography are a typical recent deposit containing valuable Industrial Minerals. These minerals include titanium bearing minerals, zircon, monazite, and sillimanite. Garnet is free from these red sediments and quartz is present as gangue mineral. At first instance recovery of individual industrial heavy minerals is made by desliming the sample followed by spiral concentrator to reject the quartz minerals. The heavy minerals sand subjected to wet magnetic separation where ilmenite and monazite are recovered as magnetic minerals. These magnetic minerals further subjected to dry electromagnetic separator to recover ilmenite as conducting mineral and monazite as non conducting mineral. The grades of these minerals are around 98–99%. The non-magnetic minerals are again subjected to gravity separator using wet gravity tables to reject the excess of quartz present in the non-magnetic fraction. The non-magnetic heavy minerals contain sillimanite, rutile, and zircon. Sillimanite was recovered by flotation. The flotation tailings contain zircon and rutile which were separated by electrical conducting separation where rutile as conducting fraction and the zircon as non conducting fraction. The grades of these minerals are around 98–99%. Recoveries of individual mineral flowsheets with mass balance are presented in the present investigation. Except monazite, all minerals are characterized. The value addition process of ilmenite to titania and iron slag by microwave heating source, sillimanite to mullite using microwave heat energy and zircon grindability using planetary mill are presented here.

Keywords Red sediments · Badlands topography · Ilmenite · Zircon · Monazite · Rutile · Sillimanite · Mullite · Titania · Iron Slag · Microwave heat energy

---

R. B. Rao (✉)
Formerly CSIR-IMMT, Bhubaneswar, India
e-mail: dr.r.bhimarao@gmail.com

Mineral Processing Division, Formerly CSIR Institute of Minerals and Materials Technology, Bhubaneswar, India

K. Randive et al. (eds.), *Current Trends in Mineral-Based Products and Utilization of Wastes: Recent Studies from India*, Springer Proceedings in Earth and Environmental Sciences, https://doi.org/10.1007/978-3-031-50262-0_17

1 Introduction

India has the largest coast both eastern and western side along with commercially exploitable industrial minerals. Red sediments are also potential resources for industrial minerals. These sediments existed all along the coast in different parts of the country specially Tamilnadu, Andhraprasesh and Odisha. In Tamilnadu it is known as "*teri sands*" as where in Andhra Pradesh it is known as "*erra dibbalu,*" and in Odisha it is known as "*Nalli sediments*". Since ilmenite contains iron along with titanium, it is highly magnetic, can be separated by magnetic separators from other minerals. Ilmenite is mainly used for synthetic rutile. The other uses are for paints, pigments, paper and plastics and fertilizer industries. The review of literature reveals that a few researchers have studied on estimation of reserves, characterisation, and recovery of ilmenite and value addition [1–14]. Recoveries of individual mineral flowsheets with mass balance have not been attempted. The value addition process of ilmenite to titania and iron slag by microwave heating source, sillimanite to mullite using microwave heat energy and zircon grindability using planetary mill are also not attempted by many researchers. The present paper reviews' the recent contributions in this field.

2 Experimental

2.1 Materials

Representative samples were collected all along the coastlines of red sediments from Andhra Pradesh, Odisha, and Tamil Nadu (Fig. 1). Figure 2 shows the topography of red sediments and its size and mineral assemblage. The beneficiation studies have been carried out to recover ilmenite and sillimanite and subsequent value addition of these two minerals.

2.2 Characterization

The size, sink-float and mineralogical analyses were carried out on the representative samples. Total industrial minerals were estimated using organic liquids of 2.89 specific gravity. Mineralogical modal analysis was carried out using a Lecia petrological optical microscope. X-ray diffraction (XRD) for assessing minerals has been used in the present investigation.

Fig. 1 Occurrence of red sediments all along the eastern coast line of India

2.3 Recovery of Placer Heavy Minerals

At first instance, the sample was thoroughly subjected for attrition and separated slimes in a hydro-cyclone. The sand was fed to spiral concentrator to recover total heavy minerals. The heavy minerals sand subjected to wet magnetic separation where ilmenite and monazite were recovered as magnetic minerals. These magnetic minerals were further subjected to dry electrical separator to recover ilmenite as conducting mineral and monazite as non conducting mineral. The non-magnetic minerals are again subjected to gravity separator using wet gravity tables to reject the excess of quartz present in the non-magnetic fraction. The non-magnetic heavy minerals contain sillimanite, rutile, and zircon. Since the sillimanite is in large quantity, flotation of sillimanite was carried to recover it in preparation of flotation froth fraction. The value addition process, ilmenite will be used for making titania, iron slag sillimanite bricks, and mullite by using microwave heating source energy. The plan of experiments is shown in Fig. 3.

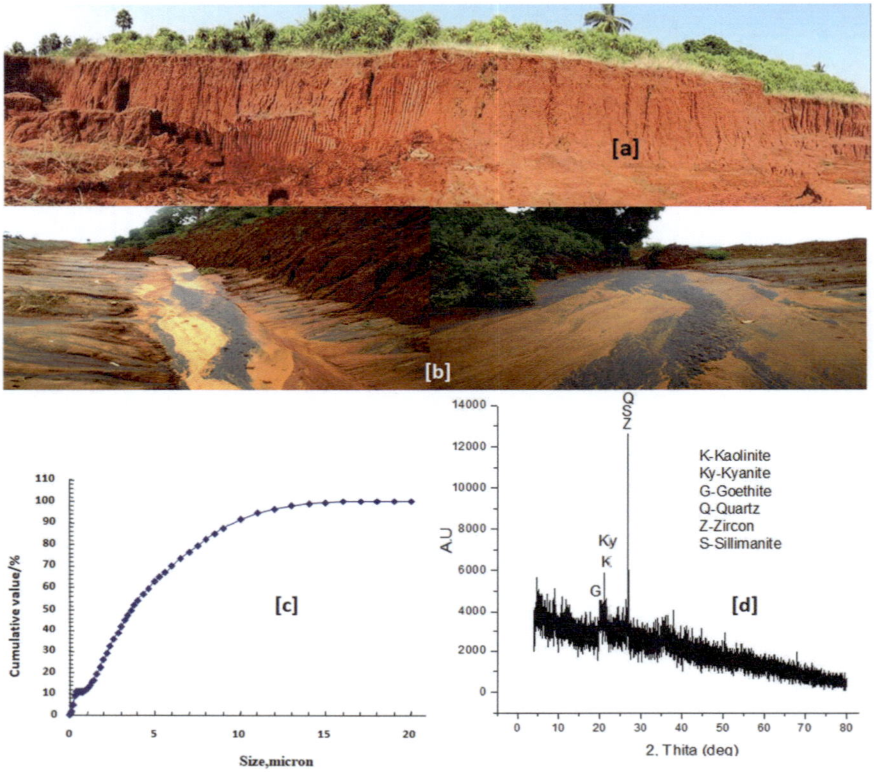

Fig. 2 Red sediments topography and its physical and mineralogical characters. **a** section showing red sediments; **b** release of placer minerals during rain; **c** size analysis of red sediments slimes; **d** XRD data of red sediments

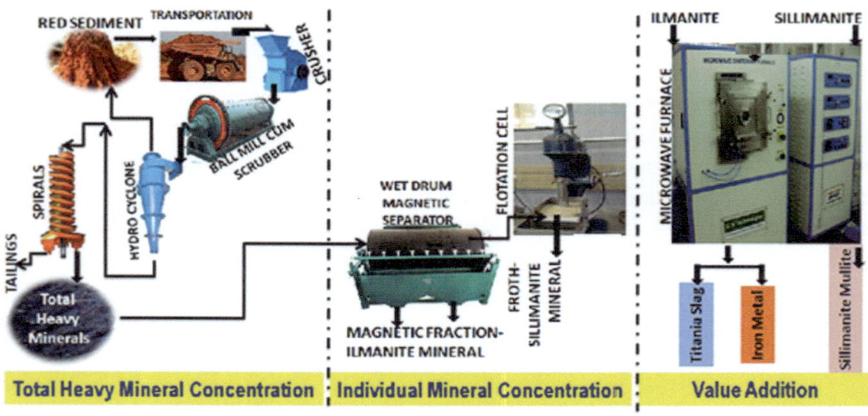

Fig. 3 Schematic diagram exhibiting red sediments from mine to metal

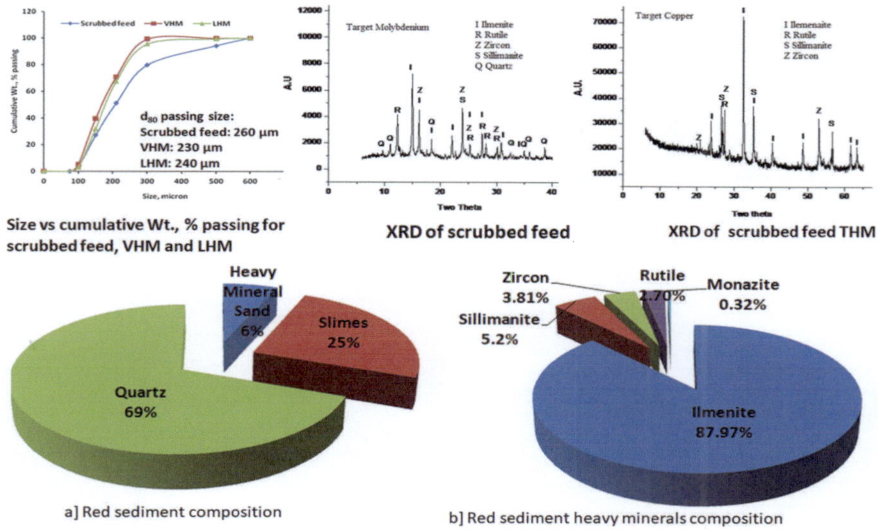

Fig. 4 Typical textural, XRD and mineralogical mode analysis of red sediments of India

3 Results and Discussion

3.1 Characterization Sample

The deslimed feed, very heavy minerals, and light heavy minerals, XRD of scrubbed feed, total industrial minerals, and mineralogical modal analysis of the sample are shown in Fig. 4. The data indicate that the slimes were accounting to 25% in which 6% were heavy minerals. It was further observed that 95% very heavy minerals [VHM] and 5% light heavy minerals [LHM] were present in the 6% of total heavy minerals. It was also noted that the total magnetic minerals were 88% and the rest were non-magnetic minerals. The size analysis of the scrubbed feed indicates that the d80 passing size for scrubbed feed is 260 μm, where as VHM exhibits 230 μm and for LHM is 240 μm. The mineralogical modal analysis data indicate that the sample contain 87.97% ilmenite. The other heavy minerals include 5.2% sillimanite, 3.81% zircon, and 2.7% rutile.

3.2 Ilmenite Recovery

The THM was recovered from multi stage spirals contain 98.2% THM with 88.6% recovery and 5.5% yield (Fig. 5).

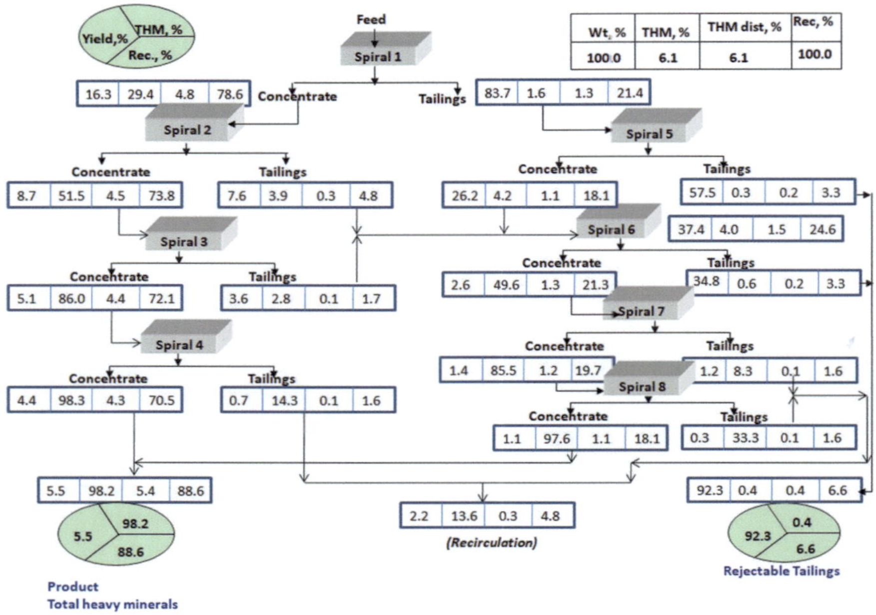

Fig. 5 Flow sheet with material balance on recovery of total heavy minerals from red sediments

By using judicious combinations of spiral, WHIS and HTS the ilmenite has been recovered (Fig. 6).

It can be seen from the flowsheet presented in Fig. 6 that by using spirals, a product recovered contains 98.2% total industrial minerals out of which ilmenite is 87.97% and 88.6% recovery. The spiral product was subjected to WHIMS. The results of wet magnetic separator reveal that a magnetic product contains 98.6% ilmenite with 99.6% recovery. In order to enrich the ilmenite content, the ilmenite concentrate has been further subjected to HTS. The ilmenite concentrate obtained from HTS contains 99.3% grade and 99.5% recovery.

3.3 *Beneficiation Studies for Recovery of Sillimanite*

In any mineral separation plant, sillimanite is recovered as by product of ilmenite concentrate. In this process, the ilmenite is recovered by using spirals, magnetic and electro-static roller separators. The non conducting fraction contains zircon and sillmanite. The sillimanite is being recovered by flotation process. Recovery of sillimanite is shown in flowsheet development with material balance in Fig. 7. The product obtained contains 53.4% sillimanite grade has been subjected to gravity

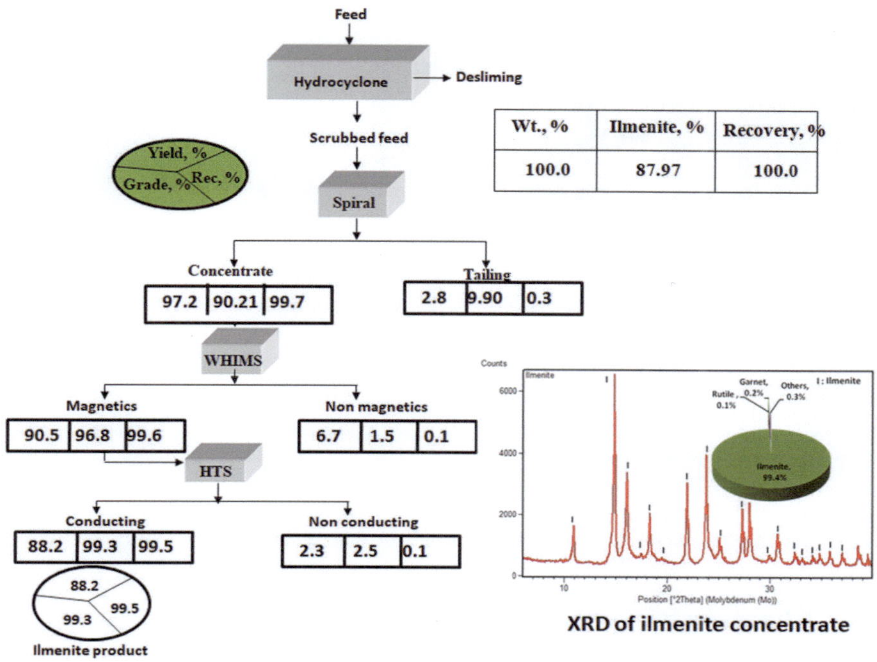

Fig. 6 Flow sheet with material balance on recovery of ilmenite from red sediments

tables followed by flotation. The sillimanite concentrate obtained achieved 96.9% grade and 67.1% recovery.

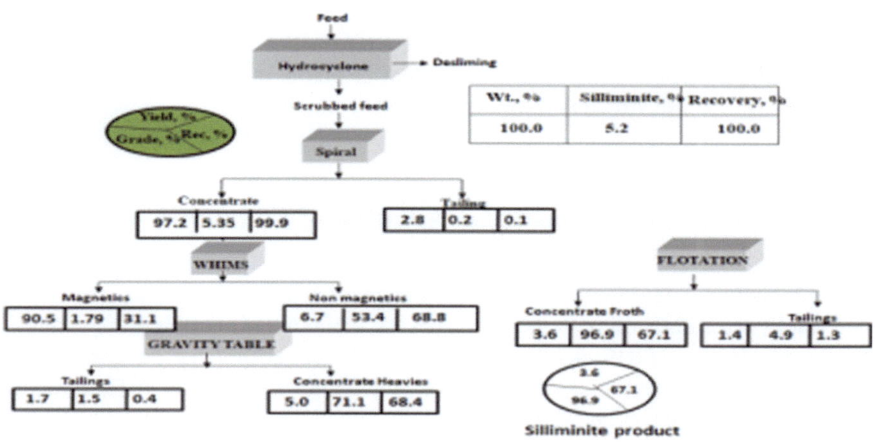

Fig. 7 Flow sheet with material balance on recovery of sillimanite from red sediments

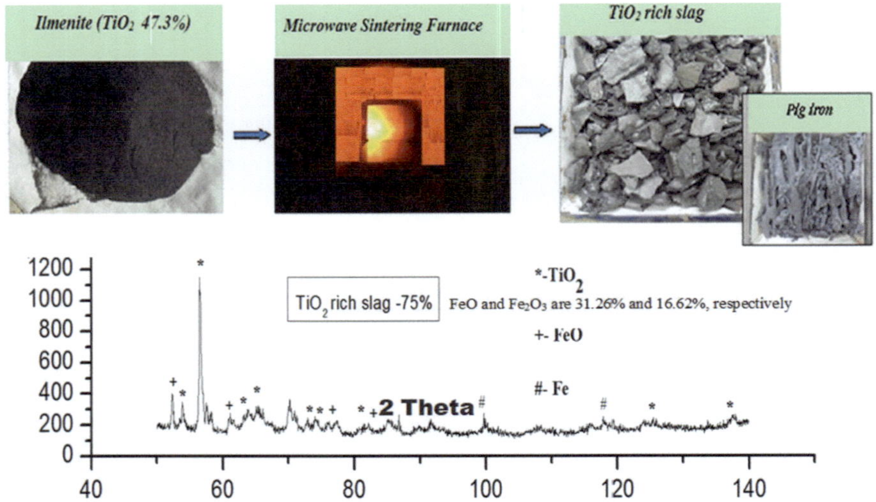

Fig. 8 Summary of microwave heat treated ilmenite for value addition

3.4 Characterization and Value Addition of Ilmenite

The complete mineralogical data of recovered ilmenite and the XRD data of ilmenite are shown in Fig. 6. The ilmenite has been pre oxidised at 900 °C for 3 h in conventional furnace and the same subjected to muffle furnace as well as in microwave sintered furnace. The best results obtained with microwave furnace at very low temperature and in few minutes. The products obtained are titania slag and iron from ilmenite concentrate characterised. These products are shown in Fig. 8.

It must be noted that the chemical analysis of red sediment ilmenite contains 47.0% TiO2. The FeO is 31.3% and Fe_2O_3 is 16.6%. The coastal sand contains in general, more than 50% TiO2. The low content of titania in red sediments is due to unweathered ilmenite which is coveted with red sediments.

3.5 Characterization and Value Addition of Sillimanite

Physical, chemical, and thermo-mechanical properties are evaluated and the developed bricks are compared with the standards for their suitability in industrial applications. The Sillimanite bricks made from red sediment sillimanite are comparable to standard sillimanite bricks in all aspects and found to be suitable for application in ceramic industry. The bricks made from beneficiated sillimanite (Fig. 9) can be used in high alumina refractories. The Refractory bricks made from sintered mullite (Fig. 9) find applications in different industries such as metallurgical, glass and ceramics (kiln furniture). The XRD graphs for mullites obtained by conventional furnace and Microwave heat furnace (Fig. 10). It can be seen from the data that mullite can be formed at very low temperature and very less time by using microwave heat furnace than a conventional furnace. Such studies with sillimanite and other minerals are further elaborated by [15, 16].

Thus, the innovative products developed in the present investigation are more economical and flexible to set up not only in upcoming red sediment placer deposit industries but also in the existing beach sand mineral industries to recover individual

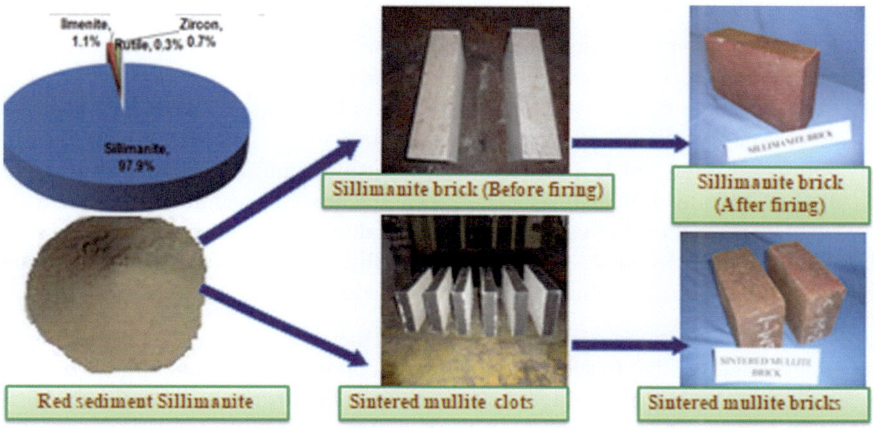

Fig. 9 Preparation of bricks and mullites using sillimanite mineral

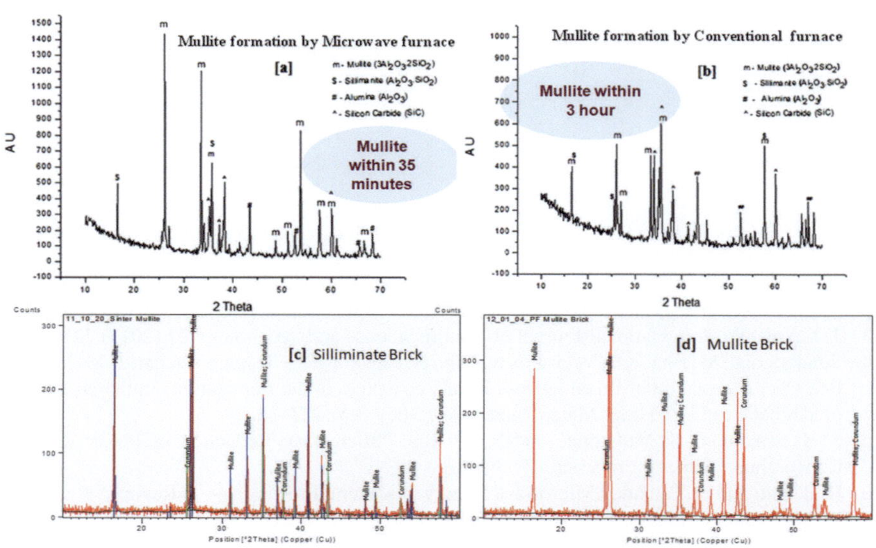

Fig. 10 Value added products from sillimanite recovered from red mud

industrial heavy minerals from red sediments of badlands topography for different industrial applications.

4 Summary

The conclusions drawn from the study on recovery of placer minerals from the of Badlands topography and its value addition are as follows.

1. The red sediments contain 6.1% total heavy mineral, in which ilmenite is 87.97%. The ilmenite recovered contains 99.3% grade and 99.5% recovery.
2. The sillimanite mineral which is a biproduct of ilmenite recovered by flotation process contain 96.9% grade and 67.1% recovery
3. Sillimanite brick made from red sediment sillimanite is comparable to standard sillimanite brick in all aspects and parameters. Sillimanite brick made from beneficiated sillimanite can be used in high alumina refractories.
4. Refractory bricks made from sintered mullite find applications in different industries such as metallurgical, glass and ceramics (kiln furniture).

Thus, the innovative products developed in the present investigation are more economical and flexible to set up not only in upcoming red sediment placer deposit industries but also in the existing beach sand mineral industries to recover individual industrial heavy minerals from red sediments of badlands topography for different industrial applications.

Acknowledgements The author wishes to thank all his research students for their support to write this paper.

References

1. Rao Gajapathi R, Shoo P, Panda NK. Heavy mineral sand deposits of Orissa. Explor Res Atom Miner 2001; 13:23–52.
2. Babu N, Vasumathi N, Rao Bhima R. Recovery of ilmenite and other heavy minerals from terisands (red sands) of Tamil Nadu. India J Miner Mater Charact Eng 2009;8(2):149–159.
3. T. Laxmi et al./International Journal of Mining Science and Technology 23 (2013) 725–731.
4. Sundararajan M, Bhat KH, Velusamy S, Babu N, Janaki MEK, Sasibhooshanan S, Mohan Das PN. Characterization of ilmenite from Kerala coastline, India: implications in the production of synthetic rutile. J Miner Mater Charact Eng 2009;8(6):427–38.
5. S. S. Srikant, P. S. Mukherjee and R. B. Rao, "Microwave Reduction of Placer Ilmenite Concentrate," Mineral Processing Technology, 2011.
6. R. B. Rao and N. Patnaik, "Microwave Energy in Mineral Processing—A Review," Journal of the Institution of Engineers (India)—Mining, Vol. 84, No. 2, 2004, pp. 56–61.
7. C. A. Pickles, "Microwaves in Extractive Metallurgy: Part 1—A Review of Applications," Minerals Engineering, Vol. 22, No. 13, 2009, pp. 1102-1111. https://doi.org/10.1016/j.mineng.2009.02.015

8. E. T. Thostenson and T.W. Chou, "Microwave Processing: Fundamentals and Applications," Composites Part A: Applied Science and Manufacturing, Vo 1055–1071. l. 30, No. 9, 1999, pp. https://doi.org/10.1016/S1359-835X(99)00020-2
9. Chen TT, Dutrizac JE, Haque KE, Wyslouzil W and Kashyap S. 1984. The Relative Transparency of Minerals to Microwave Radiation. Canadian metallurgical quarterly, 23, 1:349–351.
10. Bhima Rao R and Patnaik N. 2004. Microwave Energy in Mineral Processing—a Review. IE(I) Mineral Journal, 84, 1:56–61.
11. Pickles C A. 2009. Microwaves in extractive metallurgy: Part 1—A Review of Fundamental. Elsevier Journal: https://doi.org/10.1016/j.mineng.2009.02.015, 1:1102–1111.
12. Kelly R M and Rawson N A. 1998. Microwave Reduction of oxidised ilmenite concentrates. Minerals engineering Jr, 11, 1:1427–1438.
13. Haque K E.1999. Microwave energy for mineral treatment processes—a brief review. International Journal of Mineral Processing, 57, 1:1–24.
14. Xia D K and Pickles CA.1997. Applications of microwave energy in extractive metallurgy—a review, CIM Bull. 90, 1:96–107
15. S. Tripathi and G. Banerjee, "Synthesis and Mechanical Properties of Mullite from Beach sand Sillimanite: Effect of TiO2," European Journal Ceramic Society, Vol. 18, No. 14, pp. 2081–2087.
16. Sundarajan, M., Bhat, K.H., Velusamy, S., Janaki, M. E. K. and Sashibhooshanan, S., "Investigation on physical and chemical characterization on Indian sillimanite and preparation of Aluminium-Silicon master alloy", World applied sciences journal, 2010, 8(11), 1394–1397.

Innovative Approach to Transform Mining Waste into Value Added Products

Pratik Godbole, Kaustubh Deshpande, Sanjeevani Jawadand, Pranay Meshram, M. L. Dora, Rajkumar Meshram, Tushar Meshram, Ashok Kumar Nandi, and Kirtikumar Randive

Abstract The global economy heavily relies on the mining industry for the extraction of valuable minerals and resources. Nonetheless, this extraction process results in a substantial volume of waste, presenting environmental challenges and sustainability issues. Within this waste, valuable elements and minerals often lie dormant, awaiting proper extraction and processing to be converted into valuable products. This not only reduces the environmental impact but also taps into the economic potential of mining waste. In recent times, there has been a burgeoning interest in exploring creative solutions for the conversion of mining waste into products with added value. In this chapter, we delve into the prospect of harnessing value from mining waste through inventive methods. It delves into diverse innovative strategies, including advanced separation methodologies, bioleaching, and hydrometallurgical processes, all of which exhibit potential in recovering precious materials from mining waste. By addressing these themes, this chapter makes a valuable contribution to the continual drive for a mining industry that is more sustainable and environmentally conscious.

Keywords Value added products · Innovative technologies · Mining · Mining waste · Environmental impact · Tailings

P. Godbole · K. Deshpande · P. Meshram · K. Randive (✉)
Department of Geology, RTM Nagpur University, Nagpur, India
e-mail: randive101@yahoo.co.in

S. Jawadand
Shri Mathuradas Mohota College of Science, Nagpur, India

M. L. Dora
Geological Survey of India, Hyderabad 500068, India

R. Meshram
Regional Petrology Division, Geological Survey of India, CR, Nagpur, India

T. Meshram
Central Region, Geological Survey of India, Seminary Hills, Nagpur, Maharashtra, India

A. K. Nandi
Mineral Information & Development Centre (I) Pvt. Ltd., Nagpur, India

K. Randive et al. (eds.), *Current Trends in Mineral-Based Products and Utilization of Wastes: Recent Studies from India*, Springer Proceedings in Earth and Environmental Sciences, https://doi.org/10.1007/978-3-031-50262-0_18

1 Introduction

Mining plays a significant role in resource-rich nations, as it serves as a wellspring of both raw materials and revenue. These contributions have the potential to significantly bolster economic growth. The global surge in mining productivity highlights the critical concern of managing mineral wastes in an environmentally sustainable manner, given their substantial potential for causing pollution [1]. Depositing waste on land typically leads to the deformation of the natural terrain, eradication of plant life, and the depletion of forests. This practice also results in the emission of greenhouse gases, a decline in biodiversity, and the deterioration of the visual appeal of the nearby environment. Furthermore, it diminishes the productivity of the land, alters the overall landscape, jeopardizes the stability of slopes, and consequently heightens the susceptibility to soil erosion caused by wind and stormwater runoff [2].

Hence, improper management of mine waste can lead to persistent and at times, severe environmental deterioration. Given the existing mining techniques, mine waste is not expected to diminish; rather, it is anticipated to accumulate more rapidly due to the decline in primary ore quality and the growing needs of society. Nonetheless, mine waste possesses the potential to become a valuable societal resource. It can serve as an alternative reservoir of partially critical raw materials (CRMs), supplementing primary reserves and offering a chance to reduce supply vulnerabilities. This approach supports sustainability, helps manage geopolitical pressures, and ensures a stable supply [3].

The global scarcity of raw materials has spurred a growing interest in utilizing mineral waste materials as a promising reservoir of valuable metals and minerals [4]. So, effective waste management approaches are vital to ensure the streamlined recycling of substantial volumes of mining byproducts produced annually from numerous mining activities spanning the nation [5]. Hence, considering waste as a supplementary resource for achieving economic balance rather than a liability is the immediate requirement.

2 Mining Waste

India possesses a rich endowment of mineral resources, encompassing a total of 95 minerals. These can be categorized into 4 fuel minerals, 10 metallic minerals, 23non-metallic minerals, 3 atomic and 55 minor minerals (including building and other materials) highlighting the country's abundant and diverse mineral wealth [6]. To meet the escalating needs of a growing population, the focus on exploration and mining has intensified, serving as a critical pillar for strengthening the domestic economy. However, this pursuit of extracting valuable metalliferous, non-metalliferous, and radioactive minerals to attain profitability also gives rise to a substantial volume of mining waste. The rapid increase in the accumulation of mining waste surpasses the growth rate of the Gross Domestic Product (GDP) [7]. The origin of mining waste stems from the excavation process, encompassing both chemical and physical mechanisms, which result in the segregation of the desired value-based materials from various accompanying by-products [5]. As outlined in the Minerals, Metals, and Sustainable Development report (MMSD), the practices of mining and mineral processing are required to uphold the physical and chemical stability of disposed waste materials. Additionally, these practices should strive to minimize water consumption and prevent the discharge of residual chemicals and metals into the environment, thereby ensuring environmental integrity and sustainability [8].

Mining waste serves as a repository for potential gangue minerals and hazardous substances, necessitating our unwavering focus and concern. The presence of these elements calls for thorough examination and effective management to prevent environmental degradation and health risks [5]. Characterizing mining waste into distinct categories based on mineralogical and chemical attributes holds significant importance in predicting its geotechnical properties. This characterization process aids in forecasting key geotechnical parameters such as bulk density, dry density, shear strength, particle size distribution, and plasticity, among others. This knowledge is crucial for understanding the behavior and potential impacts of mining waste on the surrounding environment and infrastructure [9]

The diversity and quantity of mining waste differ across various countries, influenced by factors such as the abundance of natural resources and the economic worth of minerals driven by market demand. Consequently, mining waste can range from minimal amounts to substantial proportions. These variations underscore the intricate interplay between local resource availability, economic dynamics, and environmental considerations in shaping the extent of mining waste generated in different regions [10]. Addressing the challenges associated with mining waste is crucial for advancing towards sustainable development. This waste generated during mining operations holds the potential to contain valuable heavy metals and minerals. However, the storage and management of such waste can lead to negative environmental consequences. Therefore, it becomes imperative to focus on proper utilization and safe disposal methods for mining waste. By doing so, the detrimental impacts on the environment can be mitigated, while simultaneously unlocking the potential value of the contained precious metals and minerals [5]. Figure 1 shows various types of mining waste.

The types of mining waste mentioned above can be broadly classified into primary waste which forms directly during the extraction and secondary waste which forms during further refining of the ore [14]. Indeed, several types of the previously mentioned mining waste harbor toxic substances and heavy metals that pose a significant threat to the environment and ecosystems. Mining waste generated during the extraction of uranium and thorium contains substantial amounts of radionuclides. These radionuclides are extremely hazardous and can have severe repercussions on various aspects of the environment, as well as anthropological, zoological, and botanical resources. Exposure to such radioactive materials can lead to adverse impacts, including contamination of soil, water, and air. This contamination can disrupt the delicate balance of ecosystems, harm wildlife and plant life, and ultimately have far-reaching consequences for the health and well-being of both human and non-human

Fig. 1 Various types of mining waste (modified after [7, 11–13])

entities. Given the potential dangers associated with these materials, it is crucial to implement rigorous measures to properly manage, contain, and safely dispose of mining waste containing toxic substances, heavy metals, and radionuclides to prevent irreversible environmental damage [15]. Radioactive metals present a significant risk as potential carcinogens to human health, particularly when they enter the food chain. Managing radioactive waste is a formidable challenge in the context of achieving sustainable development [16]. To alleviate financial burdens, embracing recycling and extracting value-added products from the waste emerges as an ecologically sound strategy [17]. However, certain challenges must be discussed which are commonly faced during mining waste management. They are discussed in the next section respectively.

3 Challenges in Mining Waste Management and Environmental Mining Practices

The management of mining waste has evolved into a worldwide concern, impacting both environmental integrity and a country's financial well-being. Across all mining locations, there are significant challenges related to waste management that necessitate solutions. These challenges can be effectively addressed through the adoption of the Best Available Technology (BAT). Utilizing BAT presents a proactive approach to combat the pressing waste management issues associated with mining sites, ultimately safeguarding both the environment and economic interests [18]. Effectively managing mining waste poses substantial challenges, spanning five key domains: economic, environmental, social and governance, legal, and technical [19]. Undertaking mining activities within forested regions presents formidable hurdles, given its potential to trigger deforestation and disrupt the delicate soil cover. The accumulation of substantial rock waste heaps directly impacts soil microorganisms, and managing the disposal of such waste becomes a complex task, influencing the surrounding environment. Throughout mineral processing, the noise and drilling involved can perturb the local wildlife, while the generated waste can become accessible to wildlife, posing a direct threat to their survival [20]. A significant obstacle in the management of mining waste revolves around the appropriate disposal of toxic constituents. The occurrence of toxic elements and metalloids such as Cadmium (Cd), Arsenic (As), Mercury (Hg), Chromium (Cr), and Lead (Pb) varies, yet their notable presence can disrupt the flourishing of plant life and degrade the water quality within the vicinity [9] (Fig. 2).

Depositing non-hazardous waste at disposal sites poses a risk to the groundwater quality. The potential for contamination at these sites evolves over time due to factors such as hydraulic conductivity, soluble constituents within the freshly mined rock, air permeability, and the presence of geochemically unstable elements like sulfides. These factors can lead to the eventual release of primary and secondary contaminants.

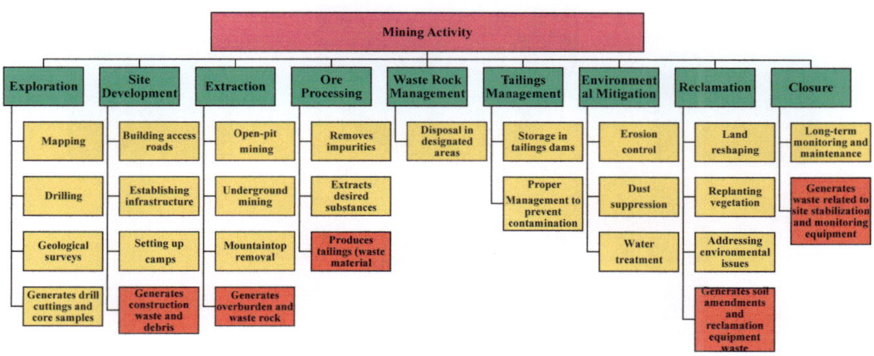

Fig. 2 Steps involved in mining waste formation [13]

Additionally, external elements such as hydrological, hydrogeochemical, meteorological, and anthropological influences further complicate waste management at these disposal sites, presenting a multifaceted challenge [10] (Fig. 3).

Navigating through the laws and regulations set forth by the government presents its own set of challenges in the management of mining waste. The Mines and Minerals (Development and Regulation) Act of 1957 stipulates specific considerations that must be adhered to during waste management. Section4A (2) of this act underscores the obligation for all mining operations to employ effective methodologies for waste management. Chapter V of the Mineral Conservation and Development Rules 2017 is dedicated to addressing the aspects of sustainability related to mining and the management of mining waste. This section includes various provisions (Fig. 4) aimed at promoting responsible practices within the mining sector and ensuring the proper handling of mining waste [21]. Taking into consideration all the stipulated provisions, the management of mining waste emerges as a delicate and highly significant facet of environmental well-being. Within this management framework, considering the geochemistry of the materials becomes a pivotal factor in ensuring their secure disposal. This is especially crucial as these materials can potentially harbor hazardous contaminants, including toxic substances and radioactive

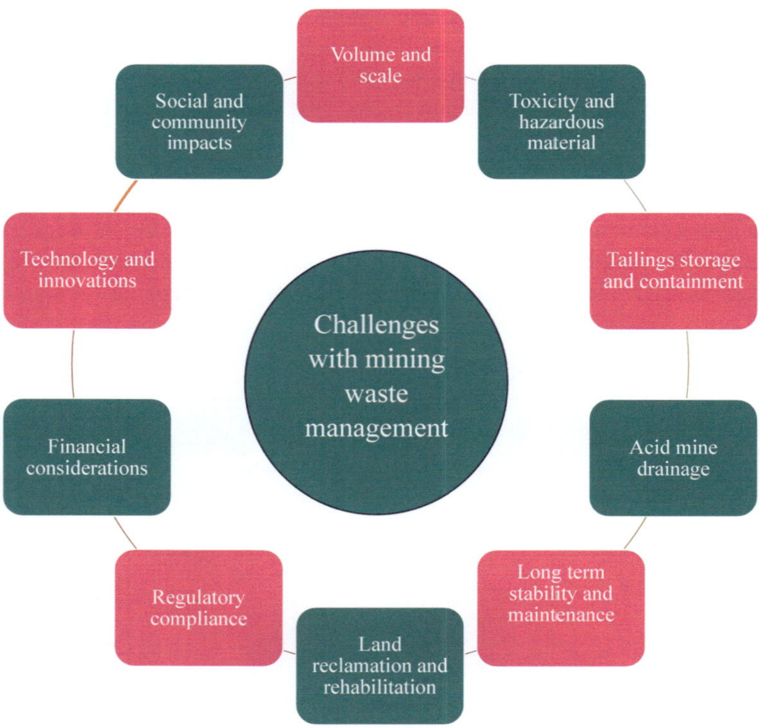

Fig. 3 Challenges associated with mining waste

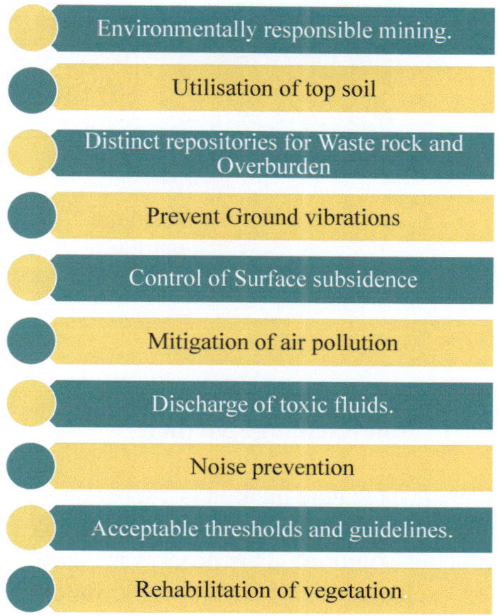

Fig. 4 Provisions for environmentally friendly mining waste management [22]

elements. Consequently, mining waste presents a major concern that requires meticulous consideration. However, within this challenge lies an innovative economic opportunity for the mining industry—the reutilization and reprocessing of existing waste. This strategy has the potential to not only enhance profit margins for mining enterprises but also to tap into untapped markets. Moreover, it can establish an alternative revenue stream for miners, while simultaneously enabling the transformation of waste into value-added products [7].

4 Innovative Technologies for Value Addition

4.1 Extraction of Rare Earth Elements (REEs) from Tailings

Rare earth elements (REEs), have indispensable applications in electronics, renewable energy systems, and defence systems. Though REES are extracted from their primary ores, the increasing demand and environmental constraints have paved the way for focusing on secondary ways of REE extraction emphasizing its recycling and extraction from tailings and abandoned mines.

Following are some methods for procurement of REEs from tailings (Fig. 5).

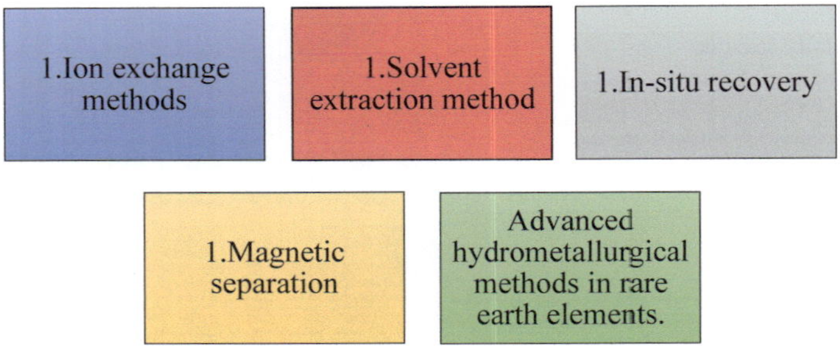

Fig. 5 Methods of REE recovery

4.1.1 Ion Exchange Method

The ion exchange method is a separation and purification technique that involves the reversible exchange of ions between solid materials (resins) and liquid (usually an aqueous solution). This process is based on the principle that certain materials i.e., ion exchange resins have specific functional groups on their surface which attract, adsorb, and bind ions of opposite charge from surrounding materials. The adsorbed ion can later be released from the resins by introducing another solution with a high ion concentration or with a change in pH [23]. This technique is profound for treating REEs (Fig. 6).

Because of their trivalent nature, REEs strongly interact with hard bases like carboxylate, carbonate, fluoride, sulphate, phosphate, and hydroxyl anion to form inner sphere complexes. They also pair with soft bases like nitrate, chloride, and sulphate anion with the formation of weak outer sphere complexes. Hence, the leachant solution mainly consists of HNO_3, Citric acid, H_2SO_4, HCl etc. [23, 24]

Solid materials (resins) that are used as adsorbents are mainly of two types depending on the charge of groups viz.

1. Anion exchange resins consist of secondary, tertiary, and quaternary ammonium groups (weak base resin WBA and Strong base resins SBA) e.g., WBA: Purolite RUA21207 (Target ion- Sc); SBA: Dowex1 X-8 (TI–Sc, Y, La, Ce, Pr, Nd, Eu, Sm, Lu, Yb, Er, Ho, Dy, Gd) [24].
2. Cationic exchange resins consist of strong acid cation resin SAC (Sulfonic acid as a functional group), and weak cation exchanger WAC (carboxylic acid as a functional group). SAC: Purolite C150 (Target ion TI -Nd), Dowex 50WX8 (TI: Y, La, Ce, Nd, Dy, Gd) [24] WAC: CYBBER LX 280, Amberlite IRC 50 [25].

Although the ion exchange method is a very distinct extraction method for REEs and other valuable metals which is effective for small metal concentrations, it is a temperature-dependent method that is quite expensive for resins and chelating separation material.

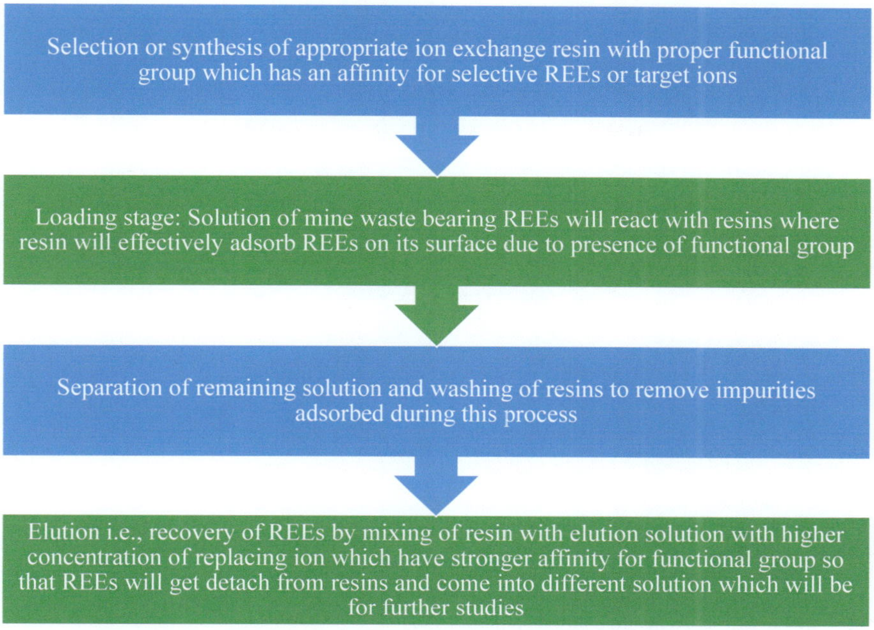

Fig. 6 Steps for extraction of REEs by ion exchange method [23, 24]

4.1.2 Solvent Extraction Techniques

The solvent extraction method is a preferred method for the separation and purification of REEs after extraction from mine wastes like red mud (bauxite and lateritic mine waste), Coal fly ash, phosphatic mine waste, etc. [26] (Fig. 7).

This technique is cost-effective as it does not require expensive equipment and reagents, where nitric acid leaching results in the separation of iron (III) hydroxide which precipitates REEs in the form of oxalates or hydrates [28]. Sometimes there could be a risk of leakage of poisonous gases, but it is replaceable. This method is also applicable for the extraction of REEs from coal fly ash by leaching and purification using Di-2-Ethyl hexylphosphonic acid (P_2O_4). This technique can also be blended with fractional precipitation, crystallization, and redox membranes [29]

4.1.3 In-Situ Recovery

The in-situ recovery method is different from other conventional mining methods like open pit or underground mine, as it avoids excavation and disturbance of habitat and environment. In this method, a series of injections and extraction wells are drilled within the ore body which forms a network to allow leaching solution injection for dissolution of REEs. [30]. This leaching solution further flows, spreads into the entire

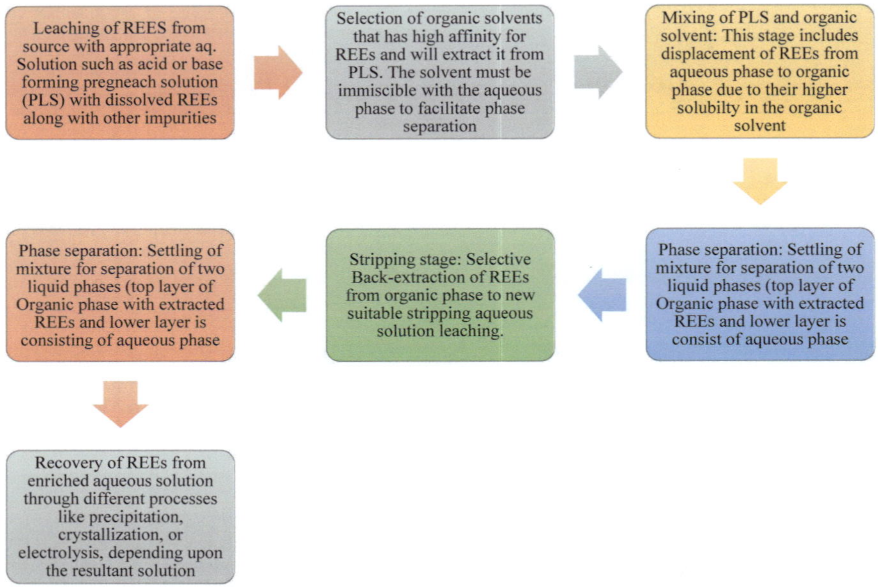

Fig. 7 Steps of Ion exchange method technique [27]

ore body, and leaches out the REEs. This REEs enriched solution then pumps back to the surface through the extraction wells. The remaining solution is reconditioned and recycled back into the ore body for continuous recovery [31]. Apart from REEs, elements like Cu, Au, Ag, Zn, Cd, Pd, Mn, Li, Mo, Se, V, Sc, Y, In, Be, Cr, Ni etc. can also be extracted via this method. This method is suitable for ore deposits of lower grade which are at deeper parts and hence prove useful in ore extraction, grinding, and less tailings [32].

4.1.4 Magnetic Separation Techniques

This technique is based on the magnetic character of elements. Some oxides of REEs having elements like Nd, Dy, and Gd exhibit magnetic character when the magnetic field is applied to a mixture of magnetic and non-magnetic materials (Fig. 8).

4.1.5 Froth Flotation Method

The froth flotation method is further adopted where sodium oleate and hydroxamic acid work as collectors and sodium silicate and starch act as a depressant. Rare earth oxides are obtained from the concentration of flotation debris. During the

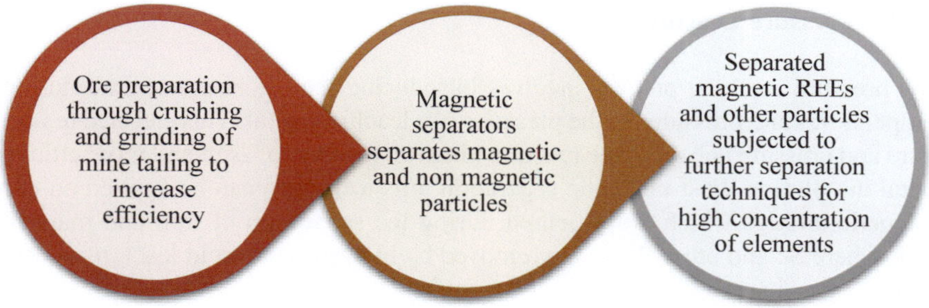

Fig. 8 Showing steps of magnetic separation [33]

management of iron oxide silicate-rich tailings, rare earth elements are recovered as a by-product [34].

4.2 Advanced Hydrometallurgical Methods in Rare Earth Elements.

Advanced hydrometallurgical methods pertain to intricate approaches applied for the recovery and refinement of metals from ores or alternative origins through the utilization of aqueous-based solutions (processes involving water-based solutions). These methods are now proliferated to address the problems associated with the mining industry's sustainability, environmental impact, and efficiency. These processes aim to facilitate higher metal recovery, minimize waste production reduce energy consumption, and utilize more eco-friendly reagents.

Some examples of advanced hydrometallurgical processes include (Fig. 9).

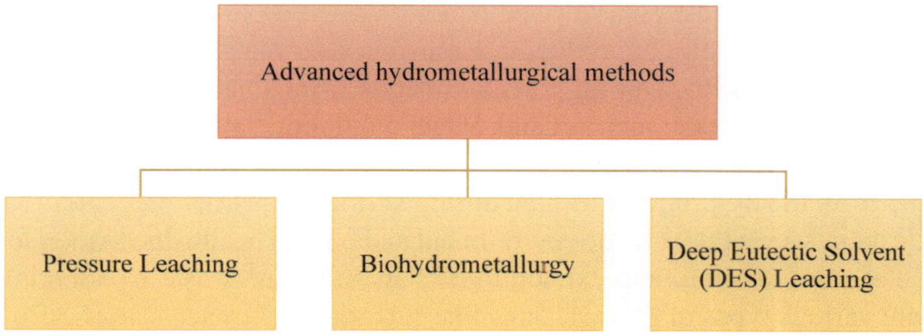

Fig. 9 Advanced hydrometallurgical methods [35]

4.2.1 Pressure Leaching

The pressure leaching process involves introducing the ore or concentrate to high temperatures and pressures in the presence of a leaching agent. This elevated temperature and pressure enhances the reaction kinetics, leading to faster and more efficient metal dissolution. Post magnetic separation, this technique can be applied on non-magnetic tailings which are procured during the separation of iron and magnetic REE where Sc and other REEs are removed by high-pressure acid leaching process [33].

4.2.2 Biohydrometallurgy

Biohydrometallurgy is a subset of hydrometallurgy that utilizes microorganisms, such as bacteria and archaea, to aid in metal extraction from ores or concentrates. Microorganisms catalyse reactions that enhance the metal solubilization thereby making the process more environmental friendly and cost-effective [36].

4.2.3 Deep Eutectic Solvent (DES) Leaching:

Deep Eutectic Solvent (DES) Leaching: DES leaching is a green solvent extraction technique where deep eutectic solvents are used to dissolve and extract metals from ores or concentrates. DES is composed of a eutectic mixture of two or more components, often formed by combining a hydrogen bond donor and an acceptor. DES leaching can be an eco-friendly alternative to traditional acidic or alkaline leaching processes [35].

4.2.4 Recovery of Precious Metals from Slag

Slag is the glass-like solid byproduct that remains after metals have been separated from their ores [37]. Recovering precious metals from slag is a process aimed at extracting valuable metals, such as Gold, Silver, Platinum, Palladium, Rhodium [38], Titanium [39], Zinc, Nickel, Chromium, Iron, Aluminium, Vanadium, etc. They are extracted from the by-products formed during various industrial processes [37].

Toxic metals are a threat to the environment and human health due to their toxicity hence, cleaning, refining, and extraction of slag and mine waste is essential [40]. There are various types of furnace slags like Electric arc furnace slag (EAF), Basic oxygen furnace (BOF) slag, Linz-Donawitz processed slag formed in Steel industry [37], Blast furnace slag from Iron industry [39], copper matte smelting slag, iron silicate slag [38].

Prior to the recovery process, slag needs some microanalysis such as EPMA, XRD, ICP analysis, OCP analysis, etc. to determine the concentration of precious metals. This slag is further crushed and ground to a suitable size to increase the surface area. This step is essential to expose the precious metal particles that are trapped in the slag [38]. Various extraction methods including magnetic separation, roasting, pyrometallurgical, hydrometallurgical, and bio-hydrometallurgical techniques are used to procure precious metal constituents from slag [37].

For separating phases that are dependent on density differences, the gravity method involves the use of spirals, shaking tables, jigs, etc. Jigging is used on coarse fractions and tabling is used on fine fractions of slag [41]. Nearly 90% of Chromium has been recovered from the granulated slag by magnetic and jigging techniques. The method of Froth floatation is also effective for recovering Copper and other hydrophobic sulfides. Various floating agents are used along with water such as fatty acids, thiocarbamate, dithiophosphates, xanthates, etc. Magnetic separation technique is profound to separate magnetic and non-magnetic slag through drum magnetic technique [37].

Pyrometallurgical techniques convert precious metal slag into its elemental form by high-temperature chemical reactions within the furnace. It is done in several steps like calcination, roasting, smelting, drossing, refining and sometimes carbothermic reduction is also performed in association with pyrometallurgy [37]. This pyrometallurgical technique is performed through the direct reduction method (metal recovery using a reducing agent like coal) [40] and the selective reduction method (selective used of reductants and typically used for recovery of Cu, Cr, and Ni) [37].

Hydrometallurgical processes have been done in two steps e.g., Leaching and Recovery of metals and used for recovery of non-ferrous metal from ferroalloy slags or converter slags. Leaching includes the reaction of lixivants i.e., leaching agents such as acids, and salts to recover and extract metals into an aqueous solution [24]. In the final stage, metal can be recovered through precipitation processes such as electrowinning, electrolysis and solvent extraction, ion exchange methods, etc. [37].

4.3 Utilization of Mine Waste in Construction Industries

The sustainable utilization of mine waste in construction materials is an approach aimed at turning mining byproducts and waste materials into valuable resources. Incorporation of mine waste into construction materials, such as concrete, bricks, and aggregates has several benefits. This industrial waste such as ground granulated blast furnace slag, fly ash, and mine tailings can be used in this regard [42].

Nowadays mine waste is one of the secondary resources for extracting metals but it can also act as construction material in road and building construction industries. This is possible with a special synthesized inorganic material called geopolymer which is used for manufacturing construction products like geopolymer bricks, tiles, and concrete [43]. Geopolymers are mainly composed of two components i.e., source materials and alkaline liquids. Aluminosilicate material with a SiO_2/Al_2O_3 ratio of 1.6 to 5 is used as raw material. The main source material for geopolymers is rich in silicon and aluminum, coming from natural minerals such as kaolinite, clays, or byproduct material like fly ash, ground granulated blast furnace slag, silica fumes, slag, rice husk ash, red mud, etc. Alkaline liquids are a combination of soluble alkali metals like a combination of sodium hydroxide or potassium hydroxide and sodium silicate or potassium silicate. Unlike ordinary cement, it will not form C–S–H gel(calcium-silicate-hydrates), on the other hand, it uses polycondensation of silica and alumina and high alkali content. [43, 44]. C–A–S–H gel is formed in high calcium systems while N–A–S–H gel is formed in low calcium systems [44]. It is an excellent replacement for cement and brick formed by conventional methods because brick is manufactured with iron ore tailings and red mud (bauxite mine). Geo-polymerized brick attains a compressive strength of 50 MPa whereas traditional bricks made of red mud and GGBs bricks have a compressive strength of 7.56 MPa. In addition to this, if concrete blocks are coated with red mud fly ash-based geopolymers it will increase the strength, durability, and thermal stability of the blocks. [45] (Fig. 10 and Table 1).

Copper mine tailings, low calcium slag, iron mine tailing, and blast furnace slags can be used in road construction materials due to their high compressive strength and

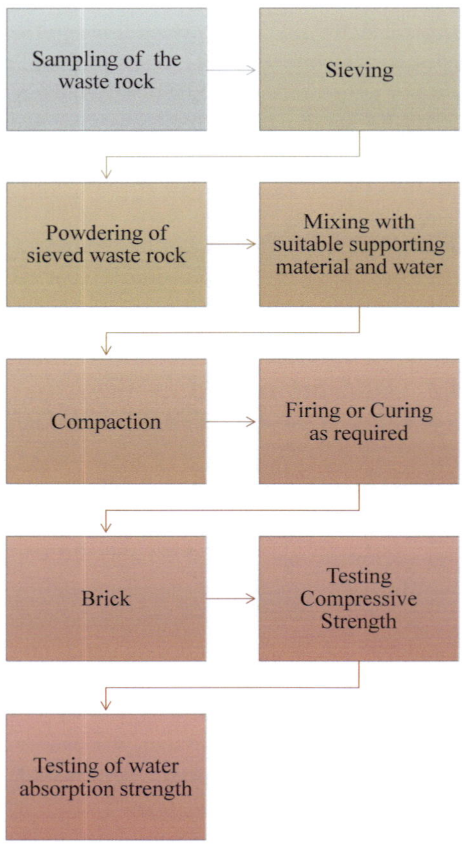

Fig. 10 Manufacturing of brick from mine waste rock (modified after [46–49])

Table 1 Manufacturing of bricks using different types of waste

Waste material	Supporting material	Compaction pressure	Firing/curing	Waste absorption in 24 h (%)	Achieved flexural strength (MPa)	References
Gold mine waste rock (80%)	Natural clays (20%)	–	1000–1050 °C	< 20	> 2.5	[47]
Iron ore waste (< 91%)	Fly ash and cement (> 9%)	20 MPa	Curing in sunlight for 28 days	6.20	3.5	[48]
Copper mine tailings	15 M NaOH solution with deionized water (8–18%)	15 MPa	90 °C	3.2	21	[49]
Aluminum tailing i.e., red mud (45%)	Fly ash (40%), Lime (10%), Gypsum (5%)	15 MPa	Cured with Steam (95% humidity; Temperature 80 °C	Low	4.4	[50]

high mechanical properties. Along with the addition of geopolymers, the mechanical properties can be improved by introducing natural and synthetic fibers such as Polypropylene and glass [42]. In steel industries, blast furnace slag is known for the exploitation of precious metals as well as for aggregates in construction for asphalt pavements, roads, concrete, and waterways. Electric arc furnace slag is also known for its strong frictional and abrasion-resistant qualities which is why it is employed for building crossroads, industrial roads, and parking lots [37]. Mining wastes are used as an alternative to dry compacted road materials and embankments having properties such as grading, impact, and water resistance e.g., Coal mine waste rock and phosphate mine waste rock) and used for backfilling and dumps [51] (Fig. 11).

4.4 Bioleaching

Bioleaching is a microbial-based process where different types of microorganisms are employed to extract elements from ore or mine tailings. This process is well known with respect to mine tailings and low-grade ores because mine tailings contain various microorganisms that use atmospheric oxygen to oxidize and solubilize sulfides and iron-containing ores which leads to the extraction of metals from the solid matrix.

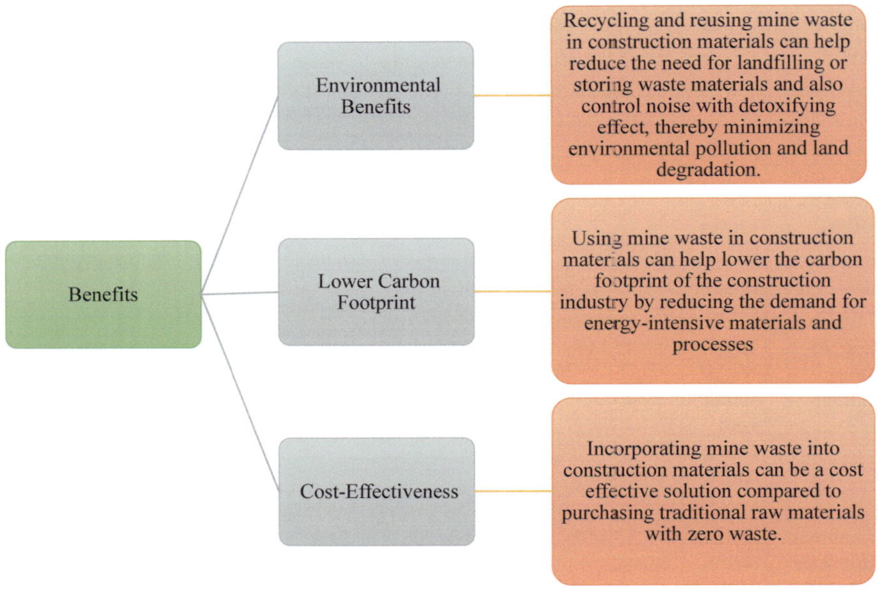

Fig. 11 Benefits of re-using mine waste [43]

This also leads to polluted effluent called acid mine drainage. So, for tailing remediation and removal and recovery of other metals in an eco-friendly and sustainable way, bioleaching is a much better way of extraction though it has low extraction rates. To accelerate the reaction efficiently it needs to mix with catalytic agents, and biosurfactants, apply light illumination or weak magnetic fields, and mutating bioleaching bacteria by UV irradiation [52].

Bacteria and fungi show mineralytic effects on minerals based on acidolysis, complexolysis, and redoxolysis. For metal mobilization, microbes go through steps like proton formation (formation of organic and inorganic acids), ligand formation (excretion of complexing agents), and oxidation and reduction reactions. Leaching can be direct or indirect. Direct leaching is the process of a cell immediately adhering to the surface of a mineral and oxidizing metal sulfides where electrons are acquired directly from reduced minerals; whereas microbial oxidation of metal ion [e.g., Fe (II) to Fe (III))] which act as an oxidant to oxidize metal sulfides without direct contact with metal and microbes is indirect leaching [53] (Fig. 12).

Suitable environmental conditions are essential for successful bioleaching. Factors like temperature, pH, oxygen availability, and nutrient levels must be optimized to support the growth and activity of the microorganisms. This method is used for the recovery of Cu, U, Mn, Ni, Au, PGE, REEs, etc. by the activity of chemolithotrophic bacteria such as *Thiobacillus ferrooxidants, T. thiooxidans, T. prosperus, Leptospirillium ferroxidans, Acidianus brierleyi, Sulfolobus, Sulfobacillus thermosulfidooxidans and Fungi such as genera Aspergillus and Penicillium* etc. (for Mn leaching).

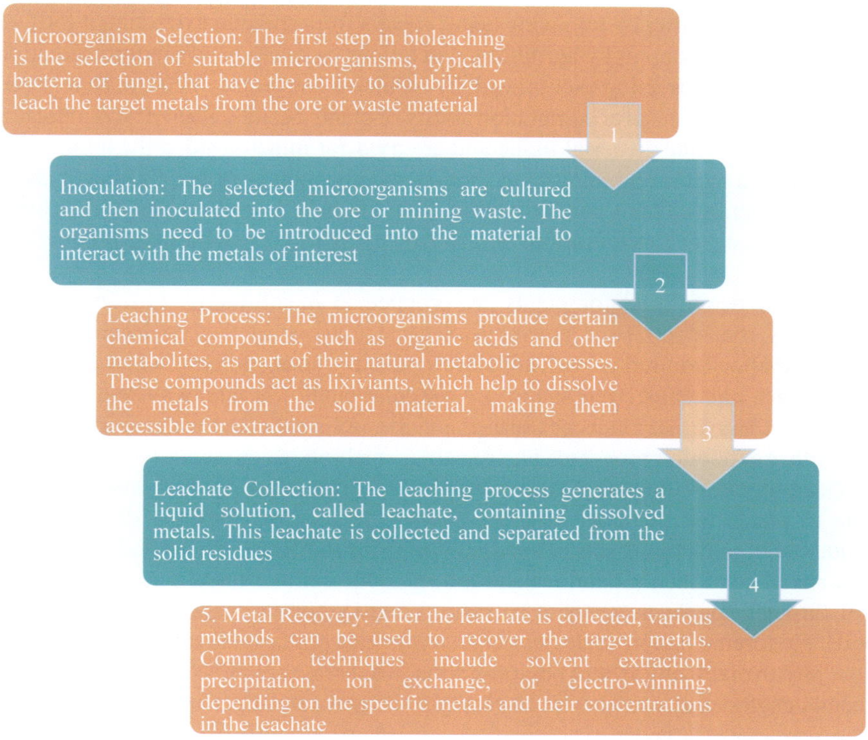

Fig. 12 Step by step process of Bioleaching [53, 54]

This technique is captivating through Percolator, submerged Column, and industrial leaching where the tailing dump, heap, underground, and tank leaching can be utilized. Bacterial leaching is carried out in an acid environment at pH values between 1.5 and 3 at which most metal ions remain in solution. But, some thiobacilli species grow only at higher pH values to oxidize sulfur and sulfides. Metal recovery from sulfide minerals is based on the activity of bacteria that convert insoluble metal sulfides into soluble metal sulfates. Non-sulfide ore can be treated by heterotrophic bacteria and fungi because metal is extracted due to the production of organic acids and chelating and complexing compounds excreted into the environment [54].

4.5 *Phytomining*

Nowadays there is a need to extract elements through environment-friendly methods. Phytomining is one of the methods where plants are used to extract metals through their metabolic and physiological activities [55]. This phytomining method might be used as a detoxifier by removing heavy metals from soil and extracting it [56], as another financial support by agromining on tailing, or as a type of phytoremediation of mine land by extracting excess partially unknown elements which can be restored for future use. The metal accumulation ability of plants decides whether it is indicator (accumulated metals in aerial tissue and are used for mineralized soil exploration), excluder, or a hyperaccumulator (accumulated high amounts of metals in their shoot without suffering). Plants used in phytomining might be hyperaccumulators (exceptionally high metal concentration) or fast-growing plants (great biomass production) [57].

These hyperaccumulator plants are cropped in metalliferous soil under toxic conditions (but don't have any physiological toxicity in this environment) and harvested after required growth and development. The absorbed mineral and metal translocated and transferred as root → Vascular bundle → Shoot system like axil, leaves, etc. Harvested plants get burned and the ash is collected with carbon dioxide for further extraction processes like roasting, smelting, electrowinning, etc. [55].

Generally, heavy metals mostly exist in insoluble form in soil which is not bioavailable to plants but plant increase their bioavailability by releasing different root exudates. It can change the pH and solubility of the Rhizosphere and then metal gets absorbed at the root surface and moves to the cellular membrane into the root cell through apoplastic (passive diffusion) and symplastic pathway (active transport). Phytoremediation strategies generally include phytostabilization, phytoextraction, phytovolatilization, and phytofiltration. Sometimes plant-associated microbes, genetic engineering can also be employed to improve plant performance [56].

4.6 Waste to Energy

Continuous use of fossil fuel can increase the carbon dioxide level in the atmosphere but this carbon be a source of renewable energy source through biodiesel production because of its biodegradability and less CO_2 and NO_2 emission. Abandoned area of mine (coal mine) can be the best site for algal or microorganisms' cultivation for renewable energy production. Algae has high photosynthetic efficiency and a great ability to fix carbon dioxide. This fixed carbon dioxide is then rapidly converted into lipids such as triglycerides by algae which is important for biodiesel production. Some algal species such *Oedogonium* and *Spirogyra* can produce such a high lipid/oil content through which nearly 15,000 gallons of oils/acre/year can be produced. Though there is no standard lipid extraction method but Bligh-Dyer (Chloroform + methanol) method, chemical extraction using different solvents and sonication, direct saponification (ether + KOH), and supercritical CO_2 extraction. With lipid extraction, the carbohydrate content of algae can also be fermented into bioethanol or biobutanol [58, 59].

Biomass is one of the important contributors to the world economy and an energy source completely CO_2 neutral after burning (assumed that CO_2 released in the atmosphere is matched with CO_2 used in production). Direct combustion will produce steam and turn to the turbine, turbine drives a generator with the production of electricity. For the conversion of biomass to combustible biogas gasifier will be used. During surface mining, short-term restored land can be useful for planting fast-growing energy crops which will have much higher growth with shorter cutting shifts that produce large amounts of biomass [60]. There is another method of generating electricity from coal-containing or other -carbon-containing mine waste, crushed biofuel through direct carbon fuel cells (DCFCs), and hybrid DCFCs (HDCFCs). There are two electrolytes i.e., solid oxide and molten carbonate. Through solid oxide membrane oxygen ions enter an anode molten carbonate oxidizes carbon and electron flow from the anode to the cathode generates electrical power. There is the conversion of chemical energy of biofuel into electricity. The complete process of electrochemical conversion of biofuel energy is divided into two stages depending on the presence and absence of free carbon. The first stage is the gasifier cell (free carbon) and the second stage is the HDCFC fuel cell (without free carbon) which has already an ordinary solid oxide fuel cell that runs on synthesis gas [61].

5 Summary

Mining waste encompasses the residual materials produced during mineral and ore extraction and processing. Frequently, these waste substances contain valuable elements or minerals that have the potential to be reclaimed and transformed into products with added value. The concept of extracting value from mining waste revolves around the application of inventive technologies and processes to recover and utilize these valuable components, thereby reducing waste and optimizing resource utilization. The creation of value-added products from mining waste offers numerous advantages and opportunities. Nevertheless, it also presents challenges, including the technical intricacies of extracting and processing valuable elements, the economic feasibility of recovery processes, and the regulatory and environmental considerations linked to waste management and resource utilization. To tackle these challenges, continuous research and innovation are imperative. Advanced technologies like sophisticated separation methods, bioleaching, and hydrometallurgical processes are under exploration to enhance the efficiency and efficacy of recovering valuable materials from mining waste. These innovations mark the beginning of both environmentally responsible practices and the potential for additional revenue streams in the future.

References

1. Shengo, L.M. Review of Practices in the Managements of Mineral Wastes: The Case of Waste Rocks and Mine Tailings. *Water Air Soil Pollut* **232**, 273 (2021).
2. J.P. Castro-Gomes; Abílio P. Silva; Rafael P. Cano; J. Durán Suarez; A. Albuquerque (2012). *Potential for reuse of tungsten mining waste-rock in technical-artistic value added prod ucts., 25(none), 34–41.* https://doi.org/10.1016/j.jclepro.2011.11.064
3. Glen T. Nwaila, Yousef Ghorbani, Steven E. Zhang, Leon C.K. Tolmay, Derek H. Rose, Phumzile C. Nwaila, Julie E. Bourdeau, Hartwig E. Frimmel, Valorisation of mine waste - Part II: Resource evaluation for consolidated and mineralised mine waste using the Central African Copperbelt as an example, Journal of Environmental Management, Volume 299, 2021, 113553, ISSN 0301-4797, https://doi.org/10.1016/j.jenvman.2021.113553
4. Silin, I., Rubinstein, Y., Volobayev, I. *et al.* Valorisation of Abandoned Low-Grade Tailings in Albania for Recovery of Metal Concentrates and Mineral Products. *Russ. J. Non-ferrous Metals* **62**, 483–494 (2021). https://doi.org/10.3103/S106782122105014X
5. Jawadand, S., Randive, K. (2021). A Sustainable Approach to Transforming Mining Waste into Value-Added Products. In: Randive, K., Pingle, S., Agnihotri, A. (eds) Innovations in Sustainable Mining. Earth and Environmental Sciences Library. Springer, Cham. https://doi.org/10.1007/978-3-030-73796-2_1
6. Annual report, Ministry of Mines, pp 11, (2018–19), https://mines.gov.in/admin/storage/app/uploads/6433dea77b42d1681120935.pdf
7. Randive K., and Jawadand S., (2020), Mineral economics an Indian perspective, Nova Science Publishers Inc., 213–230 pp.
8. Daniel M. Franks; David V. Boger; Claire M. Côte; David R. Mulligan (2011). Sustainable development principles for the disposal of mining and mineral processing wastes, 36(2), 114–122 https://doi.org/10.1016/j.resourpol.2010.12.001
9. Bian, Z.; Miao, X.; Lei, S.; Chen, S.-e.; Wang, W.; Struthers, S. (2012). *The Challenges of Reusing Mining and Mineral-Processing Wastes. Science, 337(6095), 702–703.* https://doi.org/10.1126/science.1224757
10. Irena Twardowska, Jadwiga Szczepańska, III.7—Coal combustion waste, Editor(s): Irena Twardowska, Waste Management Series, Elsevier, Volume 4, 2004, Pages 387–449.
11. Collins, R.J., Miller, R.H. Utilization of mining and mineral processing wastes in the United States. *Minerals and the Environment* **1**, 8–19 (1979). https://doi.org/10.1007/BF02010592
12. Shom, S., Mhaske, Dr. S., Pathak, Dr. K., & Tiwari, Dr. M. (2021). Mine Waste as Resource: Indian Mining Scenario of Coal and Non-Coal Mining Sector. In The International Journal of Recent Technology and Engineering (IJRTE) (Vol. 9, Issue 6, pp. 250–252). Blue Eyes Intelligence Engineering and Sciences Engineering and Sciences Publication—BEIESP. https://doi.org/10.35940/ijrte.f5396.039621
13. Pal, B.K., Khanda, D.K., & Dey, S. (2012). Problems of mining wastes management in India and its suggestive measures—case studies.
14. Rankin J (2015) Towards zero waste. Aus IMM Bull 32–37.
15. Randive K., Godbole P., Jawadand S., Chopra V., Dora M. L., Dhoble S., (2023) Radioactive waste management in India: Present status and future perspective, Editor(s): Nishikant A. Raut, Dadasaheb M. Kokare, Bharat A. Bhanvase, Kirtikumar R. Randive, Sanjay J. Dhoble, 360-Degree Waste Management, Volume 2, Elsevier, 2023, Pages 273–298, https://doi.org/10.1016/B978-0-323-90909-9.00005-8
16. M. Deb and S.C. Sarkar, (2017) Minerals and Allied Natural Resources and their Sustainable Development, Springer Geology. https://doi.org/10.1007/978-981-10-4564-6_11
17. Deshpande VP, Shekdar AV. Sustainable waste management in the Indian mining industry. *Waste Management & Research.* 2005;23(4):343–355. https://doi.org/10.1177/0734242X05056994

18. ISSN 1691-5402 Environment. Technology. Resources Proceedings of the 9th International Scientific and Practical Conference. Volume 1 © Rēzeknes Augstskola, Rēzekne, RA Izdevniecība, 2013, Improvement of Technologies for Mining Waste Management Veiko Karu, Angela Notton, Julia Gulevitš, Ingo Valgma, Tiit Rahe.
19. Kaczan, W.; Wirth, H.; Kudełko, J.; Sroga, C. (2021) Challenges in the Management of Mining Waste. Environ. Sci. Proc., 9, 28. https://doi.org/10.3390/environsciproc2021009028
20. Deb, M., Sarkar, S.C. (2017). Issues of Sustainable Development in the Mines and Minerals Sector in India. In: Minerals and Allied Natural Resources and their Sustainable Development. Springer Geology. Springer, Singapore. https://doi.org/10.1007/978-981-10-4564-6_11
21. https://ibm.gov.in/writereaddata/files/04192017182323MCDR_2017.pdf (mineral conservation and development rules, 2017) Indian bureau of mines. 2017.
22. https://ibm.gov.in/writereaddata/files/09012016174152Ch8.pdf (Environmental and forest legislation related to mines, Indian bureau of Mines).
23. Hérès X., Blet V., Natale P., Quaattou A., Mazouz H., Dhiba D., Cuer F., Selective Extraction of Rare Earth Elements from Phosphoric Acid by Ion Exchange Resins, 2018.
24. Ouardi Y., Virolainen S., Mouele E., Laatikainen M., Repo E., Laatikainen K., The recent progress of ion exchange for the separation of rare earths from secondary resources-A review, 2023, https://doi.org/10.1016/j.hydromet.2023.106047
25. Vladimir Rychkov, Evgeny Kirillov, Sergey Kirillov, Grigory Bunkov, Maxim Botalov, Vladimir Semenishchev, Denis Smyshlyaev, Andrey Malyshev, Aslan Taukin & Ata Akcil (2022) Rare Earth Element Preconcentration from Various Primary and Secondary Sources by Polymeric Ion Exchange Resins, Separation & Purification Reviews, 51:4, 468–483, https://doi.org/10.1080/15422119.2021.1993255
26. Pyrgaki K. Gemeni V., Karkalis C. Koukouzas N., Koutsovitis P., Petrounias P., Geochemical occurrence of rare earth elements in mining waste and mine water: A Review, 2021, https://doi.org/10.3390/min11080860
27. Lane D., Cook N., Grano S., Ehrig K., Selective leaching of penalty elements from copper concentrates: A review, 2016, https://doi.org/10.1016/j.mineng.2016.08.006
28. Cherkasova T., Cherkasova E., Tikhomirova A., Golovachev A., Dormeshkin O., Study of the Processes of Rare and Rare-Earth Metal Extraction from Ash and Slag Materials, 2020, https://doi.org/10.1051/e3sconf/202017402031
29. Pan, J.; Zhao, X.; Zhou, C.; Yang, F.; Ji, W. Study on Solvent Extraction of Rare Earth Elements from Leaching Solution of Coal Fly Ash by P204. Minerals 2022, 12, 1547. https://doi.org/10.3390/min1212154
30. Seredkin, Maxim, Zabolotsky, Alexander, Jeffress, Graham, In situ recovery, an alternative to conventional methods of mining: exploration, resource estimation, environmental issues, project evaluation and economics, Ore Geology Reviews (2016), https://doi.org/10.1016/j.oregeorev.2016.06.016
31. Gorman G. Michaelis H. Olson G., Novel In-situ metal and mineral extraction technology, 2004, DE-FG26-03NT4193
32. Echeverry-Vargas, L.; Ocampo-Carmona, L.M. Recovery of Rare Earth Elements from Mining Tailings: A Case Study for Generating Wealth from Waste. Minerals 2022, 12, 948. https://doi.org/10.3390/min1208094
33. Zinoveev, D.; Grudinsky, P.; Zhiltsova, E.; Grigoreva, D.; Volkov, A.; Dyubanov, V.; Petelin, A. Research on High-Pressure Hydrochloric Acid Leaching of Scandium, Aluminum and Other Valuable Components from the Non-Magnetic Tailings Obtained from Red Mud after Iron Removal. Metals 2021, 11, 469. https://doi.org/10.3390/met11030469
34. G.B. Abaka-Wood et al., The upgrading of rare earth oxides from iron-oxide silicate rich tailings: Flotation performance using sodium oleate and hydroxamic acid as collectors, Advanced Powder Technology (2018), https://doi.org/10.1016/j.apt.2018.08.01
35. Karan R., Sreenivas T., Kumar M., Singh D., Recovery of rare earth elements from coal flyash using deep eutectic solvents as leachants and precipitating as oxalate or fluoride, 2022, https://doi.org/10.1016/j.hydromet.2022.105952

36. Watling H., Review: Microbiological advances in biohydrometallurgy, 2016, https://doi.org/10.3390/min6020049
37. Samanta N., Das P., Dhara S., Purkait M., An overview of precious metal recovery from steel industry slag recovery strategy and utilization, 2023, https://doi.org/10.1021/acs.iecr.3c00604 Ind. Eng. Chem. Res. 2023, 62, 9006−9031.
38. Avarmaa K., Johto H., Taskinen P., Distributionn of Precious metals (Ag, Au, Pd, Pt and Rh) between copper matte and iron silicate slag, 2015, https://doi.org/10.1007/s11663-015-0498-4
39. Valighazvini F., Rashchi F., Nekouei R., Recovery of Titanium from blast furnace slag, 2013, https://doi.org/10.1021/ie301837mlInd. Eng. Chem. Res. 2013, 52, 1723−1730.
40. Maweja K., Mukongo T., Mutombo I., Cleaning of copper matte smelting slag from water-jacket furnace by direct reduction of heavy metals, 2008, https://doi.org/10.1016/j.jhazmat.2008.08.107
41. Ramachandra Rao, S. Chapter 8 Metallurgical Slags, Dust and Fumes. Waste Manag. Ser. 2006, 7 (C), 269−327, https://doi.org/10.1016/S0713-2743(06)80093-0
42. Krishna R., Faiz S., Mishra J., Lazorenko G., Kasprzhitskii A., Review article: Mine tailings-based geopolymers: Properties, applications and industrial prospects, 2021, https://doi.org/10.1016/j.ceramint.2021.03.180
43. Raghunandan Kumar, Pranab Das, M. Beulah and H.R. Arjun, Chapter: Geopolymer Bricks Using Iron Ore Tailings, Slag Sand, Ground Granular Blast Furnace Slag and Fly Ash, Geopolymers and Other Geosynthetics, https://doi.org/10.5772/intechopen.81748
44. Marboum S., Moukanna S., Machi A., Taha Y., Benzaazoua M., Hakkou R., (2020), Mine waste based geopolymers: A critical review, Cleaner engineering and technology, vol. 1, 100014, https://doi.org/10.1016/j.clet.2020.100014
45. Beulah M., Sudhir M., Mohan M., Gayathri G., Jain D., Research Article: Mine Waste-Based Next Generation Bricks: A Case Study of Iron Ore Tailings, Red Mudand GGBS Utilization in Bricks, 2021, https://doi.org/10.1155/2021/9499613
46. Murmu A., and Patel A., (2018), Towards sustainable bricks production: An overview, Construction and Building materials, 165, 112–125 pp., https://doi.org/10.1016/j.conbuildmat.2018.01.038
47. Benahsina A., Taha Y., Bouacher R., Elomari M., Bennouna M, (2021), Manufacture and Characterization of fired bricks from Gold mine waste rocks, Minerals, 11, 695, https://doi.org/10.3390/min11070695
48. Lamani S., Mangalpady A., Vardha H., (2016), Investigating the utility of Iron ore waste in preparing non-fired bricks, J. Inst. Eng. India, Ser. D., https://doi.org/10.1007/s40033-016-0129-5
49. Ahmari S., and Zhang L., (2012), Production of Ecofriendly bricks from copper mine tailings through geopolymerization, Construction and Building materials, 29. 323–331 pp., https://doi.org/10.1016/j.conbuildmat.2011.10.048
50. Mandal A., (2021) Production of Lime stabilized thermal insulation bricks using Aluminum mine tailings, J. Inst. Eng. India Ser. A., 102(1): 249–258 pp., https://doi.org/10.1007/s40030-020-00497-7
51. Ally, A.N., Blanche, M.M., Nana, U.J.P., Grâce, M.M., François, N. and Pettang, C. (2021) Recovery of Mining Wastes in Building Materials: A Review. Open Journal of Civil Engineering, 11, 379–397. https://doi.org/10.4236/ojce.2021.114022
52. Hernández I., Díaz H., Morales F., Romero L., Camacho J., Bioleaching of metal polluted mine tailings aided by ultrasound irradiation pretreatment, 2023, https://doi.org/10.1016/j.eti.2023.103192
53. Mishra D., Rhee Y., Kim D., Bioleaching: A Microbial Process of Metal Recovery; A Review, Article in Metals and Materials International · June 2005 https://doi.org/10.1007/BF03027450
54. Klaus Bosecker, Bioleaching: metal solubilization by microorganisms, FEMS Microbiology Reviews, Volume 20, Issues 3–4,1997, Pages 591–604, ISSN 0168-6445, https://doi.org/10.1016/S0168-6445(97)00036-3
55. Kumari P., Kumar P., Kumar T., An Overview Of Phytomining: A Metal Extraction Process From Plant Species, © 2019 JETIR January 2019, Volume 6, Issue 1.

56. Yan A, Wang Y, Tan SN, Mohd Yusof ML, Ghosh S and Chen Z (2020) Phytoremediation: A Promising Approach for Revegetation of Heavy Metal-Polluted Land. Front. Plant Sci. 11:359. https://doi.org/10.3389/fpls.2020.00359
57. Dinh T., Dobo Z., Kovacs H., Phytomining of noble metals—A review, 2021, https://doi.org/10.1016/j.chemosphere.2021.131805
58. Singh D., Nikhil K. Algae for Lipid as Renewable Energy Source in Coal Mining Area: A Critical ReviewInternational Journal of Engineering and Technical Research (IJETR).
59. Hossain A., Salleh A., Boyce A., Chowdhary P., Naqiuddin M. Biodiesel Fuel Production from Algae as Renewable Energy, American Journal of Biochemistry and Biotechnology 4 (3):250–254, 2008 ISSN 1553-3468 © 2008 Science Publication.
60. Menendez J. Energia H., Loredo J., Engergia H. Capture of CO2 emissions by reforestation with energy crops in closured mining areas, 2018.
61. Zhauk A. Ivanov P, Kiseleva E., Modeling of the Electrochemical Transformation of Chemical Energy of Biofuel to Electricity, Capture of CO_2 emissions by reforestation with energy crops in closured mining areas https://doi.org/10.1134/S0018151X2002022

Commercial and Technical Prospects of Unused Mineral Values of Aluminium Industry

Mohamed Najar, Pravin Bhukte, Upendra Singh, Manoj Nimje, and Anupam Agnihotri

Abstract In general, the overall requirement of aluminium metal is met by primary aluminium production involving the Bayer chemical process clubbed with Hall–the Héroult electrolytic process and secondary aluminium production based on the recycling of old and new scrap metal. The process of recycling undergoes remelting or refining to metal based on the type of scarp. Both primary and secondary production of aluminium is left with a variety of solid rejects which are useful materials for productive applications. These material rejects appear in the form of unprocessed solid materials of various kinds having differential chemical and mineralogical compositions containing aluminium, iron, and silica as major components such as mining rejects. Low-grade Bauxite (LB), Overburden Laterite (OBL), Kaolinitic Khondalite (KK), Partially Lateritic Khondalite (PLK), lithomargic clay known as Saprolite (SL), etc. are some of the major unprocessed resources suitable for commercial exploitation. Processed rejects of aluminium production mainly represent the Red Mud (RM), Aluminium Dross (AD), and Spent Pot-lining Materials (SPL). In the present paper, the potential of the processed and unprocessed reject as resources for preferred applications are discussed along with technical hints of value addition for commercial exploitation.

Keywords Aluminium industrial rejects · Process options · Value addition · Commercial exploitation · Market and customer demand

1 Introduction

Over a period of one century, aluminum has evolved as the most used processed metal after steel. This non-ferrous metal is largely in use for a variety of alloys, castings, forging and various products ranging from extrusions, wires, cables, and pipes as

M. Najar (✉) · P. Bhukte · U. Singh · M. Nimje · A. Agnihotri
Analytical Division, Jawaharlal Nehru Aluminium Research Development and Design Centre, Amravati Road, Nagpur 440023, India
e-mail: najar@jnarddc.gov.in

K. Randive et al. (eds.), *Current Trends in Mineral-Based Products and Utilization of Wastes: Recent Studies from India*, Springer Proceedings in Earth and Environmental Sciences, https://doi.org/10.1007/978-3-031-50262-0_19

well as flat rolled products for industrial and domestic uses. Further, in the modern world, the use of aluminium is inevitable for essential infrastructure facilities, agricultural, engineering, chemicals, and packaging industries. Production of primary aluminum (PA) begins with the mining of bauxite and alumina is produced by further crushing and grinding the bauxite ore followed by digestion in sodium hydroxide. The overall upstream activities of the metal production generate a wide variety of solid rejects which are broadly classified into two major types; processed and unprocessed. Most of the processed rejects are byproducts of the metallurgical process and contains added materials to have different chemical composition of parent materials. On the other hand, natural rejects are leftover materials of low quality and possess physical and chemical properties unsuitable for economic utilization. The production of recycled aluminium (secondary aluminium) contributes to the large input of aluminium metal in the overall aluminium consumption. Secondary aluminium (SA) production involves comparatively less energy consumption for the recycling of aluminium scrap. Compared to PA, the production of SA entails approx. 15 times less energy [1]. Secondary aluminum can be equally used in many of the applications where primary aluminum is utilized. However, the inflow of a significant amount of impurity elements in SA makes it unusable for specific applications such as foil for capacitors and disk blanks, etc. [2]. Largely, the process of aluminium recycling essentially resolves the demand for aluminum metal and the environmental issue associated with scrap accumulation. The major disadvantage of SA production is the generation of environmentally hazardous salt slag which demands treatment/utilization of the highly alkaline slag [3] in view of environmental concerns.

The disposal of industrial waste in open lands is inhibited by its rising cost, restrictions in transportation, and various norms relevant to pollution control. Consequently, industries are best trying to reduce the amount of waste being produced and for adequate ways of recycling or reusing the rejects. Thus, the choice for transforming industrial discards into value-added resources for supplementary utilization is largely in demand [4, 5].

The inherent properties of aluminium largely influenced the overall growth in aluminium demand. The Technical properties such as an optimum strength-to-weight ratio, low melting point, corrosion resistance, better electrical and thermal conductivity, and better recyclability provide better usage of aluminium over other ferrous and nonferrous metals. According to a Niti Aayog report, per capita consumption of aluminium in India is 2.5 kg which is far below the global average of 11 kg. This gap presents ample opportunity for the Indian aluminium industry to expand and grow across various end-use segments. Equally, it highlighted the need to address new usage of aluminium in relevant sectors. More usage of PA and SA metal indirectly provides more scope for exploration and exploiting the unused mineral values in associated rejects as products and resources.

2 Unused Resources of Aluminium Industry

In aluminum industry, several wastes are generated during the conversion of bauxite to aluminium and its alloys. The major solid rejects generated in primary aluminium production are bauxite residue known as red mud, dross, and spent pot lining (SPL). During the refining of alumina from bauxite by the Bayer Process, red mud is generated as waste. This is a highly alkaline solid waste that cannot be disposed of in the environment. Red mud generated at the refinery is stored in a protected pond lined with an alkali-resistant impervious layer which does not allow any soaking of liquid from the pond. Nearly the annual generation of bauxite residue (red mud) in India is five million tonnes. The residue holds several oxides such as Al_2O_3, Fe_2O_3, TiO_2, SiO_2, and Na_2O and they are present in complex form. Another bye product generated from aluminium industries is dross. India produces around eighty thousand tonnes of dross per annum. Dross is generated from the smelter and downstream areas. The aluminium content in dross varies from 10 to 70% and it is classified into four different categories based on aluminium content. The lowest grade dross contains around 10–12% aluminium which is often found chemically bonded and termed black dross. It is practically impossible to separate aluminium physically from black dross because it is chemically bonded. Similarly, during the smelting of aluminium another solid waste i.e. spent pot lining (SPL) is generated. Although SPL predominantly has two fractions (i) carbon part and (ii) refractory part, it also contains certain toxic materials. Among them, cyanide is one of them in SPL which needs to be addressed in order to find some suitable applications. India produces around 35,000 tonnes SPL per annum which is stored by primary aluminium producers in a specially covered manner so that it can be kept away from direct contact with soil.

The major source of raw materials for secondary aluminium production comes from two different routes. The first and preferred source of raw materials are industrial waste and manufacturing scraps such as off-cuts and waste streams from aluminium smelters are directly transferred as feed for secondary smelting. In contrast, end-of-life products of various uses and applications which are generally stated as urban mines form the other source of raw materials for secondary aluminium production. The raw materials from both sources are treated similarly for recycling and the process steps involve melting of scrap in rotary or reverberatory furnaces under a bath of molten salt. The salt is typically a mixture of sodium and potassium chlorides comprising low ranges of fluorides (cryolite) and possesses lower melting points compared to the constituents in the mixture. The melted aluminium metal and floating salts are tapped out of the system. The end cycles of the process result inflow of residual aluminium metal (around 5%) and various oxides of aluminium in the salts and the solidified mixture is termed "salt slag". The salt slag known as aluminium salt cake formed in the secondary aluminium smelters contains 12–30% aluminium oxide, 28–57% sodium chloride, 12–32% potassium chloride, 4–8% metallic aluminium, and other nonmetallic impurities such as carbides, nitrides, sulfide, phosphides, etc. Depending on the type and quality of the raw mix, the range of production of salt slag varies from 200 to 500 kg per tonne of metal [6]. About

40% of salt slag by weight is the ideal original salt mix used in the melting process. Salt slag generated secondary aluminum production are hazardous leftover materials and possess adverse impacts on plant and human life. Therefore, the valuable mineral resource needs to be recycled/treated for reclaiming potassium and sodium chloride contents back to the smelting process as well as preparation of value-added products from the final nonmetallic residue.

3 Process Options for Value-Addition

Conservation of natural resources and saving of energy are two important aspects on which the modern developing world is giving utmost importance in view of depleting resources, high demand for products, and short supply of fossil fuels. Worldwide, the large availability and refined forms of different types of industry rejects have received focused attention in the recent past and considerable efforts have been initiated to utilize the industry discards in place of natural resources. Maximum consumption of rejects based on the least process steps and low energy usage leading to zero waste generation is most highlighted based on the circular economy concept. The scarcity of natural raw materials and growing demand for finished products in high volume triggered ample scope of new R&D initiatives for refining and beneficiating various kind of industry rejects into fast raw materials suitable for specific applications. The newly found demand for reject materials outside the parent industry provides much-needed relief in the storage, maintenance cost, and usage of large land areas. This overall change leads to the focus on maximizing the use of existing technologies for the affable treatment of industrial solid rejects. Unfortunately, for various industrial wastes, an adequate way of recycling is still not available. Disposal by landfill is inhibited by its rising cost, restrictions in transportation, and implementation of norms relevant to pollution control. Thus, transforming industrial wastes into valuable materials or resources for further utilization is most amenable and desirable. Utilizing processed and untreated rejects as a source for value-added products attracts attention. The dealing out of reject materials depends on the category and compositions. Both chemical and physical methods with or without hydrothermal treatments are generally in use for waste processing and value-addition.

4 Research and Developments, Outcome and Scopes

Jawaharlal Nehru Aluminium Research Development and Design Centre (JNARDDC) executed successful research activities to establish practical process options to the reject materials of aluminium industry. Extensive studies have been carried out for recovering iron values of red mud for steel making and left-over lighter fractions rich of alumina and silica into insulating materials for engineering applications to achieve complete utilization of red mud based on process byproduct utilization to achieve zero waste.

In order to achieve bulk and large-scale use of red mud, building and construction materials were developed for various engineering applications. Changes in physical properties were achieved in the product range for utilizing the blocks and bricks for different construction applications. The fired bricks and blocks developed from red mud (with or without fly ash and other additives) comprised hard blocks, lightweight blocks, and foamed lightweight blocks. For hard and stabilized brick making, red mud was used along with mineralizer, soapstone $Mg_6[Si_8O_{20}](OH)_4$, and highly siliceous plastic clay available locally under different combinations. The clay material used in the process was predominantly siliceous material with 40–50% SiO_2 and 20–30% Al_2O_3 content. Mineralizer was also used for binding and vitrification. Under optimum conditions, it was observed that 60–80% of red mud could be used in the process along with the clay and mineralizer. The fired lightweight foamed bricks (LWFB) were tested as per IS: 1077: 1992 for common burnt clay building bricks specification. Apart from this test, a number of additional tests were carried out to characterize the fired bricks as per the test specified in IS 1077: 1992 which includes the determination of compressive strength, water absorption, and efflorescence. Additionally, the dry density, bulk density, and porosity were also determined to characterize the product. Caustic leachability was tested to ensure the absence of free alkali. The fired bricks were tested in accordance with the procedure laid down in IS 3495 (Part 1): 1992 for the compressive strength that compares the minimum average compressive strength for various classes of building bricks. From the compressive strength data as per IS 3495 (Part 1):1992, the bricks qualified for the class 20–25. Accordingly, bricks can be produced from red mud with a cost of rupees three (Rs. 3/-) per brick against the market price of rupees six to eight (Rs. 6–8). Additionally, the strength of red mud bricks is three times higher as compared to conventional bricks (Fig. 1).

In another attempt, value-added products from red mud reject were established in the direct conversion of red mud and fly ash mixture into lightweight and foamed

Fig. 1 Construction of red mud brick wall using Portland cement binder

Fig. 2 Lightweight foamed blocks and bricks made of red mud-fly ash mix design

products [7], building block tiles, paver blocks, and bricks based on the heat treatment process. Accordingly, an experimental model has been designed for the development of lightweight foamed bricks (LWFB) from red mud involving minimum processing steps. A two-component raw material consisting of red mud and fly ash was mixed in a high-speed mixer for homogenization. The homogenized mass was made into a slurry with water and foaming agents. Subsequently, the slurry was transferred into molds for drying and fired at elevated temperature in an electrically controlled furnace for 1–2 h. An ideal composition range of 30–60% red mud and 0.25–1% foaming agents in LWFB were derived from the experimental trials (Fig. 2).

Through tiny air-filled pores in the blocks and tiles support the acoustic performance and the light weight makes it suitable for constructing partition walls, floor screens/roofing, and panel material in the auditorium. The developed material exhibited very low thermal conductivity (0.15 Wm^{-1} K^{-1} at 30.4 °C for 6.3 mm compared to the vespel standard). The thermal resistance is proportional to the thickness of a layer of the construction and inversely proportional to its conductivity. A construction layer with a high thermal resistance (e.g. rock wool), is a good insulator; one with a low thermal resistance (e.g. concrete) is a bad insulator and the LWFB made of red mud and fly ash proved as one of the best materials for making partition walls in a high-rise building to achieve low load bearing and construction cost for achieving good market and entrepreneur's attention towards setting up of production units in line with primary aluminium producers in India.

Alumino siliceous reject mixture comprising a combination of red mud, fly ash, and glass-forming substance was used for developing glass ceramic tiles for engineering applications. The mixture consists of 40–80% red mud, and 0–35% fly ash (or similar siliceous material) with few other oxides and coloring agents. The laboratory studies confirmed the formation of glass–ceramic phases in the mixture after heat treatment and annealing process steps. It was observed, that the tiles produced from the solid rejects hold an excellent glossy finish, good mechanical strength, and abrasion resistance. The color of glass ceramic tiles varied by modifying the material

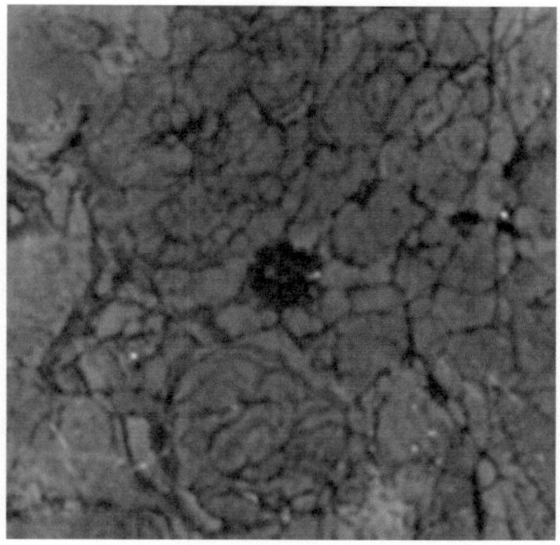

Fig. 3 Glass ceramic tiles made of red mud-fly ash-glass former mix design

mix composition or by adding coloring agents (metallic salts). Testing and characterization of the tiles confined, the properties are at par with marble and granite-based products. Glass ceramics tiles may be a cost-effective substitute as compared to decorative wall tiles. Conversion of red mud into glass ceramic tiles of comparable quality with marble and granite based on vitrification was enabled [8] (Fig. 3).

Spent pot-lining materials (SPL) generated as rejects at aluminium smelter plants contain toxic materials, such as cyanide, and fluoride which are susceptible to leaching in soil and groundwater. Due to the presence of toxic leachable cyanide in SPL, the material is treated as hazardous waste. As per EPA (Environmental Protection Agency) guidelines, material containing total cyanide > 560 ppm, and leachable cyanide > 30 ppm is considered hazardous waste and cannot be dumped in landfilling. Since the presence of toxic (leachable) cyanide is more than the safe limit of EPA guidelines, SPL is considered hazardous waste generated in aluminium industry and it requires special attention for its safe disposal. To study the heat treatment process, the SPL sample (carbon portion) was crushed and classified into different-size fractions. For easy handling, a sample size of − 1.0 mm + 0.5 mm was selected and homogenized for experimental work. With the aim of destruction of leachable cyanide, the heat treatment studies were carried out on representative SPL samples in a specially fabricated electrically heated rotary furnace under controlled conditions and below the ignition temperature of carbon. Under optimum conditions, around 98% of the leachable cyanide was removed from heat treated mass. Some preliminary experiments were also carried out to remove and recover soluble sodium & fluorides from the heat-treated samples while leaching the same with water. It is found that most of the soluble sodium and fluoride can be extracted from heat-treated SPL material. However, with the removal of 98% of leachable cyanide, the SPL is now safe for processing for further applications [9, 10] (Fig. 4).

The recovery of metal from dross is carried out by conventional metallurgical processes all over the world. In the process, the remaining portion of dross after

Fig. 4 Spent pot lining materials (SPL) before and after heat treatment

recovery of metal is a problem for disposal. Furthermore, this dross still contains aluminium metal in finer size which cannot be recovered by conventional metallurgical process. The main constituents of dross are aluminium metal and alumina, metal varies from 10 to 80% depending upon grade. Apart from aluminium and alumina, dross also contains other chemical compounds. Considerable research has been carried out to explore the possibility of extracting value-added products from waste-grade dross and developed methodologies for the preparation of alumino-ferric alum with minimal iron content (hence called low ferric alum) for wastewater treatment purposes [11]. Further studies were also carried out on the utilization of leftover process byproducts into castables to support the concept of zero waste utilization [12] (Fig. 5).

Utilization of mining rejects such as Partially Lateritic Khondalite (PLK) for making ceramic proppant [13] (Fig. 6) was attempted to establish suitable industrial utilization other than recovery of alumina for metallurgical application for the unused resource/reject. Ceramic proppant is useful for providing a high-conductivity pathway for hydrocarbons in oil wells and the current requirement is met by commercial materials produced from high-grade bauxite or similar entities.

Similarly, successful research work was executed for enriching alumina and silica after reducing the concentration of iron oxide (Fe_2O_3) by chemical treatment of PLK at moderate temperature (80–99 °C). Subsequently, the recovered alumina was

Fig. 5 Alum cake obtained from aluminium dross

Fig. 6 Green and ceramic (fired) proppant made of PLK

converted into alpha alumina [14] as well as silica into nano-silicates for suitable commercial exploitation as value-added materials. Industries such as semiconductors, automotive, and medical, are also driving the world market for high-purity alumina. If the alumina separated from PLK is purified to 3N level, the resource could be useful as a natural source for a precursor for making High Pure Alumina (HPA) preparation. Fused alumina is the most demanding application of alumina and there exists the possibility of preparing fused alumina from pure alumina isolated from PLK. Also, highly pure alumina (HPA) nano-particles (HPA-NP) are in great demand in the Indian context due to the non-availability of in-house resources and more research work is needed to establish the process route for commercial use (Fig. 7).

Development of low iron and high-silica and alumina fractions from PLK for making white ceramics and paint industry as well as preparation of high-temperature refractory aggregate [15] from lithomargic clay known as saprolite meet the increasing demand for refractory castable as genuine substitute for calcined clay

Fig. 7 High silica residue and alpha alumina made from PLK

based refractory materials. The bauxite mining industry is well associated with saprolite resources and the resource is readily available (without exploration) for utilization because it is found just below the bauxite zone and is easy to mine due to its soft nature.

Further, the concept of applying physico-chemical treatment for saprolite followed by chemical precipitation of silica by sol–gel method for value addition revealed the possibilities of recovering low iron silica from saprolite for making value-added nano-silicate for industrial utilization. It is expected that this work will promote the value addition and recovery of silica's mineral values in saprolite, overburden laterite, PLK, KK, and similar reject materials containing silica, which are plentiful for appropriate commercial exploitation (Fig. 8).

The rejects and discards of processed and unprocessed materials of aluminum industry are alumino-siliceous in nature. It contains a good amount of alumina and silica in amorphous, hydrated, and combined forms. The chemical nature of these materials is well useful for making geopolymer-based building materials as received or in combination with rejects from other industry origins such as the steel industry, biomass, power industry, or mining rejects of non-ferrous metals such as Mn, Zn, Pb, etc. JNARDDC conducted thorough studies on the synergistic utilization of about 75 industry rejects in combination with aluminium industry rejects for making geopolymer products for building and construction. Lightweight, foam type [16], sandwich type (layered), and hard and high-strength products were developed by the non-fired route for suitable commercial exploitation (Fig. 9).

The studies identified good synergy among aluminum industry rejects—steel industry rejects as well as biomass-based rejects such as rice husk ash, bagasse ash and corncob ash, etc., for the development of green building materials based on geopolymerization. The maximum crushing strength obtained for blocks prepared with materials from aluminium industry and steel industry was 23.94 and 40.80 MPa respectively, which confirmed the specified Indian Standard. The amount of soda leaching on the surface of bricks had less than moderate rating of efflorescence as well as less than 20% water absorption. Blocks prepared with single, double, and multi constituents in the mix design were also found promising in view of curing time,

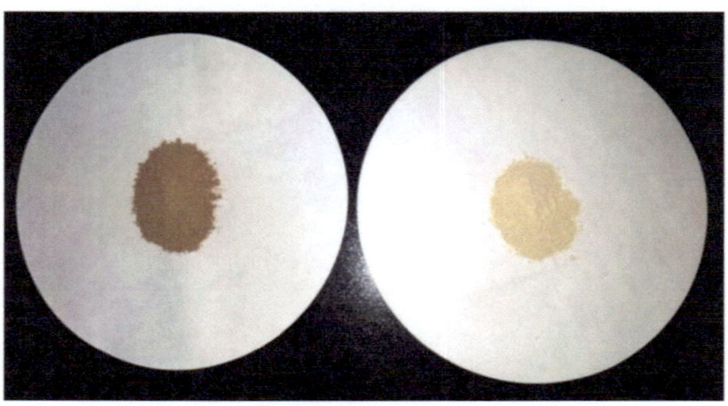

Fig. 8 PLK before and after acid leaching

Fig. 9 Geopolymer-based engineering materials made of red mud-Fly ash-GBFS mix designs

rate of rejection, and overall workability. The bricks, when tested for water absorption in accordance with IS: 3495 (Part-2):1976, were found less than 20% and efflorescence with IS: 3495 (Part3):1976 reported moderate for mix designs comprising aluminium industrial rejects and zero efflorescence for steel industrial rejects. Most of the rejects from the steel industry have proven good drying properties during geopolymer formation at room temperature. Unprocessed blast furnace slag, MSP slag, plant sludge, and granulated blast furnace slag showed good geopolymerization process as the compressive strength of blocks prepared with these raw materials reached up to 40 MPa. A mini-pilot plant for the production of 300 bricks of standard dimension per cycle has been installed and is fully operational for feasibility trials and cost estimation. A similar route was employed for making a geopolymer-based flower pot from a red mud-fly ash mixture (Fig. 10).

Fig. 10 Geopolymer-based flower pot made of red mud-fly ash-GBFS mix design

5 Conclusions

The scope of utilization highlighted on the reject materials is prominent research work executed at JNARDDC in the area of recovery and value-added utilization of processed and unprocessed solid rejects of aluminium industry. The applied R&D activities executed as in-house R&D and sponsored projects for primary aluminium industries in the country as well as the Ministry of Mines, Gov. of India were intended for opening pathways of value addition of rejects which remains unutilized in the vicinity of industrial units for long. The proven R&D efforts and quality products developed at the laboratory are still looking for appropriate scaleup activities, pilot plants, and setting up of industrial units for transferring the technology from lab to field. A consensus of government agencies, aluminium producers (primary and secondary), and entrepreneurs need to be established as working models for developing appropriate business opportunities. Instead of considering the industrial rejects as a major concern, its utilization is our responsibility and essential for supporting development activities in the country. Appropriate action plan leading investable solutions and longstanding partnership needs to be created for business models. Material efficiency utilization services supported by safe and hassle-free transportation, continuous availability with an acceptable range in the change of materials composition, and incentives and motivations for startup establishment with relevant inputs of material and quality criteria of end-use industries such as ceramic, alloys, paint, castable, building & construction, metal recycling and recovery (steel, aluminium, titanium, REs, etc.) has to be established.

References

1. Gaustada G, Olivetti F, Kirchain R (2012) Improving aluminium recycling: A survey of sorting and impurity removal technologies, Resources, Conservation and Recycling, 58, 79–87.
2. Kondo M, Maeda H, Mizuguchi M (1990) The production of high-purity aluminum in Japan, JOM 42, 36–37.
3. Padamata SK, Yasinskiy A, Polyakov P (2021) A Review of Secondary Aluminum Production and Its Byproducts; Recovery, Sorting, and Processing of Secondary Aluminum, JOM, 73, 2603–2614.
4. Aguirre NV, Montes-Morán MA, Ania CO, Canals-Batlle C, González-Suárez V, Martin MJ (2009) Environmental applications of steel industry solid wastes, 1st Spanish National, Conference on Advances in Materials Recycling and Eco—Energy Madrid, 12–13 November 2009.
5. Smith KM, Fowler GD, Pullket S, Graham NJD (2009) Sewage sludge-based adsorbents—A review of their production, properties and use in water treatment applications, Water Research, 43, 2569–2597.
6. Tsakiridis PE (2012) Aluminium salt slag characterization and utilization—A review; Journal of Hazardous Materials, 217–218, 1–10.
7. Najar M, Nimje MT, Chaddha MJ, Mukhopadhyay J, Satpathy BK (2018) Process for preparation of lightweight foamed bricks LWFB) utilizing red mud and fly ash admixtures, Patent No **292333**.

8. Nimje MT, Najar M, Chaddha MJ, Mukhopadhyay J, Satpathy BK (2019) A process for preparing of glass ceramic tiles, Patent No. 326108.
9. Nimje MT, Najar M, Patra B, Satpathy BK (2018) Controlled heat treatment process for destruction of leachable toxic cyanide in first cut aluminium spent pot lining materials, Patent No. 301579.
10. Singh U, Mukhopadhyay J, Patra B, Bandopadhyay P (2022) Process for preparation of low ferric alum from waste or low-grade aluminium dross, Patent No. 406384.
11. Singh U, Nimje MT, Ansari SM, Agnihotri A, Mishra ASP (2017) Development of process for revitalization of high alumina castables from residual Aluminium dross, Application No/ 201721037172.
12. Nimje MT, Najar M, Agnihotri, A., Mishra A.S.P., and Satpathy, B.K. (2022): A process for conversion of hazardous aluminium spent pot-lining (SPL) waste into non-hazardous material and recovery of sodium, fluoride and carbon value, Patent No. 407276.
13. Bhukte PG, Puttewar SP, Chaddha MJ, Agnihotri A, Pradhan B, Bandopadhyay P (2018) Partially lateritised khondalite based ceramic proppant and its preparation, Patent Application No. 201821003354.
14. Najar M, Karn A, Nageshwar P, Chaddha M, Agnihotri A (2020) A Process for selective separation and recovery of alpha alumina and silica, Patent Application No. 2021022813.
15. Bhukte PG, Puttewar SP, Nimje MT, Chaddha MJ, Najar M, Singh U, Agnihotri, A. (2015): Development of process for conversion of saprolite into refractory aggregate, Patent Application No. 3585/MUM/2014.
16. Najar M, Numanuddin MA, Madatali WS, Chaddha M, Bhukte PG, Puttewar SP, Agnihotri, A, Jain S (2022) A lightweight foamed geopolymer (LWFGEOP) and its preparation, Patent No. 409005.

Potential Application of Red Mud in Cement: An Indian Perspective

Suchita Rai, M. J. Chaddha, Prachi Pradhan, K. J. Kulkarni, M. Panchal, and A. Agnihotri

Abstract The insoluble residue remaining after digestion of bauxite with NaOH at high T and P is commonly referred to as "red mud" or "bauxite residue" in the Bayer process of obtaining alumina from bauxite ore. Red mud is a large-scale waste, and because of its alkaline nature and fine granulometry, poses a grave threat to the ecosystem. The construction and building industry have the potential to consume red mud in large quantities and hence its use in the production of cement is being explored. Red mud contains oxides of iron, aluminium, and silica, which are important ingredients of cement, therefore the red mud finds application in the cement industry. The qualities, availability, and characteristics of red mud and cement have been explained along with their Indian and global situation. The paper provides information about key aspects of red mud's potential use in the cement industry. The anticipated technical and financial difficulties of employing red mud in cement have been thoroughly examined. Utilising red mud in the cement industry will support a circular economy and environmentally sound red mud management, enhancing the metal industry's sustainability.

Keywords Red mud · Environmental problem · Cement production · Circular economy · Challenges

1 Introduction

By digesting pulverised bauxite in a caustic soda solution at a high temperature, the Bayer's process produces aluminium oxide (Al_2O_3), also known as alumina. The majority of the bauxite's alumina content enters the solution during digestion to form sodium aluminate, while contaminants including iron, titania, silica, undigested alumina, etc. remain in suspension. After digestion, the resultant slurry is separated

S. Rai (✉) · M. J. Chaddha · P. Pradhan · K. J. Kulkarni · M. Panchal · A. Agnihotri
Jawaharlal Nehru Aluminium Research Development and Design Centre (JNARDDC), Wadi, Amravati Road, Nagpur, Maharashtra 440023, India
e-mail: suchitarai@jnarddc.gov.in

K. Randive et al. (eds.), *Current Trends in Mineral-Based Products and Utilization of Wastes: Recent Studies from India*, Springer Proceedings in Earth and Environmental Sciences, https://doi.org/10.1007/978-3-031-50262-0_20

by sedimentation or filtration into clear sodium aluminate solution and solids which are washed and finally stored in a disposal area. This solid waste of an alumina refinery is called 'bauxite residue' or 'red mud.'

The quality of the bauxite and the parameters of the digestion process are the two key determinants of red mud generation. The generation of red mud will be lower with good-quality of bauxite in terms of higher alumina content and suitable digestion technology whereas the same will be higher with low-quality of bauxite and inferior digestion technology. Per tonne of alumina, 1.0–1.5 tonnes of red mud are produced, depending on the same. Red mud consists of unextracted alumina, unreacted titanium and iron, sodium aluminosilicates along with caustic soda and can potentially be used in the production of cement based on the quality of limestone and pet coke used in the cement industry. The research has addressed the potential use of red mud in cement manufacturing while keeping considerations from the Indian perspective. Table 1 is a brief summary of the literature on the use of red mud in cement.

Table 1 Use red mud in cement

Studies carried on	References
Development of a process to produce hydraulic cement consisting of red mud as a major constituent	[1]
Study on the possibilities of cement preparation from red mud which may have an edge over normal cement when applied to briquetting of fine grain iron ore	[2]
Investigation on how the addition of red mud to lime affects the strength of cement and other characteristics	[3]
Cement clinkers are made by heating a mixture of red mud, limestone, and marl to 300 °C in a rotary kiln	[4]
The process of making cement using red mud with high iron content	[5]
Development of a process to remove the alkaline part of red mud and its utilization as a source of iron for the cement industry	[6]
Investigation on the specific process for using red mud in cement	[7]
Research on using red mud and lime in the context of the region to make cement in the absence of appropriate raw materials	[8]
Red mud from HINDALCO, Renukoot, India, was utilised for producing cement. The best conditions for preparing cement were raw mix iron oxide to alumina (F/A) ratios of 0.8 to 1.2 and a fire temperature of 1250 °C for 1.0–1.5 h	[9]
Investigated the viability of making cement using gypsum, bauxite, red mud, and lime	[10]
Alkali-slag-red mud cement development. The Shan Dong alumina refining plant produced red mud	[11]
Review of three approaches for employing red mud in cement production: making cement clinkers, composite cement, and alkali-activated cement	[12]

(continued)

Table 1 (continued)

Studies carried on	References
Cement is produced through the combination of 30–35% of HINDALCO red mud, 15–20% bauxite, 7.5–10% gypsum, and 45–50% lime, depending on the type of cement required	[13]
Alkali-activated cement was made using a significant amount of red mud, low basic aluminosilicate compounds with a glass phase (blast-furnace slag), and additives chosen from high basic Ca-containing cement with a crystalline structure (Portland cement)	[14]
Use of red mud as a binder in making hydraulic cement through an elevated temperature process that leads to cement clinker	[15]

2 Cement Industry in India

As the government focuses on infrastructure and housing development for all by 2024, the housing and real estate sectors are the major drivers of cement demand. 502 million tonnes of cement are produced annually at the installed capacity. According to the Indian Brand Equity Foundation 2021, India produced 329 million tonnes (MT) of cement in FY20 and 294.4 million tonnes (MT) in FY21. India produced 7.8% more cement in February 2020 than in February 2019. India manufactured 262 million tonnes (MT) of cement total in FY21 (through February 2021) (Indian Brand Equity Foundation 2021) and is the second-largest cement producer in the world. The share of demand for cement in various fields is shown in the pie chart below in Fig. 1.

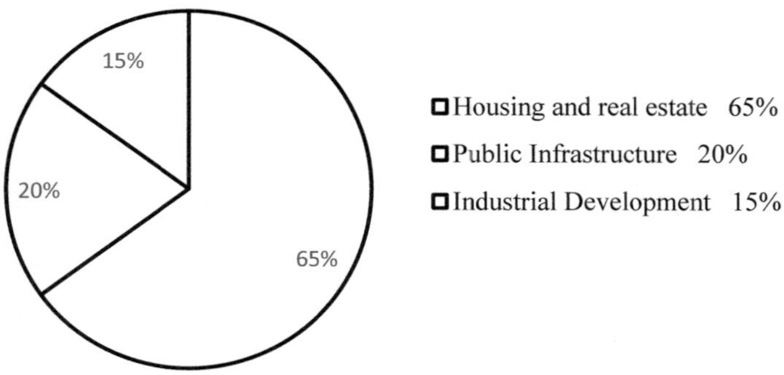

Fig. 1 Demand for cement in various sectors in India

3 Red Mud Generation: Global and Indian Scenario

150 million tonnes of bauxite residue is generated globally each year [16]. Red mud is typically stored in huge red mud ponds and is currently generated by six alumina refineries in India. These Indian alumina refineries generate more than 9.0 million tonnes of red mud yearly.

With the expansion of alumina production capacity by Vedanta and starting up of ANRAK Aluminium, red mud generation is bound to increase by approx. 3.5 million tonnes. These figures are likely to be increased further with the setting up of new alumina refineries and decreasing bauxite quality in India as well as the world over. Between 95 and 100 million tonnes of red mud are being kept in mud ponds in India, but 2.5 billion tonnes of red mud are kept in land deposits around the world [17].

Figures 2 and 3 illustrate the red mud production in India and globally, respectively. Although red mud's use in cement is being mentioned here, it is very important to know the locations of these alumina plants. In Fig. 2, the quantity shown is collectively from all the 4 alumina refineries of HINDALCO situated at Muri in Jharkhand, Renukoot in Uttar Pradesh, Belgavi in Karnataka, and Utkal in Orissa. Vedanta's plant at Lanjigarh and NALCO's plant at Damanjodi is also situated in Orissa.

Since the demand for aluminium metal is projected to double between 2010 and 2050, or 160–200 million t/y of metallurgical alumina + 8 million t/y of chemical grade alumina, the red mud generation would be over 210–250 million t/y by the year 2050 in the world.

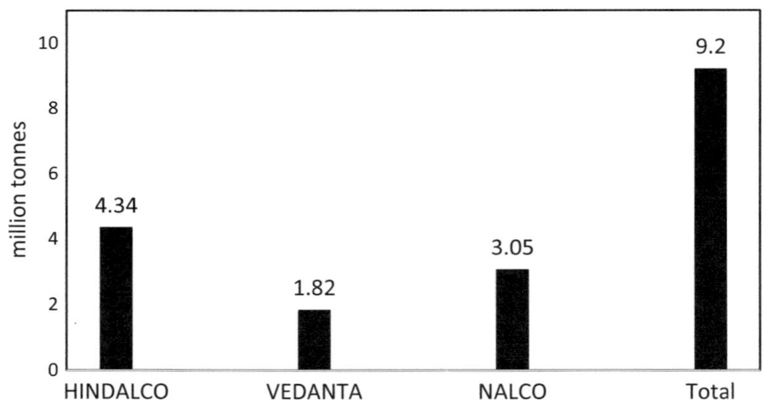

Fig. 2 Red mud generation in India

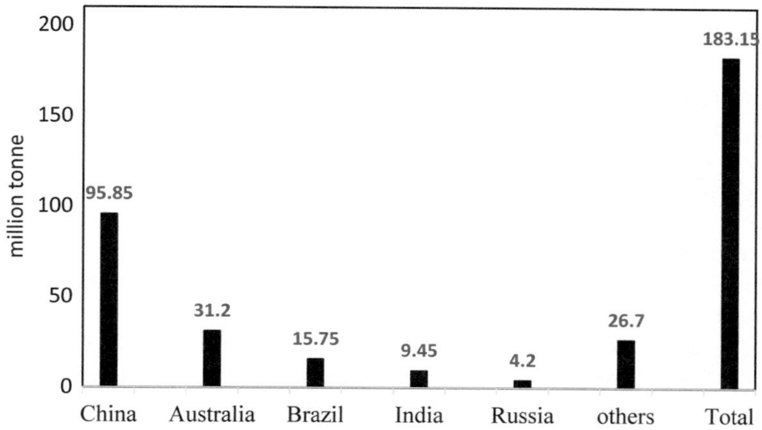

Fig. 3 Red mud generation in the world

4 Red Mud Utilization in Cement

i. Bauxite residue is used to make cement in factories in China, Ukraine, India, Russia, Georgia, Moldova, and Greece.
ii. Over 260,000 tonnes of red mud from the Nikolayev alumina refinery in Ukraine are used annually in the cement industries of Ukraine, Russia, Georgia, Moldova, and Belarus.
iii. A cement company in Milaki was planned to produce about 180,000 tonnes of red mud per year, however it has since stopped because of financial difficulties.
iv. Utilization of bauxite residue in small quantity in the Tulcea was observed in a local Romanian cement plant.
v. After a lot of pilot plant trials with nearby cement industries, 14 cement factories receive 35% of the bauxite residue that HINDALCO Belgavi produces each day.

Red mud is increasingly being used to produce cement because it can consume enormous amounts of waste when used for the manufacturing of construction and building products. Nearly all the red mud produced each year can currently be used, given the capacity of cement manufacturing, if 3–4% of it is employed in the making of cement. Red mud from Hindalco is currently being used in UltraTech cement plants where it has successfully replaced up to 3% of the volume of the clinker raw mix, replacing mined resources. Knowing the differences between the two is crucial for the use of red mud in the manufacturing of cement. These differences are described in the following sections.

4.1 Characteristics of Red Mud

High in alkalinity, fine-sized, irregular, and aggregated is red mud. It has weak soil-mechanical properties and is thixotropic. Red mud has a wide range of chemical and mineralogical compositions depending on the bauxite source and the process variables. It is found to be an extremely complex substance that contains six major oxides, including Al_2O_3, Fe_2O_3, TiO_2, SiO_2, Na_2O, and CaO, as well as several minor/trace element oxides, including V, P, Ga, Cr, Zr, U, Th, Mg, Sr, Ba, Li, K, Pb, Cu, Mn, Ni, Zn, Co, rare earths, etc. Table 2 displays the normal chemical contents of Indian red mud.

In addition to these, red mud is composed of 15–20 different mineral phases, the number of which depends on the source of the bauxite and the conditions under which it is processed. Mineral phases such as gibbsite, boehmite, diaspore, alumogoethite, haematite, magnetite, goethite, rutile, anatase, quartz and kaolinite have their origin in the bauxite from which red mud is formed. On the other hand, phases such as sodium aluminosilicate (NAS), also known as sodalites, cancrinite, calcium aluminosilicate (CAS), sodium titanate, calcium titanate, calcite, etc., are created throughout the process. The multiplicity of mineral phases, each with a unique set of properties, is what gives red mud its complicated and unpredictable behaviour. As a result, they have an impact on both the disposal strategy and the possible application. The physio-chemical characteristics of red mud are given below:

Table 2 Chemical content of red mud in India

Components	Weight %
Al_2O_3	16–23
Fe_2O_3	38–62
SiO_2	4–10
TiO_2	5–16
CaO	0.8–2.6
Na_2O	3–6.5
LOI	8.5–13

Table 3 Chemical composition of cement [18]

Components	Weight %
CaO	60–70
Al_2O_3	5–10
SiO_2	20–25
MgO	< 5
Fe_2O_3	2–3

1. Alkalinity: pH 12–13
2. Particle size: $d_{50} < 10\ \mu m$
3. specific surface: 14–25 m^2/g (depends upon process technology)
4. real density: 3–3.8 g/cm^3.

Based on the data above, Indian red mud can be broadly divided into two categories:

(i) Red mud with a high iron content but low titanium content, that is produced by the NALCO alumina refinery from bauxite from the east coast.
(ii) Red mud with a moderate iron content and a high titanium content, that is produced at the Hindalco, Renukoot, and Muri alumina plants from bauxites from Central India. Belgaum Plant's red mud quality is in between the first and second group.

4.2 *Characteristics of Cement*

Cement is a construction material that sets, hardens, and binds other materials such as sand and gravel together. It binds coarse and fine aggregates to form concrete. It is inorganic and is lime or calcium-silicate-based. Its capacity to set in the presence of water determines whether it is classified as hydraulic or non-hydraulic. Limestone, clay, bauxite, and gypsum are the raw materials used for cement production. Limestone is first burned to produce lime (calcium oxide) liberating carbon. Calcium silicate and tricalcium silicate are produced when lime reacts with silicon dioxide. The lime reacts with aluminium oxide to form tricalcium aluminate. Finally, the reaction between calcium oxide, aluminium oxide, and ferric oxide produces cement. The following reactions occur during cement production:

$$CaCO_3 \rightarrow CaO + CO_2$$

$$2CaO + SiO_2 \rightarrow 2CaO.SiO_2$$

$$3CaO + SiO_2 \rightarrow 3CaO.SiO_2$$

$$3CaO + Al_2O_3 \rightarrow 3CaOAl_2O_3$$

$$4CaO + Al_2O_3 + Fe_2O_3 \rightarrow 4CaO.Al_2O_3.Fe_2O_3$$

5 Key Aspects Required for the Use of Red Mud in Cement Production

As red mud consists of Fe_2O_3, Al_2O_3, and SiO_2 (as seen in Table 2), it meets the raw material demand of limestone, clay, bauxite, and gypsum needed for cement production. Red mud's alumina and iron oxide content make up for limestone's lack of these elements, which is the main raw material used to make cement. The presence of caustic soda in the red mud, when used to produce clinker, balances the sulphur level of the pet coke used to burn the clinker-enrooted cement and improves the setting properties of the cement. Red mud, which contains dicalcium silicate, is another essential component of cement clinker. The following are the main factors for using red mud in cement manufacturing:

1. A red mud dry disposal system can easily be used to achieve red mud's comparatively low moisture content, which is between 20 and 25%.
2. It has been claimed that a value of 2.5% Na_2O is adequate, although this will depend on the other raw materials' chemical composition. Levels of sodium are relatively low.
3. According to one study, the best results are achieved by using aluminium oxide and iron oxide in appropriate proportions. However, some companies utilise the bauxite residue to boost the iron level while others use it to increase the alumina content.
4. Reasonable proximity to a cement plant—A distance between the cement plant and the location of the red mud pond must be both practical and reasonable.

6 Challenges

The use of red mud as a raw material for cement production comes with several technical and economic constraints. Based on the following criteria, prospective use of red mud in the cement industry needs to be assessed:

6.1 Technical Challenges

1. Inherent moisture, alkaline and hygroscopic characteristics make red mud moist, sticky, and difficult to handle. To overcome these associated problems, the industry must go for a dry disposal system that can reduce these problems to a certain extent. Some Indian alumina refineries have a dry disposal system of disposing of red mud.
2. Usage of red mud is restricted to non-monsoon months owing to the above problems.

3. The utilization of red mud needs to be addressed plant-wise since each cement plant has a different raw mix design and the requirement of iron and alumina in a raw mix depends on the limestone characteristics and fuel used for firing.
4. Quality inconsistencies in red mud led to blocking and jamming during pyro-processing and are detrimental to cement quality. The quality of red mud is a very important criterion because each cluster has got a different configuration of limestone which could be one of the challenges regarding its utilization. To overcome these problems the red mud generated from a fixed source of bauxite should be used to maintain quality consistency.
5. The quality of limestone and the type of fuel being used in the cement industry are the main controlling factors for the utilization of red mud.
6. The compatibility studies of red mud composition with the cement raw mix will vary with each cement plant and a more realistic assessment is required for its utilization.
7. Potential role of TiO_2 in affecting clinker chemistry at high concentrations needs to be studied with red mud generated using Central Indian bauxite having higher TiO_2 content (12–14%).

6.2 Economic Challenges

1. The cost of transportation of red mud to the cement manufacturer should be mutually decided as the transportation cost may be more compared to that of red mud disposal cost.
2. The location of the alumina refinery and cement plant manufacturer should be within a 300–400 km radius. The possibility of exploring transportation through the rail at some concessional tariff should be explored.
3. The cost of transportation of red mud increases many folds (7–8 times) if the moisture content is not regulated.
4. Red mud generation is limited to one cluster (Eastern), whereas Cement Industry is located far off from there. Logistics and movement of red mud to western/southern/Northern/NE clusters of cement plants need to be addressed. Freight charges should be taken into consideration.
5. Red mud should be made commercially viable for the cement industry for it to be considered as an alternative option. Red mud of consistent quality should be provided to the cement plants. Cost-effective alternatives should be adopted to maintain a certain quantity of alumina and iron in the red mud.

7 Summary

The chemical constituents of red mud make it potentially suitable for utilization in cement production. Technical and economic suitability are the most important things to be investigated for the utilization of any waste and the basic economic viability model should be followed for the use of red mud in cement. The utilization of red mud in cement is location specific. Red mud properties should be altered and made marketable to utilize. The aluminum industry must start looking at how the requirement of the concerned cement industry can be met. The concerns of both sides (the alumina and cement industry) must be properly addressed and the best solution should be implemented. Red mud utilisation will encourage a circular economy and lessen the cement industry's reliance on natural resources. Red mud might be effectively recycled on a wide scale while simultaneously lowering clinker emissions if used in the manufacturing of cement.

References

1. Shiro, S., Yuichiro, I. and Hideo, K (1963): Chemical Abstracts.
2. Istvan, S., (1960): Epitoanyag: 28–37.
3. Shunzo, K. and Yoshikazu, H. (1963): Rika Gaku Kenkyusho Hokoku: 224–225.
4. Stanislaw, S., Artur, K., Wladyslaw, S., Henryk, W., Marian, B. and Jerzy, B. (1967): Chemical Abstracts.
5. Wargalla, G., (1973): Chemical Abstracts.
6. Youth, C C., (1973): Kuang Yeh: 25–27.
7. Shimano, S. and Koga, T. (1980): Light Metals: 53–80.
8. Allik, A R., (1980): Chemical Abstracts.
9. Singh, M., Upadhyay, S N. and Prasad, P M. (1996): Preparation of Special cements from red mud, Waste Management 16(8): 665–670.
10. Singh, M., Upadhayay, S N. and Prasad, P M. (1997): Preparation of iron-rich cements using red mud, Cement and Concrete Research 27(7): 1037–1046.
11. Zhihua, P., Li, D., Yu, J. and Yang, N. (2003): Properties and Microstructure of the hardened alkali activated red mud slag cementitious material, Cement and Concrete Research 33(9): 1437–1441.
12. Xiaoming, L. and Zhang, Na. (2011): Utilization of red mud in cement: A review, https://doi.org/10.1177/0734242X11407653.
13. Samal, S., Ray, Ajoy, K. and Bandopadhyay, A. (2012): Proposal for resources, utilization, and processes of red mud in India—A review, international journal of Mineral Processing 118, 43–55.
14. Krivenko, P., Kovalchuk, O., Pasko, A., Croymans, T., Hult, M. and Lutter, G. (2017): Development of alkali-activated cements and concrete mixture design with high volumes of red mud, Construction and Building Materials 151, 819–826.
15. Pontikes, Y., David, A-M. and Hertel, T. (2019): Novel binders incorporating high volumes of bauxite residue: Calcium sulfo-ferroaluminate and ferrite-rich cements as well as Fe-rich inorganic polymers. In proceedings of 7th International conference and Exhibition on Aluminium. Bhubaneswar, India.
16. K. Evans et al. (2016): The history, challenges and new developments in the management and use of bauxite residue, Journal of Sustainable Metallurgy **2**, 316–331. https://doi.org/10.1007/s40831-016-0060-x.
17. V. Denton et al. (2014): Environmental sustainability of the alumina industry in western Europe, Sustainability **6**, 9477–9493. https://doi.org/10.3390/su6129477.
18. Y. Pontikes, G.N. Angelopoulos, Bauxite residue in cement and cementitious applications Current status and a possible way forward, Resources, Conservation and Recycling, Volume 73, 2013, Pages 53–63, ISSN 0921-3449, https://doi.org/10.1016/j.resconrec.2013.01.005.

Value Added Bauxite Products and Beneficiation Opportunities

Ashok Kumar Nandi and Hiresh Shrirame

Abstract In the world about 10–15% bauxite are used in the non-metallurgical applications including value-added products, particularly for refractory, abrasive, proppants, alumina cement, ceramics, and scores of chemicals. These applications, in general, require more stringent specifications compared to metallurgical industry, particularly with reference to alumina, iron and titania contents and specific physical properties. China, India, Guyana, and Brazil are the major producers of these value-added bauxite products. Some of the bauxite and special alumina products are similar, however, there is a vast difference in their quality, product specification, industrial applications, and prices. The natural white bauxites are rapidly depleting in the world. In order to produce high alumina bauxite, it is necessary to beneficiate widely available metallurgical grade ore. In this context, a large bauxite deposits of Eastern Ghats belt of India are identified as potential raw material target as this has low level of impurities and any process which can significantly decrease iron content can make this chemically suitable for refractory and abrasive application. As the value of the calcined bauxite is comparable to alumina, 3–4 stages of physical/chemical/bio beneficiation techniques can be applied, which may lead to produce low iron and high alumina bauxite.

Keywords Refractory · Abrasive · Beneficiation · Non-metallurgical · Calcination

1 Introduction

Total world bauxite production is around 300 million tons per year. About 80–85% of bauxite mined are processed to produce alumina and aluminium metal. Remaining 10–15% of crude bauxite is used in the non-metallurgical industries, however, all of them are not value-added products [1]. In USA about 73% of the bauxite undergoes the Bayer's process to obtain alumina or aluminium hydroxide. The remaining

A. K. Nandi (✉) · H. Shrirame
International Bauxite, Alumina and Aluminium Society (IBAAS), IndiaNagpur
e-mail: ashok.nandi@ibaas.info; info@ibaas.info

K. Randive et al. (eds.), *Current Trends in Mineral-Based Products and Utilization of Wastes: Recent Studies from India*, Springer Proceedings in Earth and Environmental Sciences, https://doi.org/10.1007/978-3-031-50262-0_21

non-metallurgical bauxite is used in the production of abrasive, cements, chemicals, proppants, refractories, and its application in steel mills as sludge adjuster. The production of high-quality calcined Bauxite for refractories, castables, ramming mixes, abrasive grains, mortars, and gun mixes require high alumina and low iron and titania bauxite. This is also used in the production of sintered bauxite (Proppants), widely applied in oil and gas wells for hydrofracturing. Although there are several small value-added bauxite products, the major applications are in the production of high alumina refractory and abrasives. A large quantity of laterites and low-grade bauxites are used as additive in cement industry; however, they cannot be considered value added products.

There are two basic lines—first non-metallurgical products based on high quality bauxite and second special alumina hydrates and calcined alumina. There is also some product like Zeolites, produced from digested bauxite slurry, which are almost half-way of alumina hydrate production. Some of the bauxite and alumina-based products are similar, however, there is a vast difference in their quality and product specifications. For example, alum and calcium alumina cement can be produced from bauxite as well from alumina powder and they have large difference in quality, prices, and applications. Some products like AlF_3 are traditionally produced from alumina, however, a new process technology has been developed in Australia to produce this directly from bauxite also.

High quality calcined and sintered bauxites with alumina levels of 93% coupled with low impurities levels of silica, iron and alkalis can partly replace alumina and ideal for production of quality refractory bricks and castable. Currently Guyana is prominently recognized as a rich source of high-grade white bauxite, which is exceptionally well suited for refractory and abrasive industries. During the process of calcination, the material is subjected to high temperature just below its melting point which expels both moisture and chemically bound water. The loss of water can be both from moisture and chemically bound water that cannot endure such intense heat. Hydrates, carbonates, and other elements break down, resulting in the release of volatiles after high temperature treatment. As the temperature surpasses 1100 °C, bauxite undergoes profound change, becoming a compact, strong, and non-absorbent calcined form. calcined bauxite possesses such distinct characteristics that it cannot be replace by another material in refractory and abrasive industry.

2 Value Added Bauxite Products

A general non-metallurgical application of bauxite is shown in Fig. 1.

Here aluminium chemicals are mainly produced from the alumina trihydrate (ATH) and only selected high quality bauxites are used for some alumina chemicals. The general distribution of value-added bauxite products is depicted in Fig. 2 [1].

According to Fig. 2, about 30% of value-added bauxite in the world are used in the production of refractory, abrasive materials, and proppants. The other uses of

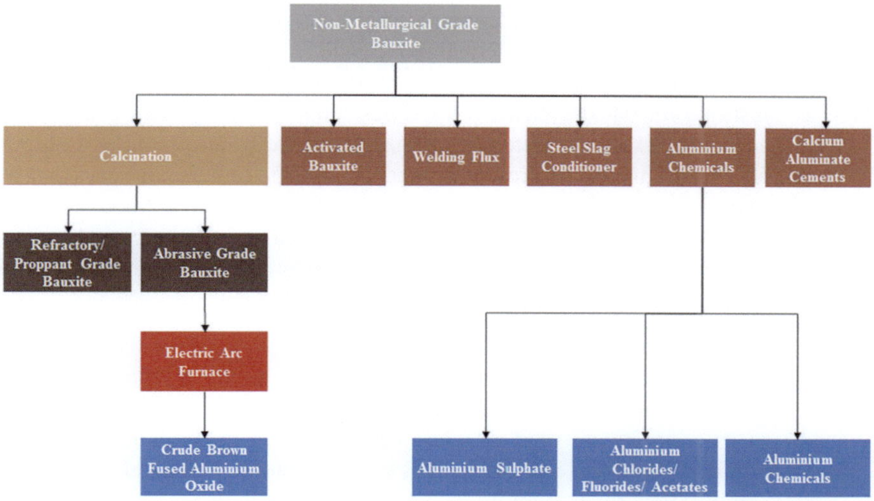

Fig. 1 Non-metallurgical uses of Bauxite

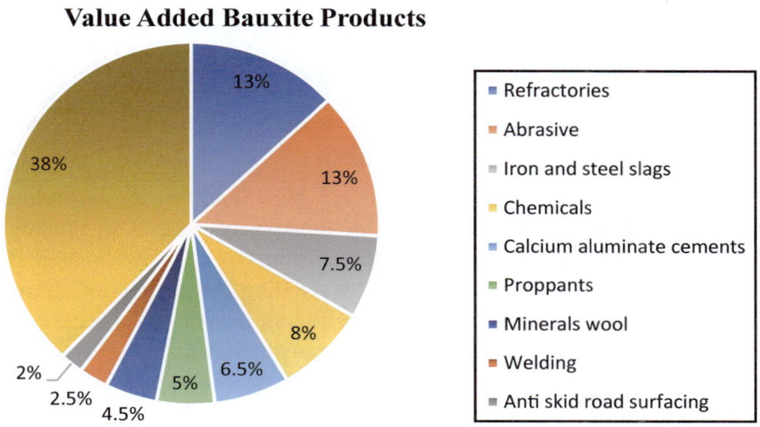

Fig. 2 Share of various value-added Bauxite products (Modified Fastmarket diagram)

bauxite (38% in above pie diagram) are in the production of various types of fluxes, glass industry, water treatment, flame retardants, ceramics, and potteries. A good quality calcined bauxite has similar market value as metallurgical grade alumina and there is a fair margin for processing metallurgical grade ore. At present, the exclusive reserves of natural high alumina and low ferric bauxite lie solely in Guyana, making the entire world, including India, reliant on this South American country. Hence it becomes important to beneficiate suitable metallurgical grade bauxite, with aims to produce bauxite with high alumina content (>55% as Al_2O_3) and low iron content (Fe_2O_3 < 3%), which is ideal for calcination purpose.

In certain countries no longer having sufficient bauxite resource, synthetic mullite, serve as substitute for bauxite-based refractories. This synthetic mullite is derived from kaolin, bauxitic kaolin, kyanite and sillimanite, providing an alternative to fulfill the refractory requirements. At present around 1.2 million tonnes per year (tpy) of calcined bauxite is used in refractories, along with abrasive industry. About 40–50% of Brown fused Aluminium oxide (BFA) are also produced from the calcined bauxite. BFA production is approximately around 1.2 m tpy in the world [1]. Calcined bauxite finds application in various other uses, including the following:

- Anti-skid Road Surfacing—The application of calcined bauxite on road surface as aggregate is relatively expensive leading to its use in high rupture zone, precisely near traffic lights and pedestrian crossings. In the UK, it has been traditionally appraised as a significant market, while other abrasive minerals are preferred in other regions. Calcined bauxite market in this context is relatively small and variable, approximately amounting to as little as 10,000 tonnes.
- The requirement of calcined product for welding purpose should be highly refined with minimal LOI and extremely low level of phosphorous, sulphur and carbon are essential. Excluding China, the projected global market demands approximately 60,000 tonnes. China is anticipated to require comparable quantity of products.

High alumina cement is broadly grouped in two categories:

1. High purity cement having Al_2O_3: 70–80%.
2. Medium purity cement with Al_2O_3: 40–50%.

High purity cement is manufactured using high quality calcined alumina or alumina hydrate (ATH) and Hydrated lime. Medium purity cements are manufactured by using raw bauxite and limestone or hydrated lime.

High Alumina Cement is manufactured by grinding the clinkers of alumina and calcareous material such as lime followed by fusing or sintering process. Initially known for its resistance to chemical attacks, it was predominantly utilized in marine construction. Its popularity surged as its advantageous properties like rapid strength gain, hardening, and ease of manufacturing were discovered and applied in various precast concreting applications. This CAC are widely used in construction, civil engineering, and refractory industries because of their ability to withstand high temperature, High resistance to chemical corrosion and serve as bonding agent when incorporated into refractory castables.

Special grades of high alumina cement are manufactured from high quality lime and alumina. Different types and grades of bauxite are also being used to produce alumina cement based on their specific application. To produce low iron- high alumina cement, it is normal to specify an Fe_2O_3 content less than 2.5%, while for the other grade, contents of 25% may be acceptable. In both cases, the silica content must be controlled, and a typical specification for cement use is Al_2O_3/SiO_2 ratio of at least 10/1, which means that a bauxite containing more than 6% SiO_2 is seldom accepted by producers of high alumina cement.

3 Required Raw Bauxite Specifications

The refractory and abrasive grade bauxite needs to meet stiff compositional standards compared to bauxite used for aluminium industry. These grades involve specially chosen, high-quality bauxite sourced from deposits for metal production, commanding significantly higher prices in the market [2]. Naturally occurring high alumina bauxite are limited even though there are extensive bauxite resources worldwide. Specifications for different categories of bauxite are given in Table 1 [3].

As shown in Table 1, the alumina content should be at least 55% or higher in raw bauxite used for calcination. Depending on mineralogy and LOI content of ore, about 1.4–1.6 tons of raw bauxite is required to produce 1 ton of calcined bauxite. The process used to manufacture the high alumina cement dictates the physical specification of bauxite.

Typical specifications of bauxites for the two types of high alumina cement in Table 2.

Table 1 Specifications of Different Grades of Bauxite by Some Industry

Grade	Major Oxides (%)			
	Al_2O_3	SiO_2	Fe_2O_3	TiO_2
Metallurgical	40–52	1.5–15	5–30	1–6
Cement	45–55	Max. 6	20–30	3
Abrasive	Min. 55	Max. 5	Max. 6	Min. 2.5
Chemical	Min. 55–58	Max. 5–12	Max. 2	0–6
Refractory	Min. 55	Max. 1.5–5.5	Max. 2	Max. 2.5
Proppants	Min. 55	Max. 6	Max. 7	Max. 8

Table 2 Specifications of two types of high alumina cement

Parameter	High Iron product	Low Iron product
Al_2O_3/SiO_2	10/1 (min.)	10/1 (min.)
Al_2O_3/Fe_2O_3	2.0–2.5/1	20/1 (min.) (preferred)
Al_2O_3/TiO_2	–	16/1 (min.)
Mineralogy	Boehmite or diaspore	Not critical
Free moisture	5% (preferred max.)	Dry preferred
Particle size	50–100 mm	Not critical

3.1 Refractory Grade Bauxite

The Characteristic of refractory grade bauxite are high alumina content and low alkali content, along with minimal impurities (Iron oxide and Titania specifically), are instrumental in providing stable temperature strength.

- High alumina content and low iron oxide content generally > 55% and 2.5% maximum after calcination.
- Maximum TiO_2 content 4% and only trace amounts of alkalis and alkaline earths.
- Silica content exceeding 10% might be permissible, as long as it predominantly exists in the form of clay mineral (as kaolinite, halloysite etc.) and not as quartz.

During the calcination process of refractory grade bauxite, temperature lies between 1600 °C and 1700 °C, at this temperature all water i.e., free and combined is eliminated. The clay's silicate minerals experience solid-state reactions and transform into mullite, an aluminium silicate represented by formula $3Al_2O_3.2SiO_2$. General alumina minerals are gibbsite, boehmite and diaspore after dehydration undergo series of phase changes. A portion of this alumina is absorbed in the reaction with silica and balance forms α-alumina (corundum). High refractoriness, good mechanical strength and moderate slag resistance are the qualities of refractory grade bauxite.

The high-temperature properties of bauxite such as refractoriness and melting point are predominantly governed by its material composition. Under rigorous operating conditions, the primary physio-chemical factor influencing the wear resistance of a refractory is density. In refractory, the diasporic bauxite is preferred choice, although mixed gibbsitic-boehmitic or boehmitic ores are also suitable options.

Currently the proven deposits of this specific bauxite for refractory applications are available only in Guyana and China. The First Bauxite, Guyana mines produces high grade bauxite for value added applications and widely used for calcination in the world. An Indian company, Dharti Minerals calcines 100% raw bauxite of Guyana and produces high quality calcined material as shown in Table 3.

3.2 Abrasive Grade Bauxite

The Calcination of abrasive grade bauxite occurs at a temperature of approximately 1100 °C, which is lower than the calcinating temperature used for refractory grade bauxite. However, it is still sufficiently high to prevent the calcined product from readily absorbing water and forming hydrated compounds. The desired alumina content should be as high as possible, typically with calcined bauxite containing over 80% Al_2O_3 (depending on mineralogy/ LOI content in raw ore Al_2O_3 equivalent to 55–60%). In the original bauxite the mineralogical form of alumina is not critical, gibbsitic bauxite typically make up the majority of feed stock. Silica is considered as main impurity as the reduction of oxide to the elemental silicon is costly in terms of

Table 3 High quality calcined bauxite based on imported raw ore

Refractory grade calcined Bauxite (Guyana origin)	
Quality certificate	
Chemical properties	
Al_2O_3	90.40%
Fe_2O_3	0.87%
SiO_2	4.41%
TiO_2	2.57%
CaO	0.28%
LOI	0.28%
Physical properties	
Bulk density	3.26 gm/cc
Apparent porosity	4.72%

energy usage and requires carbon as the reductant. Simultaneously, maintaining the silica/iron equilibrium in the resultant ferrosilicon by-product is crucial, requiring substantial iron incorporation when elevated silica content is detected in the ore. In general, calcined bauxite containing more than 7% SiO_2 is not used. Titania is important in determining the quality of the final corundum since it controls the crystal structure, which in turn controls the toughness of the alumina grains. Thus, the TiO_2 content in bauxite must be in the range of 2–4% so that the hardness of alumina grain is not impaired. The significance of iron oxide content is relatively lower, as it can act as a replacement for alumina within the bauxite feed; usually, levels are maintained below 10%. Low levels of total alkalis are specified. The presence of calcium oxide results in the generation of glassy phases in corundum and for this reason levels in excess of 0.1% may be a cause for rejecting the calcined bauxite. Finally, water, either free or lattice bound, can be hazardous as it can lead to the generation of superheated steam. Due to this consideration, the calcined bauxite's loss on ignition (LOI) should not surpass 1%.

Apart from adhering to the chemical requirements, calcined bauxite intended for abrasive uses must also meet specific physical standards. Particle size holds paramount importance, as excessive coarseness can impede calcination, while overly fine material has the potential to trigger issues with dust generation.

4 Can Bauxite Be Beneficiated to High Alumina Products?

As indicated the natural raw bauxite resources for refractory, abrasives and proppant industries in India, China and Brazil have almost depleted and these countries are looking for viable beneficiation process. For example, in India, in average about 5–10% white bauxite suitable for calcination plant is being recovered during mass mining for bauxite in Jamnagar and Kutch mines of Gujarat. Among Indian bauxite, it is feasible to bring down the iron content in the Eastern Ghats bauxite (Odisha state) as other impurities like titania, silica and calcium are quite low here. Some of the research directions for production of low iron-low titania bauxite are summarized below.

4.1 Stage 1 De-ironing of Bauxite

A typical chemical and mineralogical composition of Eastern Ghats bauxite of India, used for the beneficiation test is presented in Table 4.

In one of our experiments for two stages beneficiation of Eastern Ghats bauxite, first the ore was screened by 1 mm sieve and coarser fraction was further treated in wet high intensity magnetic separation after grinding bauxite to −200 mesh (74 micron) size. First screening has substantially improved the grade as kaolinite and goethite minerals have concentrated in the fines. The wet high intensity magnetic separation (WHIMS) work, carried out at ore dressing pilot plant of Indian Bureau of Mines, have further improved the grade appreciably and the non-magnetic fraction is found to contain alumina as high as 53%; Fe_2O_3—14.7% and titania as low as 0.33 [2]. An optimization of the wet high intensity magnetic separation process and high gradient magnetic separation (HGMS) for the fine particle separation techniques may further bring down iron content below 10% and proportionately increase the

Table 4 Chemical composition and mineralogy of typical Eastern Ghats bauxite

Chemical Composition (%)		Mineralogy	
Al_2O_3	44.3	Al_2O_3 as Gibbsite	41.18
SiO_2	1.90	Boehmite	0.42
Fe_2O_3	26.7	Diaspore	ND
TiO_2	2.70	Kaolinite	1.58
LOI	23.7	Alumo-goethite	0.81
CaO	0.01	Total	43.99%
MnO_2	0.05	SiO_2 as Kaolinite	1.86
K_2O	0.03	Quartz	ND
ZrO_2	0.05	Total	1.86
Cr_2O_3	0.05	Fe_2O_3 as Hematite	19.50
P_2O_5	0.12	Alumo-goethite	7.23
V_2O_5	0.08	Total	26.73
		TiO_2 as Anatase	2.0
		Rutile	0.7
		Total	2.7

alumina. Further physical and chemical processes are required to bring down iron content below 3% in beneficiated non-magnetic fraction.

4.2 Stage 2 De-ironing by Reduction Roasting

Laboratory developed processes for reducing iron content in bauxite involve reduction roasting followed by magnetic separation technology, specifically designed for high-iron gibbsite bauxite. It has been demonstrated that iron and aluminium can be effectively separated after reduction at specific temperature (around 500–600 °C) with the assistance of specialized additives like coke. As a result, metallic iron concentrate with high iron grade and non-magnetic products with elevated alumina content can be obtained from high-iron gibbsite ore. The metallic iron concentrate finds utility for making iron products, while alumina extraction can be carried out from the non-magnetic product. A schematic representation of the process is shown in Fig. 3 [3].

To bring down the iron content from 10 to 3% of lateritic bauxite the above process can be modified as per the requirement and bauxite quality. The other alternative is

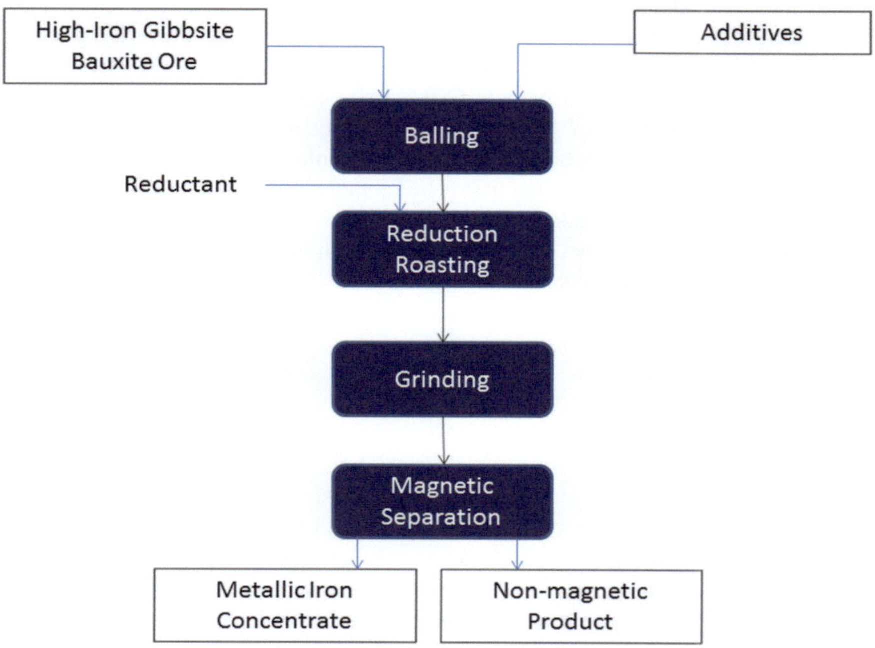

Fig. 3 Reduction roasting followed by magnetic separation flow-sheet

to convert goethite into hematite by heat treatment of bauxite at 500–550 °C, further grind to liberate hematite and again use WHIMS and/or HGMS for iron removal [3].

4.3 Stage 3 De-ironing by Chemical Method

Various tests were carried out on Indian bauxite to leach with hydrochloric acid resulting in the formation of $FeCL_3$ and produce high alumina product. A specific HCl acid concentration is required for optimum leaching and in the process the iron content substantially decreases in product. However, a trade-off was observed as there was a loss of alumina takes place during leaching process [3]. Later on, tests were carried out on heat treated bauxite at temperature of 550 °C and it is observed that the reactivity of alumina with HCl decreases thanks to phase changes in bauxite, whereas iron readily reacts with HCL. During the reaction iron oxide get converted into ferric chloride (FeCl3), which is soluble in water. Symbolic form of reaction is given below.

$$\mathbf{Fe_2O_3(s) + 6HCl(l) \rightarrow 2FeCl_3(l) + 3H_2O(l)}$$

The transition of bauxite colour from reddish-brown to a lighter shade indicates successful ore leaching [3]. The optimal results were achieved by leaching 300 micron-sized bauxite with 25% HCl concentration for 60 min at 950 °C. The recovery rate for beneficiated bauxite (No. 1) was about 62% while beneficiated bauxite (No. 2) obtained after 40 min leaching showed a 76% recovery rate. During the process, there was a substantial reduction in titania content, resulting in a total $TiO_2 + Fe_2O_3$ of only 3.61% in the first beneficiated bauxite, making it an excellent quality non-metallurgical grade bauxite [3] (Fig. 4).

Although the laboratory scale tests were carried out on raw bauxite, it is suggested to use bauxite concentrate obtained during the first stage of de-ironing tests having only 10% Fe_2O_3. In Brazil, HCl leaching of bauxite is industrially applied in the production of low iron and low titania bauxite [3].

4.4 Bio Beneficiation

Bio beneficiation is an environment friendly method that utilizes microorganisms to extract metals from ores. Bio beneficiation involves the targeted elimination of unwanted minerals from an ore through the use of microorganisms, leading to the concentration of the desired mineral in ore [4] Several microorganisms have been studied for the ability to keep aluminium unaffected and remove the iron content from ore without the loss of alumina. Example of such microorganisms are Pseudomonas, Thiobacillus, Leptospirillum, Acidophillum, Aspergillus, etc. Group of microorganisms such as Acidithiobacillus thiooxidans, Leptospirillum ferrooxidans,

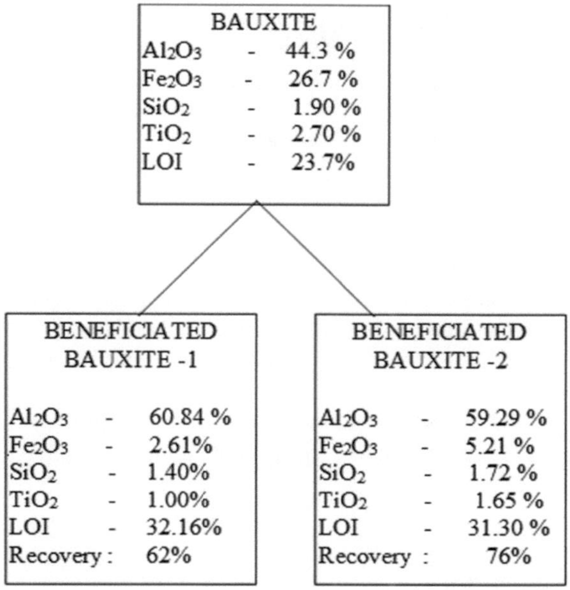

Fig. 4 Results of Bauxite after acid leaching tests

Sulpholobus spp., Acidithiobacillus ferrooxidans (A. ferrooxidans), commonly used for biochemical leaching [4, 5]. This bacterium is acidophilic that grow vigorously under low pH environment. The source of energy is iron and sulphur compound. As in chemical method acid reduces the iron similar this bacterium also reduces the iron leave alumina relatively unaffected [3].

The simplified explanation of biochemical leaching is as follow:

- A. ferrooxidans are cultured and grown in medium.
- Bauxite ore is then mixed with cultured A. ferrooxidans and nutrients rich medium.
- As the A. ferrooxidans metabolize the nutrients, they produce sulfuric acid due to which pH of solution is reduced. Due to low pH and sulphuric acid, there is detachment of iron from bauxite ore
- Alumina is resisting to this condition it remain relatively unaffected.
- After process solid residue contain alumina and liquid contain iron.

5 Summary

- The bauxite specification for the value-added products is more stringent compared to the ore used in the aluminium industry. The natural high alumina and low iron bauxite resources are fast depleting in the world and at present only Guyana produces this quality ore.

- In some bauxite mines, white bauxite occurs along with reddish brown metallurgical grade ore, and they can be separated during mining and accumulated for calcination and value addition.
- Among the value-added bauxite products, the calcined bauxite dominates, which is a raw material for high quality refractory, abrasives, castable and brown fused alumina. The best calcined bauxite is produced in India by importing high quality Guyana washed ore.
- There are opportunities to beneficiate metallurgical grade bauxite, which occurs in large quantity in the Eastern Ghats in the state of Odisha and Andhra Pradesh. A complex process involving wet high intensity magnetic separation, followed by reduction roasting and / or use of chemical / biochemical leaching may produce high alumina and low iron bauxite, suitable for calcination.
- As market value of calcined bauxite is equivalent to metallurgical alumina, complex beneficiation process can be applied to bring down the iron content and proportionately increase the alumina in gibbsitic bauxite.

References

1. https://www.fastmarkets.com/industrial-minerals/bauxite-alumina
2. Nandi A. K. and Murthy P., Production of Special Grade Bauxites and Export Quality Ore by De-Ironing Eastern Ghats Bauxite of India, 8th International Congress of ICSOBA
3. Datta B. and Nandi A. K, Bauxite Beneficiation: An Approach to value Addition in Mining, In: Innovation in Sustainable mining, Balancing Environment, Ecology and Economy, July 2021, pp. 99–114
4. Sukla L. B, Pattanaik A., Pradhan D., Advances in Beneficiation of Low-grade Bauxite, Light Metals, February 2019, pp. 3–10
5. Bahira S., Raut S., Sharma D., Bioleaching of Alumina from Low Grade Indian bauxite (41% Al_2O_3) by indigenous bacteria with reference to pH, Time and Carbon source, International Research Journal of Environmental Sciences, Vol. 7(9), September- 2018, pp 20–27

Sustainable Methods of Dewatering and Disposal of Processing Plant Tailings

K. S. Raju

Abstract Mineral processing is carried out on low-grade ore to recover valuable minerals as concentrate and reject unwanted minerals as waste, generally termed tailings. Mineral separation when done by dry processing (without using water media) either employing gravity or magnetic separation techniques, the tailings are generated in a dry state and can be transported by conveyor belts or closed tube conveyors to designated tailing dumps. In such cases, there may not be many environmental control issues except for using some dust suppression measures. But, in case mineral separation is carried out by wet methods of processing (using water media in flotation, fine gravity, and wet magnetic separations) the tailings generated are a mixture of water and solids in the form of slurry. In general practices, the tailings in the form of slurry are pumped to designated tailing ponds or tailing Dams for long storage. The disposal of tailings as a slurry with large quantities of water has many environmental concerns and is one of the major problems facing the mineral industry in maintaining environmental controls and safety standards. Impoundment of tailings into tailing dams along with large quantities of water has environmental concerns such as water conservation, and pollution of ground and surface water which are regular immediate concerns, and hazards of dam breaches and land degradation have a long-term impact. In view of the water scarcity expected in coming years, the water contained in tailings needs to be recovered and recirculated. Now several regulations are promulgated at the Government level to restrict the use of water and also to reclaim and recirculate slurry water. Secondly, make-up water to compensate for the losses of water at various stages of processing should be lower as far as possible. This depends on the type of ores processed and may vary between 10 and 30% of feed water to the plant. For this reason, permissions to establish mineral processing units are given only when these guidelines are adhered to. Recently environmental regulations also restrict the disposal of tailings as slurry into natural or artificial ponds. In this regard, it is recommended to adopt and use several methods of dewatering equipment like high capacity and efficient thickeners, vacuum or pressure filters, etc. Secondly, dry methods of processing as well as pre-concentration by dry methods

K. S. Raju (✉)
Indian Bureau of Mines, Nagpur, India
e-mail: satyamhari@hotmail.com

K. Randive et al. (eds.), *Current Trends in Mineral-Based Products and Utilization of Wastes: Recent Studies from India*, Springer Proceedings in Earth and Environmental Sciences, https://doi.org/10.1007/978-3-031-50262-0_22

are encouraged. Many mining companies in the world are now making efforts to reduce water consumption and reuse and also financing R&D projects to develop dry methods of ore process. Tailing management is considered one of the major environmental challenges for the mining industry. Under the Sustainable Development Framework (SDF), there are many new guidelines for handling and disposing of mining waste and plant tailings. More emphasis is given to the recovery of water from tailings prior to its disposal. Presently, several methods of dewatering and safe disposal of tailings have evolved in recent years and the industry is keen to adopt sustainable methods. In this paper, some of these methods are discussed.

Keywords Low-grade ore · Dry method of processing · Wet method of processing · Environmental concerns · Recovery of water · Dewatering

1 Introduction

Mineral processing is basically the separation and concentration of desired minerals or metals present in ore from unwanted minerals called gangue minerals with the purpose of producing a marketable product or for use in various mineral-based industries. During mineral processing, ROM (Run of Mine) ore is reduced to the desired size by comminution and subjected to suitable separation processes either in dry conditions or in water media. The level of size reduction and stages and methods of separation depend on the mineral composition and nature of mineralization. At the end of mineral processing, two products are generated, one concentrate and another, tailings (waste). The quantities of concentrate or tailings depend on the type of ore. In the case of industrial minerals like iron ores, limestone, rock phosphate, etc., tailings may be from 20 to 50%, whereas in the case of sulfide and gold ores, tailings may be huge from 90 to 95% [1].

Mineral processing of low-grade ores particularly, iron ores, rock phosphates, marly and kankar limestone, bentonite, coal, and talc generates tailings consisting mostly of fine clayey minerals. The clay minerals present in the tailing rejects are smectites (montmorillonite group) or kaolinite which are extremely fine in sizes below 10 microns. The slurry carrying clay minerals when pumped to settling ponds for settling of solids usually takes long periods to settle and to recover water. In such cases, detailed characterization studies as well as effective dewatering methods are required. Hence, the handling and management of tailings are one of the major challenges for the mining industry with respect to protecting the environment and controlling the costs of processing.

2 Nature of Tailings and Problems of Disposal

The tailings generated in mineral processing plants vary widely in particle sizes, mineral composition, and quantity of process water. In dry methods of separation (without water media), the tailings are mainly coarse and can be transported easily by conveyor belts to tailing stacks/ dumps [2, 3]. There may be some dust generation that can be controlled easily using dust suppression methods or by using tube conveyors. There are no major environmental issues in such cases.

Mineral processing is done in water media in many gravity methods like jigging, tabling, spirals, etc., and in wet magnetic separators and flotation. The use of water media is advantageous for easy handling of solids, better size reduction during grinding and classification, and results in best separation in gravity and flotation methods. However, the tailings are generated in the form of "slurry" (solids mixed with water). The slurry carries water with varied percentages of solids expressed as pulp density (that is the ratio of solids versus water). The disposal of tailings in slurry form has many environmental concerns faced by the mining industry.

The traditional methods of disposal of tailings are storage in ponds and specially designed tailing dams. Some of the other methods are: (i) discarding tailings after thickening into river or sea, (ii) thickened tailings are staked as a dump, used for backfilling underground and open pit mines and to raise the height of tailings dams [4]. The area occupied by tailings ponds or dams varies largely from the size of a swimming pool to thousands of acres. The disposal of tailings into dams and ponds has raised many concerns with respect to the protection of the environment and the prevention of disastrous hazards. The major environmental impacts are (i) Contamination of underground water, (ii) Generation of acid rock drainage in case tailings contain sulfide minerals, (iii) Collapse of Dam embankments and flooding of the dam, and (iv) land and soil degradation.

3 Management of Tailings

Safe and sustainable disposal of mineral processing plant tailings is also one of the key issues of the mineral industry. Impoundment of slurry tailings into tailing dams is the most common method and there are no other alternatives presently due to economic reasons. However, there are some instances of major environmental disasters caused by tailing dam failures. In order to safeguard from the adverse effects, many measures are recommended to protect the environment and prevent hazards. Some of these measures are noted below:

(i) Dewatering and thickening of slurry employing suitable dewatering techniques [3].
(ii) Removal of suspended fine particles like clay using a combination of chemicals and filters.
(iii) Recovery of water to the maximum for recirculation.

(iv) Precipitation of heavy metals before impounding to tailing dams.
(v) Neutralization of excess flotation reagents and cyanides.
(vi) Monitoring the stability of tailing dam structures periodically.
(vii) Laying impervious synthetic or clay material at the bottom of the tailing dam to prevent seepage.
(viii) Control of inflow and outflow of water to prevent flooding in the event of unexpected stormwater entering tailing dams.
(ix) Periodic dredging of material to maintain the depth of tailing dams.

In recent years, there has been tremendous public awareness to protect the environment and work for sustainable development. Amongst many issues, the safe disposal of rejects from mining, beneficiation plants, ash from thermal plants, red mud from alumina refineries, and metallurgical slag and waste from smelters gained attention. Many countries have issued guidelines and promulgated new regulations for handling and safe disposal of mining, metallurgical, and industrial wastes.

4 Sustainable Methods for Disposal of Plant Tailings

The basic requirement to plan the safe disposal of plant tailings is the characterization of tailings and determination of Batch settling tests (BST). In mineral processing, R&D studies are mainly concentrated on finding suitable processes to obtain good concentrates. Less attention is paid to studying the nature and composition of tailings and finding solutions for disposal. In some processing plant operations settlings of solid suspensions become a problem during dewatering. In such cases, it was observed that the presence of certain minerals like clay and carbonaceous minerals created havoc during settling operations identification and quantification of such minerals is a pre-requisite. Therefore, characterization and settling studies on tailings is an important step. Basically, there are four steps involved in safe and environmentally sustainable disposal of tailings:

(i) Characterization of tailings and determination of Batch settling tests.
(ii) Settling of coarse suspended solid above 0.045 mm size using hydro-cyclones.
(iii) Thickening of fine particles in Thickeners with the help of chemical coagulants and flocculants and dewatering the tailings for recirculation. Employing different thickeners such as deep cones or paste thickeners to prepare thick pulp with 60% solids.
(iv) Use of mechanical filtration equipment to prepare semisolid cake of tailings with moisture between 10 and 15%.

5 Characterization of Tailings and Determination of Batch Settling Studies

Tailings in mineral processing plants contain different minerals in different size ranges. Some minerals such as clays, carbonaceous materials, and laterite are finer in size and create problems in settling and consolidation. The following characterization studies to identify such minerals are very important.

(a) Identification of clay minerals by x-ray diffraction and Differential Thermal Analysis (DTA).
(b) Sub-micron size distribution analysis of particles using laser micro analyzers.
(c) Determination of surface charge of particles using ZETA potential equipment.

Batch settling studies called JAR Tests using different coagulants and flocculants at different PH conditions help to determine the Thickener area used in the dewatering and consolidation of suspended solids. Chemical additives such as coagulants and flocculants are a wide range of natural organic and inorganic products (namely lime, ferric chloride, alum (aluminium sulfates) to synthetic polymers). Coagulants are positively charged chemicals (cationic) that serve to reduce the negative (anionic) charge of suspended particles. This helps to reduce electrostatic repulsive force and brings particles together to form **flocs**. Flocculants are synthetic polymers of long chains that create a physical bridge to bind together flocs. Flocculants are either cationic anionic or neutral.

The batch settling tests are done using a beaker or measuring jars (1000 ml) Coagulants are added with different dosages while stirring at high speeds. Flocculation takes place by binding small particles to form big particles called “flocs” and slowly settling to the bottom. Settling studies provide data on the rate of settling and to select suitable chemicals and dosages. A view of the settling test Apparatus is shown in Fig. 1.

6 Dewatering Methods and Equipment

*Dewaterin*g is more important to recover water for recycling so as to use minimum fresh make-up water. Dewatering of concentrates to transport is always given importance. Dewatering of tailings is now becoming important due to pressures from society and environmentalists to conserve precious water. Dewatering is mainly a combination of sedimentation and filtration. Many types of dewatering equipment are available and their application depends on the nature of tailings. Each type of equipment is used either as a stand-alone or in combination depending on the requirement of moisture level in thickened solids. Many dewatering equipment manufacturers and suppliers have testing or simulation apparatus to find the suitability of a particular equipment to dewater tailings of different natures.

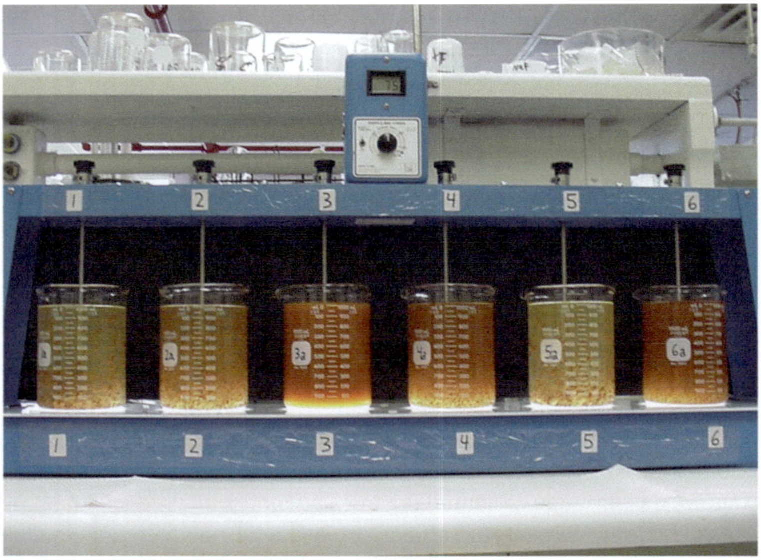

Fig. 1 Jar tests with a beaker and stirring mechanism

6.1 Gravity Sedimentation Methods

6.1.1 Drying Bed

Drying beds consist of perforated or open joint drainage pipes laid within a gravel base covered with a layer of sand. The slurry is placed on top of this sand layer and allowed to dry. Water is removed through natural evaporation and by gravity draining from the slurry mass through the supporting sand to the drainage piping. Cracks develop as the sludge dries, allowing evaporation to occur from the lower layers which accelerates the drying process. The schematic of the drying bed is given in Fig. 1.

6.1.2 Geo-textile Tubes

Geo-textile tubes are made with strong polypropylene fabric. The tubes are filled with pressurized pulp treated with flocculants. The solids are retained in the tube and allow water to flow out through the textile membrane. The advantages of geotextile tubes are static and require less mechanism, operation, and labor to install.

The main disadvantage is that the operation may require more polymer addition and the cake may not be dry to the desired extent. The geotextiles are likely to face cloth blinding resulting in slow dewatering.

6.1.3 Thickeners

One of the commonly used sedimentation equipment is Thickener. In Thickener, solids settle and concentrate at the bottom and clear water accumulates at the top and is collected for recirculation.

A thickener or clarifier is a rotary-raked vessel like a well, that accepts slurry from cyclone overflow. The dilute pulp is fed vertically downwards to the base and flocculant dosage is added to the feed in a mixing tank before being discharged to the bottom. The flocculant slowly mixes with solids and agglomerates the fine particles. The flocculated solids settle at the bottom and form a sludge blanket. The clarified water flows over a peripheral weir. The slow-rotating rake directs settled solids towards the center and it is discharged as underflow and pumped to the next dewatering equipment or tailing ponds. Thickeners handle large capacities ranging from thousands of tonnes to ten thousand tonnes per hour. Some special Thickeners, such as Deep Cone Thickeners and Paste Thickeners, are developed for special applications in place of conventional ones.

In general, thickening is done on the overflow from hydro-cyclone, and the underflow of Thickener is subjected to various filters like belts or, pressure filters. A schematic arrangement is shown in Fig. 2.

The main advantage of Conventional Thickener is to collect clean water for reuse and help as primary dewatering equipment. The disadvantage is that they occupy large space (100 ft. in diameter) and multiple units may be required. Thickener discharge is required to be treated further by other mechanical dewatering units.

6.2 *Dewatering Filter Equipment*

Some of the other commonly used filters for dewatering are described in brief below:

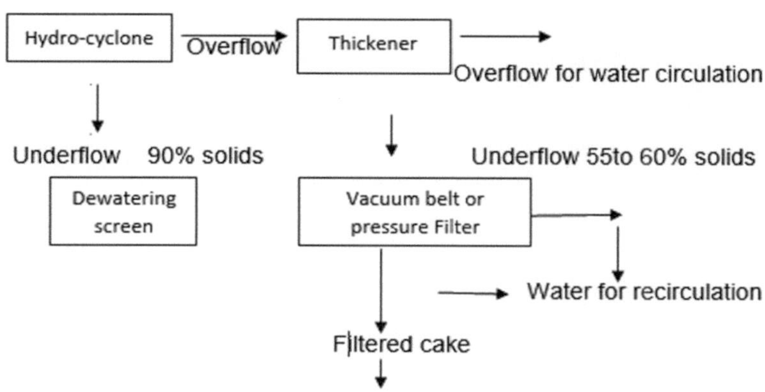

Fig. 2 A schematic arrangement of Thickener setup

Centrifuge Filters—Disc and Bowl centrifuge.
Vacuum Filters—Belt and Drum type.
Pressure Filters—Plate and Screw type.

6.2.1 Solid Bowl Centrifuge

In a solid bowl centrifuge, a high centrifugal force of up to 200G is generated. The high "g" centrifugal force takes advantage of specific gravity differences to separate solids from liquid to obtain a dry solid cake. The bowl centrifuge is always operated as a secondary step operation after primary thickening. The operation of a bowl centrifuge is a skilled mechanism and may not produce a cake with the required moisture.

Centrifuge separates solids from liquids through sedimentation and centrifugal force. Solids are fed through a stationary feed tube, accelerated through ports in the conveyor shaft, and distributed to the periphery of the bowl. The bowl, spinning at high speeds, separates the water from the solids which are compacted against the bowl walls. Solids can then be conveyed to additional centrifuge drying stages while the separated liquid is discharged continuously over adjustable weirs at the other side of the unit. The solid bowl centrifuge is a continuous operation that requires less labor. The main disadvantage is that the filter cake is not as dry as required. The operation requires skilled manpower and consumes high electric power.

6.2.2 Vacuum Drum Filter

Vacuum filters involve creating a vacuum to draw out water from solids. The filter consists of a drum over which the filtering medium is laid. The drum is set in a tank with one-quarter submerged in the cake or sludge. Valves and pipes are arranged so that a vacuum is applied to the inner side of the filter medium as the drum rotates slowly, causing water to be drawn from the sludge through the filter media usually cloth. The solids move along with the drum to the top and the cake is subsequently removed by a stationary knife blade.

Operating costs are usually higher than other processes. However, vacuum filters are independent of seasonal conditions, require a small area, and run as a continuous process. They can produce dry solids enough to eliminate the need for subsequent steps such as heat treatment before disposal.

6.2.3 Belt Filters

There are two types of filters. One is a belt pressure filter and another is a belt vacuum filter. Belt filter press applies mechanical pressure on the slurry by sand witching the slurry between two wire mesh belts which squeezes the water from the slurry. These wire mesh belts move on rolls with decreasing size in diameter. At the end of the path, filter cake is scrapped. The water squeezed is collected at various chutes and drained out.

The vacuum belt filter is known as the horizontal belt filter. The vacuum belt filter is a continuously operating filter with an endless cloth supported by a grooved carrier belt. The belt and cloth are driven by around two rollers across suction boxes. The slurry is introduced at one end and dewatered cake is collected at the other end. Vacuum belt filters can process high volumes of pulp at less cost compared to other mechanical filters. The water sucked at the bottom is collected and discharged.

6.2.4 Pressure Filters

Filter presses consist of several filter plates in series. These plates are metal frames containing large sheets of filter cloth, usually polypropylene. Filter presses are operated by compressing the plates together to make a water-tight seal, and then pumping the feed through a tube down the center of the plate as seen in Fig. 3. The slurry deposits solids on the filter cloth as the filtrate leaves the press via discharge ports. The solids continue to deposit and eventually form a cake in the spaces between the filter plates. Filter presses can be operated as batch or continuous processes. When operated as a batch process, filtering is paused to remove the filter cake when it has reached a maximum thickness. For continuous processes, the filter press is modified. An automatic filter press has the same concept as the manual filter and frame filter, except that the whole process is fully automated. It consists of larger plate and frame filter presses with mechanical "plate shifters". The function of the plate shifter is to move the plates and allow rapid discharge of the filter cakes accumulated in between the plates. Pressure filters are very efficient and are mostly used to get a filtered cake with less moisture.

6.2.5 Screw Press

The screw press is designed to dewater solids by transporting them in a helical path along the length of a tapered screw fitted in a cylindrical vessel to collect water. As solids move towards the narrow screw area, solids are compressed, water is squeezed out and drained through a filter and collected for discharge [5]. The solids move up along the screw and are collected separately. A schematic diagram is shown in Fig. 4.

7 Conclusions

Mineral processing plant tailings are generally impounded to tailing ponds and tailing dams [6]. In view of many concerns of environmental pollution and hazards, storage of tailings in ponds is discouraged. Many guidelines and policies are enforced to

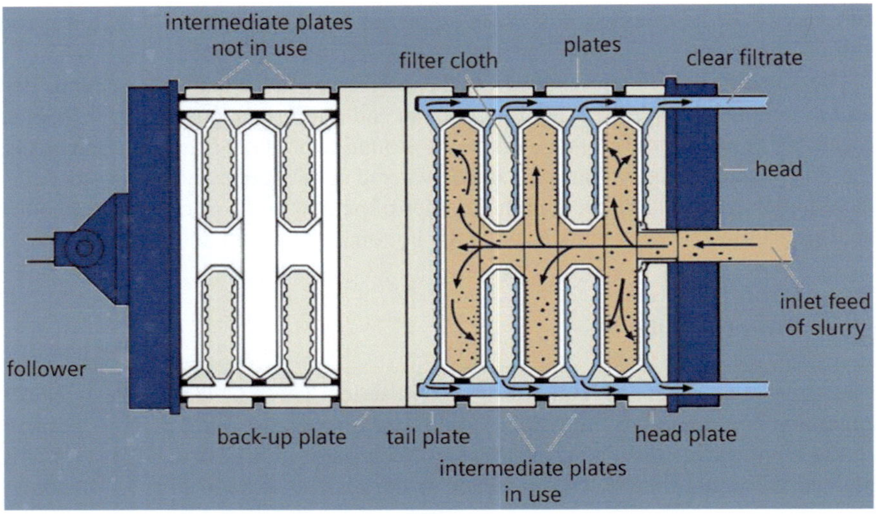

Fig. 3 View of a pressure filter

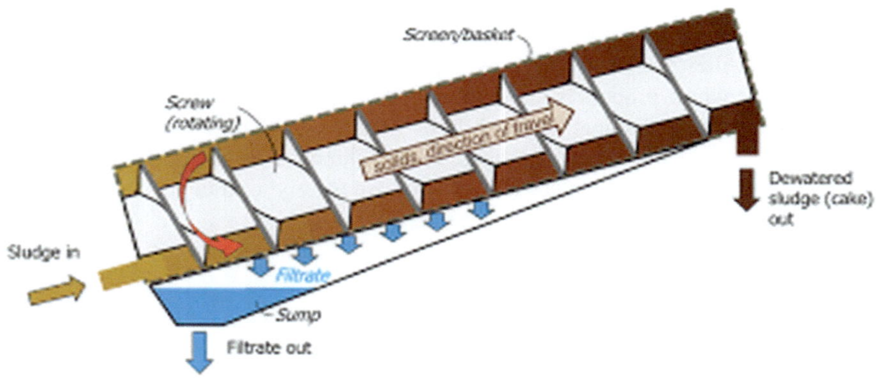

Fig. 4 Schematic view of screw press

maintain the safety of dams and undertake environmental control methods. Dewatering of tailings and circulation of water is given priority for the conservation of water. Sustainable methods for the disposal of tailings are given priority by government agencies, social groups, and environmentalists. The Ministry of Environment, Forest and Climate Change, Government of India has also issued many guidelines and instructions for the conservation of water and safe disposal of tailings and reviews critically while giving EC for setting up processing plants.

References

1. European Commission Reference Document (2009) Management of Tailings and waste rock in mining activities.
2. Holdich RG (2003) Solid-liquid separation equipment selection and modeling, Mineral Engineering. 16 (2), 75–83.
3. Tarlaton, E.S. (2007) Solid–liquid separation equipment selection and process design, Elsevier publications.
4. Stocks P (2006) Paste and Thickened Tailings, Australian Centre for Geo-mechanics.
5. Hunter DW (2006) The positive impact of polymers on sediment treatment and handling. Proceeding of the Western Dredging Association, 26th Technical Conference.
6. Wills et al. (2006) Mineral Processing Technology, Elsevier publications 7th Edition.

Utilization of Internal Waste (Fe < 35%) of Iron Ore Mines: An Initiative Towards Zero Waste Mining

Suman Krishna Sit, Kshirod Chandra Brahma, Suraj Ranjan Prasanjit, Cholagangadev Nayak, and Biswajit Jena

Abstract During mining of Iron Ore, three lithological units: Laterite/Limonite, Shale and BHJ/BHQ are being excavated as internal waste, whose Fe (Total) is less than 45%. All internal wastes are being dumped in waste dump area along with top lateritic Over Burden (OB). The heterogeneity of litho-types with differential geotechnical characteristics of it may affect the waste dump stability, as space limitation for keeping litho unit-wise waste. Banded Hematite Jasper (BHJ)/Banded Hematite Quartzite (BHQ), typical rhythmically alternate layered (Fe rich Hematite and SiO_2 rich Jasper/quartzite) and chemical precipitates of ocean sediments, is one of the intermittently occurring internal waste during iron ore mining operation. As per the Indian Bureau of Mines (IBM) classification, According to Gazette notification No. C-284/3/CMG/2017 dated 25th April 2018, the Fe < 45% BHJ/BHQ has been classified as waste rock and more than 45% Fe as sub-grade. As per recent development, waste of iron ore mines having Fe ≥ 35% also needs to be estimated and reported to the Government. BHJ is having Fe < 35%, where the hematite bands are less in comparison to jasper or quartzite bands, which gives an artistic view and the physical and mechanical properties enable it for both internal or external architectural use as well as ornamental use. Hence, utilization of such lower grade waste, whose abundance potential is substantial, is a noble attempt towards zero waste mining and improving circular economy.

Keywords Iron ore · Banded Hematite Jasper (BHJ) · Banded Hematite Quartzite (BHQ) · Over burden (OB)

S. K. Sit (✉) · K. C. Brahma · S. R. Prasanjit · C. Nayak · B. Jena
M/s Odisha Mining Corporation Ltd., Bhubaneswar 751001, India
e-mail: suman.sit@odishamining.in

K. Randive et al. (eds.), *Current Trends in Mineral-Based Products and Utilization of Wastes: Recent Studies from India*, Springer Proceedings in Earth and Environmental Sciences, https://doi.org/10.1007/978-3-031-50262-0_23

1 Introduction

Opencast mining operations involve the removal of substantial quantities of wastes dumping in designated area as well as backfilling of the excavated area. These waste materials, containing very low/no concentration of ores, are either occurring top of the ore benches (Over Burden) or within ore benches (Internal waste). In iron ore mining, three lithological units: Laterite/Limonite, Shale, and Banded Hematite Jasper (BHJ)/Banded Hematite Quartzite (BHQ) are being excavated as Internal waste and are being dumped in waste dump area along with top lateritic Over Burden (OB). The heterogeneity of litho- types with differential geotechnical characteristics of it may affect the waste dump stability, as space limitation for keeping lithological unit-wise waste.

BHJ/BHQ is a typical ocean sedimentary rock having rhythmic banding with alternate layers of Hematite (Fe-Rich) and Jasper/Quartzite (SiO_2 rich) occurring in Iron ore stratigraphic horizon and having mineralogical content of hematite and quartzite/jasper with minor amount of goethite/limonite and clay. It occurs in the iron ore mining benches either on the top of the ore or laterally with ore or within the ore, whose removal is imperative for proper development of mine benches during iron ore excavation. Based on the Fe%, as per IBM classification, Fe < 45% BHJ/BHQ has been classified as waste rock and more than 45% Fe as Sub-grade. Presently millions of tons of BHJ, BHQ are stockpiled in and around the mine site. Basing on the need of high amounts of Iron-ore to cater the need of steel industry, it was earlier proposed to use this material after upgrading the Fe% of BHJ/BHQ through beneficiation [1–8], but till date, these techniques are not properly utilized due to the lack of cost-effective beneficiation techniques for Fe separation. Odisha Mining Corporation Ltd (OMC ltd) is also examining the possibilities of beneficiation of BHJ/BHQ, whose Fe is >35%. The BHJ having Fe < 35% is difficult to beneficiate as it will be more enriched in SiO_2 and secondly, the interlocking pattern between jasper/quartz to hematite is complex, resulting high cost to grind it to smaller size to liberate the Fe mineral to make it suitable for beneficiation because of high cost. So, this material is kept in waste dumps, attracting environmental hazards as well as increase in cost of mining with respect to dump management. Hence, an attempt has been made in this paper to develop an understanding of utilization of such waste, whose abundance potential being substantial is a noble concept towards Zero Waste Mining and improving circular economy.

2 Presence of BHJ in the Mines of OMC

In OMC Iron-ore leases viz. Gandhamardan Iron Ore Mine (Block-B); Kurmitar Iron Ore Mine; Jilling-Langalotta Iron Ore Mine and Khandband Iron Ore Mine, BHJ occurs in both the grades, i.e., more than 35% Fe and less than 35% Fe. BHJ

Fig. 1 Occurrence of BHJ in mining area. (Kurmitar Iron ore mine, M/s OMC Ltd.)

Fig. 2 Different size of BHJ in dumps. (Gandhmardan Iron ore mine, M/s OMC Ltd.)

less than 35% Fe grade accounts approximately 15–20% of the total BHJ Resources, in consideration of only internal waste occurrences in mine benches (Fig. 1).

In course of mine development, the BHJ needs to be excavated and stacked separately in dumps for future purpose. BHJ can be stored in the dumps either in the form of boulders or cobbles/pebbles (Fig. 2) which are generally produced after blasting of the in-situ BHJ rocks.

2.1 Varieties of BHJ Based on Rhythmic Bands

In BHJ, the variable thickness bands of hematite and jasper/quartzite shows different pattern, which in turn give rise to a very artistic view because of the alternate colour combination. The six different varieties of BHJ observed in the field are as follows (Fig. 3 for corresponding photographs):

1. BHJ having thick bands of Japer: Jasper bands thicker than Hematite bands,
2. BHJ having equally thick Jasper and Hematite bands,
3. BHJ showing Chevron folding,
4. BHJ having almost equal bands of Jasper and Hematite,
5. BHJ having equal small thin laminations of Jasper and Hematite, and
6. BHJ having thick bands of Hematite: Hematite bands thicker than Jasper bands.

Fig. 3 Six Different varieties of BHJ (Type-1 to 6) on the basis of rhythmic band width variations. (Khondbond Iron ore mines, M/s OMC Ltd.)

The Fe < 35% BHJ has an artistic look due to the thicker band width of Jasper. Similarly, it is a massive, hard and compact rock. These properties make the BHJ as a decorative stone that can be used as building material. If these properties are utilized, then this deleterious industrial waste can be converted into a value-added product.

3 Properties of BHJ

3.1 Megascopic Properties

Megascopic (hands specimen) study unveiled the distinctive characteristics of a Banded Iron Formation consisting of few millimeters to a few centimeters thick either magnetite (Fe_3O_4) or hematite (Fe_2O_3) layer of various shades from grey to black with alternate bands of Jasper/Quartz (SiO_2), often red in colour, of similar thickness.

Table 1 Mineral % distribution in BHJ

Mineral name	% of occurrence (by volume) in BHJ
Hematite-Fe_2O_3	55–60%
Quartzite/Jasper-SiO_2	40–45%
Clay-$Al_2O_2 \cdot 2SiO_2 \cdot 2H_2O$	< 1%
Goethite-FeO(OH)/Limonite– $FeO(OH) \cdot nH_2O$	< 1%
Magnetite- Fe_3O_4	Traces

3.2 Microscopic Properties

Microscopic studies, especially their mineralogy has been studied in details. According to several researchers BHJ consists predominantly of hematite with alternate layer of quartz/jasper and sometimes minor amount of limonite or goethite and clay [9, 10]. Hematite is usually subhedral to anhedral shaped fine to medium grained and usually show platy bands alternate with silica layer. Hematite grain carries different degree of fine disseminated silica inclusion and some places hematite with silica gangue shows intergrowth/interlocking texture.

Hematite grains varies in size from 10 μm to a maximum of 1600 μm.

Occasionally, hematite is altered to Specularite (Fe_2O_3) and associated with martitised magnetite. Quartz exhibits a fine-grained texture, appearing as micro-cryptocrystalline clusters. Many of these quartz grains contain numerous tiny inclusions of iron oxide.

3.3 Distribution of Minerals

From the microscopic study, it was revealed that BHJ consists mainly of iron as well as silicate minerals along with other clay minerals [11]. The distribution of minerals in BHJ is given in Table 1.

4 Field Study

BHJ obtained from the mining faces is of mixed sizes (Boulders, Cobbles, Pebbles and Granules) with different width of layers because of the blasting and gives artistic appearances. The huge piece of BHJ boulder is placed in Shakti Sthal, Delhi gives a very beautiful look as well as encourages to think the use of this material as exterior architectures (like Rock Sculpture, path ways and paving stone etc.) as well as decorative stones. BHJ rock competency, toughness and appearance may also be suitable as building material. Thus, the physical properties are being mapped and compared with other building stones like Khondalite, Granite, Sandstone and Marble, which are being used as building material.

The blasted small pieces of BHJ are very beautiful in appearance, therefore, in order to understand their alternate utilization, a number of manual tests like cutting, polishing etc. have been done (Fig. 4).

The following properties are generally considered for evaluating any material before utilizing as building material, namely, the appearance, structure, strength, hardness, toughness, porosity and absorption, seasoning, weathering, workability and fire resistance (Table 2).

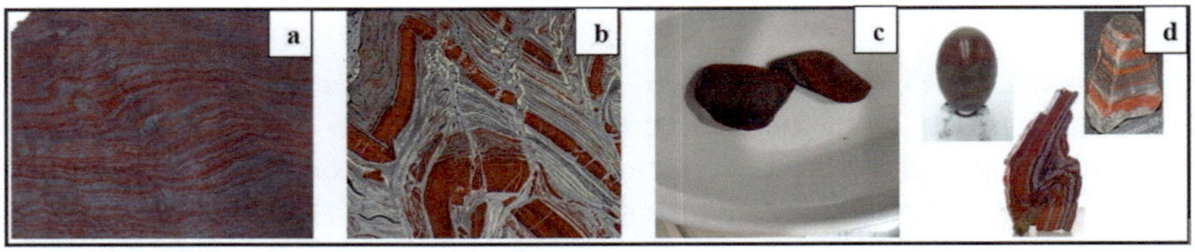

Fig. 4 Appearance of BHJ **a** BHJ after slicing, **b** BHJ after polishing, **c** BHJ colour reflectance under water, **d** Small-sized BHJ as decorative stone

Table 2 Properties of BHJ suitable for other application

Properties	Creative application	Decorative application	Ornamental application
Appearance and texture	√	√	
Colour		√	√
Thickness of bands		√	
Lustre	√	√	√
Hardness	√	√	
Grain size			√
Fracture	√		√
Strength	√	√	
Cleavage			√
Reflectance		√	
Transparency		√	√
Porosity	√		
Geotechnical properties	√	√	

Table 3 Different mode of utilization of BHJ

Mode of uses	Type of uses
Interior architecture uses	# Decorative aggregates in homes
Exterior architecture uses	# Paving stone # Rock sculpture # Garden pathways # Decorative water pathways
Other architectural uses	# Curbing material
Construction industry	# Dimension stone #Used for flooring and cladding # As stair treads, borders and window sills
Antique uses	#Artifacts
Other uses	# Touchstone # Cemetery markers # Creating artwork

5 Probable Mode of Utilization

The properties of BHJ having less than 35% Fe justify the mode of utilization for both decorative as well as for construction industries. Table 3 summarizes various ways in which BHJ can be used as a decorative stone.

6 Conclusion

The BHJ/BHQ which is a mining waste, occurs at different mining locations such as on the top of the benches or as partings within the ore body or in the waste dumps. Although, this material which typically as 45% > Fe < 35%, is considered as a waste, this material has a potential to make the value-added products. Here a case study of the Orissa Mining Company (OMC) is presented and based on this it is inferred that the BHQ/BHJ can make a suitable decorative stone, which can be commercially cut, polished and sold as a value-added material for buildings.

However, for commercialization, the following works need to be done, in consideration of availability of BHJ in four Mine locations of OMC:

- Mine location wise assessment of 35% Fe grade BHJ resources with sustainable availability.
- Performing tests to determine strength, physical properties and chemical treatment of BHJ from each location as well as conducting detailed assessment of stone properties and suitability for architectural, decorative, ornamental and associated uses based on reported characteristics from each location.
- BHJ Usage as External Architecture as well as Construction purposes, while development of in-house colony and associated infra-structures like Parks/Garden,

colony pathways and community halls etc. It is planned for on-site cutting and production of different sizes of blocks/slabs for specific usages and thereby helping in livelihood generation of local communities.

- Assessment of mass production feasibility of different sizes of BHJ blocks from mining locations and documentation of the manufacturing processes.
- Estimation of cost of production and development along with marketing strategies for finished BHJ products, which is to be used for different architectural usages.

The usage of BHJ (<35% Fe) not only reduces the waste dump space but also give the source of earning to the local people. Sale of this material can be targeted in Indian market as well as the international markets. Hence, it's a noble attempt towards Zero Waste mining and reinforcing the Circular Economy.

References

1. Beura, D., Satpathy, B. and Singh, P. (2012): Studies on Petromicrography of BIF In Eastern Limb of Bonai-Keonjhar Belt with Special Reference to Banspani-Jilling-Jajang Area of Keonjhar District, Odisha. Vistas in Geological Research. V. 11, pp. 159–167.
2. Das, B., Mishra, B.K., Prakash, S., Das, S.K., Reddy, P.S.R., and Angadi, S.I.. (2010): Magnetic and flotation studies of banded hematite quartzite (BHQ) ore for the production of pellet grade concentrate. International Journal of Minerals, Metallurgy, and Materials; 17(6):675–682.
3. Das, S.K., Prasad, R., and Singh, R.P. (2018): Characterization-assisted reduction roasting of BHJ, West Singhbhum, Jharkhand, India. Transactions of the Indian Institute of Metals; 17:1357–1362.
4. Gurulaxmi, S.N.; Ghosh, T.K. and Mukherjee, A.K. (2010): State of Art Characterization Study of Banded Hematite Jasper for Beneficiation. In Proceedings of the International Seminar on Mineral Processing Technology (MPT—2010), Jamshedpur, India, 15–17 December 2010; p. 440.
5. Mel'nik, Y.P. (1982): Precambrian Banded Iron-Formations, Physico-chemical Condition of Formation, Elsevier, 310 p.
6. Mohanty, J.K., Jena, M.S., and Paul, A.K. (2012): Integrated Mineralogical Characterisation of Banded Iron Ores of Orissa and Its Implications on Beneficiation. Journal of Minerals and Materials Characterization and Engineering; 11(12):1133–1142.
7. Roy, S. and Venkatesh, A.S. (2009): Mineralogy and geochemistry of banded iron formation and iron ores from eastern India with implications on their genesis. Jour. Earth System Sci., v. 118(6), pp. 619–641.
8. Santosh, T., Angadi, S.I, Das, Nilima, Eswaraiah, C., Tripathy, and Sunil Kumar (2019): Characterization and Comminution Studies of Low-Grade Indian Iron Ores; Mining, Metallurgy & Exploration.
9. Singh, S., Sahoo, H., Rath, S.S., Palei, B.B., and Das, B. (2015): Separation of hematite from banded hematite jasper (BHJ) by magnetic coating. Journal of Central South University; 22(2):437–444.
10. Srivastava, R. and Prasad, R. (2020): Studies on Characterization and Beneficiation of Banded Hematite Jasper of Joda Area, Eastern India. Transactions of the Indian Institute of Metals; 73(1):215–221.
11. Venkatesan, J., Ubayadullah, M., Kumar, P. Mrunmaya, Sah, Rameshwar, Kaza, Marutiram and Rao, Srinivas (2017): Use of Banded Hematite Quartzite (BHQ) in Blast Furnace for Partial Replacement of Quartzite, Trans Indian Inst Met 70(10):2529–2536, https://doi.org/10.1007/s12666-017-115-9.

Challenges and Barriers in Implementing Value Addition to Mine Waste and Its Sustainability

Kaustubh Deshpande, Pratik Godbole, Sanjeevani Jawadand, Tejashree Raut, Saili Dhok, Pradeep Kumar Jain, Krutika Jangale, and Kirtikumar Randive

Abstract The endeavor to implement value addition to mine waste while ensuring sustainability is marked by a complex array of challenges and barriers. Foremost among these is the inherent variability in the composition and characteristics of mine waste, posing difficulties in devising standardized processes for its extraction and utilization. Geological and environmental factors further compound the issue, necessitating tailored solutions adapted to diverse mining sites. Concurrently, addressing the adverse environmental consequences of traditional mining activities, including soil and water contamination, is imperative for achieving sustainable practices. Regulatory complexities and protracted permitting procedures can stymie innovation in mine waste management, amplifying the challenge. Striking a harmonious balance between economic viability and upholding environmental and social responsibility is a delicate feat. Interdisciplinary collaboration emerges as an indispensable factor, facilitating the formulation of comprehensive and sustainable strategies to transmute mine waste into valuable resources. Triumphing over these multifaceted hurdles is pivotal in harnessing the complete potential of mine waste while preserving the well-being of our planet. This chapter endeavors to elucidate the hurdles and sustainable methodologies linked to the creation of value-added products from mining waste. It delves into the prospects of revolutionizing mining waste into valuable commodities through pioneering approaches. The chapter highlights the difficulties linked to transforming mining waste into valuable commodities. It also delves into the concept of life cycle assessment in the context of achieving a circular economy for the advancement and market introduction of these value-added products.

K. Deshpande · P. Godbole · T. Raut · S. Dhok · K. Jangale · K. Randive (✉)
Department of Geology, RTM Nagpur University, Nagpur, India
e-mail: randive101@yahoo.co.in

S. Jawadand
Shri Mathuradas Mohota College of Science, Nagpur, India

P. K. Jain
Indian Bureau of Mines, Nagpur, India

K. Randive et al. (eds.), *Current Trends in Mineral-Based Products and Utilization of Wastes: Recent Studies from India*, Springer Proceedings in Earth and Environmental Sciences, https://doi.org/10.1007/978-3-031-50262-0_24

Keywords Challenges · Value added products · Life cycle assessment · Environmental footprint · Circular economy

1 Introduction

Adding value to mine waste is a complex endeavor fraught with several significant hurdles. Firstly, the diverse composition of mine waste, which can vary widely depending on the type of mining operation and the geological characteristics of the deposit, poses a substantial challenge. Developing efficient and cost-effective processes for extracting valuable materials from this heterogeneous waste is a formidable task. Environmental concerns also loom large. Traditional mining activities often result in contamination of soil and water resources, and finding sustainable methods to mitigate these environmental impacts while adding value to the waste is a critical obstacle. Moreover, achieving economic viability in mine waste valorization can be difficult. The costs associated with research, development, and implementation of value-added processes must be balanced with the potential revenue generated from the recovered materials. Regulatory and permitting issues further complicate the process. Obtaining the necessary approvals and adhering to environmental regulations can be time-consuming and costly, hampering innovation in this field. Overcoming these hurdles is crucial to unlock the untapped potential of mine waste as a valuable resource while ensuring responsible and sustainable practices in the mining industry.

2 Environmental and Economic Benefits

Heaps of historic mine waste pose both a boon and bane for society as well as the environment. If mine waste is treated and value is added to it without any negative impact on the environment then this bane of mine waste can turn out to be a boon for the society as well as the environment. Value can be added to the mine waste, with reprocessing of the waste with new innovative technologies, which was earlier dumped on the open grounds near the mine. Reprocessing of mine waste opens new doors of revenue generation and it also has environmental benefits [1, 2]. Some of the benefits of reprocessing of mine waste are discussed below.

2.1 Reduction of Environmental Footprint

Increasing demand for better human life has increased the mining and processing of minerals as well as the generation of waste. Generated waste poses threat to the environment if not handled properly. Increasing demand and increasing prices for the economic minerals have led to the reduction of cut-off grade for the minerals by mining companies. The situation is more concerning in the case of critical minerals due to their exponentially growing economic importance and limited supply chains [1]. This leads to more mining activity and eventually generates more waste. This generated waste acts as an environmental liability which leaves its impact on the environment and can be termed an environmental footprint. Geotechnical stability, and the release of dissolved metals, or suspended particles impact heavily on the environment leaving a serious environmental footprint [2].

The threatening impact of mine waste on the environment can be reduced not only with the help of reprocessing, recycling, and reuse of the mine waste but also, by adding value to the reprocessed and recycled mine waste. For this purpose, mine waste must be classified into two types viz. mineralized mine waste and non-mineralized mine waste. Mineralized mine waste which possesses a minor percentage of mineral value could be considered a potential source for the extraction of minerals in the future and can be sent to the reprocessing unit for value addition [1]. Such practice will reduce the need for additional mining activities to extract the same mineral and limit the impact of mining on the environment.

Non-mineralized mine waste can be reworked and utilized on the basis of its physical properties. This creates an opportunity to consume large volumes of non-mineralized mine waste [1]. Such non-mineralized mine waste can be considered for road construction, building construction, and brick manufacturing. But before selecting such materials for construction purposes, it should be considered that they should not pose any threat to the environment and are suitable for climatic modifications [3]. Globally, sand and gravels are the basic units for construction materials, and due to high urbanization projects, their demand has increased three folds in last two decades. This has created a problem of mining sand, leading to a decrease in groundwater level, increased water turbidity, and adverse impact on birdlife [4]. Mine waste can also counter a threat to the environment due to illegal sand mining. Processing mine overburden from coal mines into sand can reduce the environmental impact of sand mining and can be a good alternative for the river sand [5]. River sand as back-filling material can also be replaced with processed overburden from coal mines. It is a promising practice for the reduction of the environmental impact of mining and is beneficial in the case of the disposal of mined overburden [6].

With 1.4 billion population, India is the third largest CO_2 emitter in the world which is propelled by the economic growth in the twenty-first century and the consumption of energy [7]. To decrease this incorporation of CO_2 in the atmosphere, pressure is increasing at the international level. This led to the innovation of carbon capture and storage to reduce atmospheric CO_2 in silicate rocks in the form of carbonate hence it is called as carbonation process [8]. This process is combined with the production of biofuels such as biodiesel and further conversion to electricity with the help of Hybrid Direct Carbon Fuel Cells or Microbial Fuel Cells as discussed earlier, will surely reduce the part of CO_2 in the atmosphere and eventually reduce carbon footprint in the atmosphere.

Carbon footprint in the environment can also be reduced with the help of trees which can be sustained in the mining environment. Most of the mining activity is concentrated on the extraction of sulphides which results in contamination of soil by toxic heavy metals such as Cu, Zn, Pb, and Cd. Also, mining activity loosens the soil which results in soil erosion and reduction of agricultural yield. Various modern methods are applied to such contaminated environments but, when applied they are either temporary solutions or are costly. On the other hand, the treatment of such soil biologically with the help of micro-organisms, fungi, and plants along with the extraction of metals is an effective remedy for such an environment and is an environment-friendly alternative for the generation of revenue [9].

2.2 Social and Economic Benefits

The upending of world economies towards clean energy or green energy will impact more or less on the demand and supply of every mineral [10]. The abundance and scarcity of economic minerals play an important role in a nation's economy which is volatile according to the demand and supply of economic minerals. Increasing population plays an important role in increasing the demand for minerals and if an equilibrium is not maintained between the demand and supply of economic minerals then, it generates problems such as culminating production, market scarcity, and also social stress due to unemployment [11]. The imbalance between demand and supply, unavailability of employment due to lack of availability of capital in the market, and non-availability of skilled workers leads to unemployment [12]. Along with this, the increasing cost of extraction and dwindling economic returns are emerging issues due to the gradual decline in low-cost or high-grade ore. The market reacts to such an event by reducing demand or inflating prices. Profits of mining companies are eventually reduced which leads to depletion in production [13].

All these issues can be countered by the valorization of mined waste and generating revenue through adding value to the mined waste. People's ability to work can be classified into skilled and unskilled labor. Skilled labor with high formal education can directly be employed in production as well as non-production with proper training according to the demand of the industry. At the same time, the unskilled workforce can be employed for the work of production only [14]. In valorization of mine waste, work of production refers to brick making, sand production from mine waste, bioremediation, and phytoremediation of land, and also to the work where not much skill is required for the work [15]. The cost of bricks from iron ore tailings with 91% iron ore tailings and 9% Clay + Cement was around Rs. 13, which is comparable to traditional bricks (around Rs. 15 for one brick) and may reduce further on an industrial scale [16]. Agro-mining can be a good job opportunity in the field of mining sector [17]. The concept of agro-mining is the selection of metal hyperaccumulator and high biomass plant species on low-grade ore bodies or mineralized soils, wastes, and industrial sludge which is followed by harvesting and incineration of the biomass to produce a bio-ore from target metals or salts may be recovered [18]. Agro-mining can be carried for metals with high economic value such as Ni, Co, Tl, and Au [19]. There is an increase in demand for good quality ornamental plants in urban as well as in rural India. Therefore, the nursery business has been rapidly developing in recent years in our countries [20]. Demand for ornamental plants can also be fulfilled by cultivating suitable plants in the contaminated area surrounding the mine site as well as in the mine waste dump site. Ornamental plants are classified according to their uptake viz. excluders are the ones in which heavy metals are retained in the roots and not transported further to the shoot; accumulators accumulate heavy metals in above soil plant; and hyperaccumulators are the plants that accumulate heavy metals and translocate it the shoot without showing any toxic symptoms [21]. Here for nursery business, we can use excluders and accumulators having economic values. Agro-mining tends to attract metal and alloy producers for low-carbon impact metals and also Micro-Small-Medium Sized Enterprises (MSMEs) who are keen on implementing agro-mining. Agro-mining can develop sustainably by creating value and job opportunities for farmers in emerging fields for a green economy [16].

Products derived from the treatment of mine waste can be sold in the market which will generate revenue for the mining company catering to the problem of capital which can further be reinvested either in the mining or reprocessing of mined waste. It has been found that there is no difference in the strength of bricks, cement, sand, and gravels generated from the mine waste compared to traditional construction materials and hence can be promoted in the market. This will generate finance for the mining company. Mine waste-based geopolymer can serve as an alternative to cement used for construction material. Selling fuel generated from mine waste can serve as an alternative to fossil fuels [15]. As India is focusing on green fuel, the generation of biofuel from the mine waste by biotreating the mine waste can serve India as a good option. In the formulation of medicines, minerals can also be an important constituent [22]. Pyrite which is called Vimal in ayurveda is used for the treatment of chronic fever, edema, anemia, anorexia, piles, and jaundice [23]. Copper pyrite is used in the preparation of *Swarna makshika bhasma* [24]. Citrate-CuO nanohybrid materials made from the copper recovered from mine tailings showed antimicrobial response in different species of bacteria paving the way to develop low-cost antibiotics. Such practices can help to reduce the waste burden created by mines [25]. Thus, trading of products made from the processing of mine waste can lead to environment-friendly revenue generation eventually increasing the profit of mining companies [15] (Fig. 1).

3 Challenges and Barriers to Implementation

While discussing the benefits of valorization of discarded material from mining a question arises—how feasible it is to process the mine waste and create a value-added product from it? What if reprocessing is more harmful to the environment than the dumping of the material at the dumping site? These questions should be clearly answered.

Mine waste contains various types of minerals with varying proportions. Varying geochemical characteristics and varying grain sizes lead to varying metal recovery potential [26]. For example, rare earth elements are not rare but, the similarity in

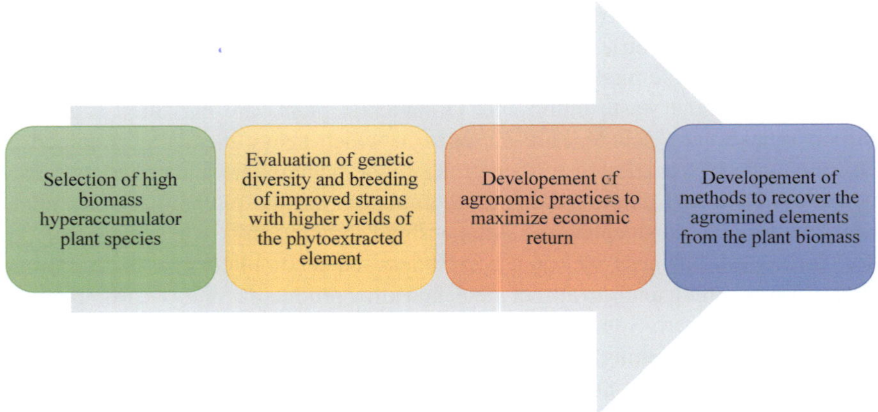

Fig. 1 Development of agromining [19]

their chemical and physical properties makes them difficult to separate from each other and hence, they are termed as rare [27]. Similar is the case of various elements making them difficult to separate. Separation of unwanted elements from desired elements poses a main challenge while dealing with mine waste. Technological and engineering developments are needed under the R&D wing for treating mine waste and focusing on a particular element.

Technological and engineering development feel to be effective while developing at a laboratory scale but, while using them practically they fail to deliver desired results. Although technology is proven to deliver desired results, industrial awareness about technology and their willingness to implement innovative technologies are the most important as industries focus more on their profit. Along with this, a problem of unwillingness for technology transfer also arises due to insecurity of the industrialists, due to increased competition and fear of losing their monopoly [28]. Such barriers create weak technical infrastructure and create a mindset that necessary technologies are not to be used [29].

Economic barriers are also created due to the dependency of commodity prices not only on consideration of demand and supply chains but also, on the profits of the mining companies [30]. Heterogeneities in the waste, uncertainty in the extracting capacities, and volatility in market situation create disbelief in the mind of stakeholders hindering initial investment decisions [31]. Investors are attracted more toward high-market-value products rather than the low-market-value products. Also, investors are more focused on less shadow-priced commodities i.e., products that improve their quality along with the better usage of material and energy from an economical, technological, and environmental point of view [32]. Refurbishing, remanufacturing, reconditioning, and recycling of the waste creates high operational costs leading to high initial investment which feels like a costly business activity to the investors [33]. Therefore, the management of mineral resources is to be combined with the overall strategy of development [16]. The government always welcomes efforts directed towards the development of new technologies for beneficiation of lower grade and finer size material for conversion of existing mineral resources into feasible economic resources. But while employing such techniques and converting small deposits into reserves, the government is not willing to neglect the safety of workers and reduce accidents. Also, the government demands proper rehabilitation, resettlement, and implementation of schemes under Pradhan Mantri Khanij Kshetra Kalyan Yojna (PMKKKY). Environmental, economic, and social considerations must be taken into account to develop a sustainable approach in the mining sector [34]. While dealing with the projects in areas of dispute, proper legislative structure, economic support, and stable government are required [35].

While discussing the social barrier, lack of seriousness, willingness, support, availability of little information, and languishing behavior of manufacturers hinders the growth of the project. Also, the release of toxic gases and chemicals during the recycling process creates fear of individual safety at the workplace [36]. Desired results are not achieved if recycling and recovering processes are wrongly implemented leading to environmental and economic loss leading to failure of the project [37].

4 Sustainability and Life Cycle Assessment

4.1 Environmental Impact Assessment (EIA)

Environmental Impact assessment focuses on the provisions for the discerning of effects of waste management and alleviation measures during the establishment, functioning, and decommissioning stages of development and contributes to waste management [38]. Environmental impact can be used as a framework for deciding the viability of different waste management methods which increases sustainability [39]. While, focusing on recycling, reprocessing, and reusing mined waste caution must be taken that, the 3Rs must impact positively on the environment and not negatively [39]. To ensure a positive impact on the environment, the influence of environmental information is a must in the designing and planning of the solutions and creativity should be employed in such a way, that the project should not be hampering the environment [40]. Commodities should be designed in such a way that the residence time of elements or metals used in them is high so that, the energy required for further mining and recycling is conserved, impacting highly positively on the environment [35].

Low-grade tailings generated from mining contain economic feasibility which can be employed if technology permits to do so as these tailings are also hazardous and act as pollutants. These tailings are an opportunity for recycling which will reduce the environmental impact [41]. Developing an increased capability of recycling from their source raw material into tradeable commodities with emerging techniques for environment-friendly economic generation is a part of the Environmental Impact Assessment [15].

4.2 Life Cycle Assessment and Circular Economy

Environmental impacts of products throughout the life of the product are studied under Life Cycle Assessment (LCA) [42]. The concept of LCA deals with the important activities in the life of a product i.e., from raw material to waste material. The implementation of LCA allows the identification of transference environmental impact from one medium to another. Non-implementation of LCA, the impact transfer may go unnoticed. Various methods of valorization of the waste are studied under LCA. Landfilling of waste has the highest impact on the environment. The recovery of valuable metals from the waste can also have a high environmental impact if toxic materials are used to extract. The manufacture of construction material has least environmental impact and also, they are economically viable. Therefore, the application of LCA hypothesizes an interesting concept for the improvement of environmental performance during the valorization of mine waste [43].

As discussed earlier, population growth will increase = demand and supply chains of raw materials and final products. The adoption of modernity by society is based on the mineral raw materials and their complete utility decides the economic growth [44]. Today luxurious goods are not only consumed by the upper class of society but also, the middle and lower classes of society. Also, stress on the environment is increased due to the increase in migration due to urbanization. Altogether resulted in demand for raw materials has increased eight-fold in the twenty-first century [45]. Ongoing decrease in the minable grade of most commodities resulting in an increase in generated waste, turning out to be a huge cost and burden for mining industries [41, 46, 47].

About 150 years ago, the Industrial Revolution introduced the concept of linear economy which is "Produce-Use-Dispose" [45, 48]. Open dumping sites or unsuitable sanitary landfills are selected for dumping the used waste [37]. The concept of linear economy does not take account of impact on society, environment, and conservation of critical resources [45]. Linear model poses to be unsustainable in the modern economy [48]. In such a situation, steps were taken to close the loop of waste materials giving birth to the concept of Circular Economy (CE) [49]. "Goods which are at the end of life are rejuvenated as a new product in CE" i.e., the residence time of raw material used in the product will be limitless [37].

Reusing previously disposed unexploited low-grade and unmineralized material at mine sites brings both economic benefits as well as reduced environmental impact [50, 51]. 95% of mine waste is the rock mass mined and due to the diversity in ore bodies, waste is also diverse and site-specific [47]. 2–10 tonnes of waste is removed for each ton of metal extracted from ore bodies which may contain the potential to become valuable as they contain mineral as well as energy resources. Adding value to waste along with environment-friendly methods such as bioleaching will generate huge finance. The process of value addition to waste is called as valorization. Valorisation of mine waste is also included in the Sustainable Development Goals (SDGs) of the United Nations [41]. Better recovery of raw materials can be attained with reprocessing of primary commodities, recovery of critical metals, and extracting reusable mineral fractions, reducing the cost of site remediation [46].

The contribution of mine companies in CE is missing, losing the opportunities for technically and economically viable material recovery [41]. The concept of value addition to mine waste seems to be good at first but, environmental impacts need to be discussed following the life cycle thinking approach including all processes [52] (Fig. 2).

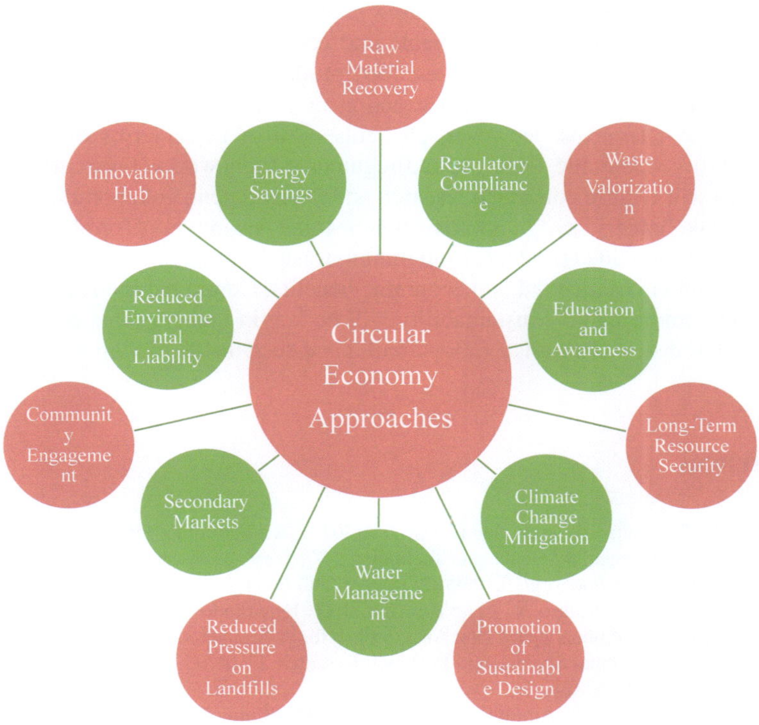

Fig. 2 Recommendations that can lead to a circular economy by the development of value-added products from mine waste

5 Summary

As the mining industry continues to face increasing pressure to minimize environmental impact and maximize resource efficiency, the development of innovative solutions for mining waste management is gaining traction. Advancements in technology and research are enabling the extraction of valuable materials from mining waste that were previously considered uneconomical or technically challenging. This opens new opportunities for creating value-added products from these waste streams. The growing focus on sustainability and circular economy principles further drives the demand for transforming mining waste into valuable resources. Governments, regulatory bodies, and stakeholders are increasingly recognizing the importance of reducing waste generation and promoting resource recovery in the mining sector. Additionally, the market for sustainable and eco-friendly products is expanding, creating a potential market for value-added products derived from mining waste. These products can range from construction materials and aggregates to metals, minerals, and chemicals. However, challenges still exist in terms of technological feasibility, economic viability, and regulatory frameworks. Overcoming these challenges will require continued research and development, collaboration between industry stakeholders, and supportive policies and incentives. Overall, the future outlook for value-added products from mining waste is optimistic, with the potential for significant environmental and economic benefits. Continued innovation, investment, and commitment to sustainability will be key in realizing this potential and driving the transition towards a more circular and sustainable mining industry.

References

1. Shaw R., Petavratzi E., and Bloodworth A., (2013), Resource recovery from mine waste. Issues in Environmental Science and Technology, 37. Waste as Resource, The Royal Society of Chemistry, https://doi.org/10.1039/9781849737883-0004
2. Dinis M., Fiuza A., Futuro A., Leite A., Martins D., Figueiredo J., Gois J., Vila M., (2020), Characterization of mine legacy site: An approach for environmental management and metals recover, Environmental Science and Pollution Research 27:10103–10114. https://doi.org/10.1007/s11356-019-06987-x
3. Segui, P.; Safhi, A.e.M.; Amrani, M.; Benzaazoua, M. Mining Wastes as Road Construction Material: A Review. Minerals 2023, 13, 90. https://doi.org/10.3390/min13010090
4. Bhatwadekar R., Singh TN., Mohammad ET., Armaghani DJ., Hasbollah D., (2020), River sand mining vis-à-vis manufactured sand for sustainability, Springer Nature Switzerland, X.-N. Bui et al. (Eds.): ISRM 2020 - Volume 1, LNCE 109, pp. 143–169, https://doi.org/10.1007/978-3-030-60839-2_8
5. Brahmam G. and Sisindree D., (2020), Experimental examination on utilization of processed mine overburden sand by substituted to river sand in construction, International Journal of Innovative Technologies and Exploring Engineering (IJITEE), Vol. 9, Issue 5, 1858–1866 pp. https://doi.org/10.35940/ijitee.E3206.039520
6. Gupta A. K. and Paul B., (2015), A review on utilization of coal mine overburden dump waste as underground mine filling material: A sustainable approach of mining

7. Arun Kumar Dwivedi & Archana Soni (2022) Drivers and critical paths of carbon emissions in India: a structural path decomposition analysis, Energy Sources, Part B: Economics, Planning, and Policy, 17:1, 2084185, https://doi.org/10.1080/15567249.2022.2084185
8. Jacobs AD., Hitch M., (2011), Experimental mineral carbonation: Approaches to accelerate CO_2 sequestration in mine waste materials, International Journal of Mining, Reclamation and Environment, 25(4): 321–331. https://doi.org/10.1080/17480930.2011.608883
9. Gupta A. K. and Paul B. (2015), Ecorestoration of Coal mine overburden dump to prevent environmental degradation: A review, Research Journal of Environmental Sciences, ISSN, 1819-3412
10. Bazilian M., (2018), The mineral foundation of Energy transition, The extractive industries and society, 5, 93–97
11. Randive K., and Jawadand S., (2020), Mineral economics an Indian perspective, Nova Science Publishers Inc., 213–230
12. Chand K., Tiwari R., and Phuyal M., (2017), Economic growth and Employment rate: An empirical study of Indian Economy, PRAGATI: Journal of Indian Economy Volume 4, Issue 2, pp. 130–137. https://doi.org/10.17492/pragati.v4i02.11468
13. Bardi U., Jakobi R., and Hettiarachchi H., (2016), Mineral resource depletion: A coming age of stockpiling, Biophys Econ Resour Qual, 1:4, 1–9
14. Falkinger J., (2002), A Theory of employment in firms, Physica Verlag Hydelberg, https://doi.org/10.1007/978-3-7908-2649-4
15. Nangulama, Horris K. (2021), Analysis of Abandoned Mine Wastes Disposal Techniques in Malawi: A Concept of Circular Economy. In: Engineering and Technology Quarterly Reviews, Vol.4, No.2, 1–9
16. Lamani S., Mangalpady A., Vardha H., (2016), Investigating the utility of Iron ore waste in preparing non-fired bricks, J. Inst. Eng. India, Ser. D., https://doi.org/10.1007/s40033-016-0129-5
17. Morel J., Simonnot M., Echevarria G., Ent A., and Baker A., (2021), Conclusions and outlook for agromining; Agromining: Farming for metals, Minerals resources reviews, pp. 485–489
18. Nkrumah P., Chaney R., and Morel J., (2021), Agronomy of 'Metal Crops' used in Agromining; Agromining: Farming for metals, Minerals resources reviews, pp. 23–46
19. Chaney R., Baker A., and Morel J., (2021), The long road to developing agromining/phytomining; Agromining: Farming for Metals, Mineral resources reviews, pp. 1–22
20. Gholap V., Benke S., and Patil S., (2022), Management and economics of ornamental nursery, International Journal for Agricultural Sciences, vol.18, Issue 1, pp. 195–201
21. Khan AHA., Kiyani A., Mirza CR., Butt TA., Barros R., Ali B., Iqbal M., Yousuf S., (2021), Ornamental plants for the phytoremediation of heavy metals: Present knowledge and future perspectives, vol. 195, 110780, https://doi.org/10.1016/j.envres.2021.110780
22. Ríos-Reyes, Carlos-Alberto; Ríos-Gutiérrez, María-Paula; Joya-Neira, Santiago, (2021), The importance of minerals in medical geology: impacts of the environment on health Archivos de Medicina (Col), vol. 21, no. 1, https://doi.org/10.30554/archmed.21.1.3589.2021
23. Sharma V., (2018), Vimal (Iron Pyrite): A medicinal mineral drug of Ayurveda- An approach to develop its mineralogical monograph, Asian Journal of Pharmaceutics, https://doi.org/10.22377/AJP.V12I04.2837
24. Meena V., Bhushan S., Chaudhary A., (2021) Pharmaceutical standardization of Swarna Makshika Bhasma (Processed Copper Pyrite): A traditional medicine of Indian Subcontinent, Asian Journal of Pharmacy and Pharmacology, 7(1), 33–37, https://doi.org/10.31024/ajpp.2021.7.1.8
25. Banerjee, A.; Ghosh, R.; Adhikari, T.; Mukhopadhyay, S.; Chattopadhyay, A.; Pal, S.K. Development of Nanomedicine from Copper Mine Tailing Waste: A Pavement towards Circular Economy with Advanced Redox Nanotechnology. Catalysts 2023, 13, 369. https://doi.org/10.3390/catal13020369
26. Komba T., (2021), Evaluation of mine waste characterization to identify opportunities for optimizing, project economics using fragmentation analysis, Mining technology, 130:4, 221–231. https://doi.org/10.1080/25726668.2021.1927587

27. Garcia AC., Latifi M., Amini A., Chaouki J., (2020), Separation of Radioactive elements from Rare-earth element bearing minerals, Metal, 10, 1524, https://doi.org/10.3390/met10111524
28. Stander HM., Broadhurst J., (2021), Understanding the opportunities, barriers, and enablers, for the commercialization and transfer of technologies for mine waste valorization: A case study of coal processing wastes in South Africa, Resources, 10(4), 35, https://doi.org/10.3390/resources10040035
29. Kazancoglu Y., Ada E., Ozturkoglu Y., and Ozbiletkin M., (2020), Analysis of the barriers to urban mining for resource meliorationnin emerging economies, Resources Policy, 68, 1–12
30. Zeng, X., Mathews, J.A., Li J., (2018), Urban mining E-waste is becoming more cost-effective than virgin mining. Environ. Sci. Technol. 52(8), 4835–4841
31. Cossu, R., Williams, I.D., 2015. Urban mining: concepts, terminology, challenges. Waste Manag. 45, 1–3. https://doi.org/10.1016/j.wasman.2015.09.040
32. Stenis, J., Hogland, W., 2014. Economic optimization of urban mining. Iran. J. Energy Environ. 5 (4), 337–344. https://doi.org/10.5829/idosi.ijee.2014.05.04.01
33. Tesfaye F., Lindberg D., Hamuyuni J., Taskinen P., Hupa L., (2019), Improving urban mining practices for optimal recovery of resources from e-waste. Miner. Eng. 111, 209–221
34. National Mineral Policy, (2019), Ministry of Mines, Government of India, pp. 1–12
35. Ali SH, Giurco D, Arndt N, Edmind N, Brown G, Demetriades A, Durrheim E, Eriquez MA, Kinnard J, Littleboy A, Meinert LD, Oberhänsli R, Salem J, Schodde R, Schneider G, Vidal O, Yakovleva N. Mineral supply for sustainable development requires resource governance. Nature 2017, 543, 367–372.
36. Arora R., Paterok K., Banerjee A., Saluja M. S., (2017), Potential and Relevance of urban mining in the context of sustainable cities. IIMB Management review, 29(3), 210–224
37. Ferronato N., Torretta V., (2019), Waste mismanagement in developing countries: a review of global issues, Int. J. Environ. Res. Publ. Health 16, 1060
38. Claassens C., Cilliers D., Retief F.,Roos C., & Alberts R., (2022) The consideration of waste management in environmental impact assessment (EIA) for developments in protected areas, Impact Assessment and Project Appraisal, 40:4, 320–330, https://doi.org/10.1080/14615517.2022.2080491
39. Mehta N., Dino GA, Passarella I., Ajmone-Marson F., Rosetti P., Anotnio de Luca D., (2020), Assessment of possible use of extractive waste coming from Abandoned mine sites: Case study in Gorno, Italy, Sustainability, 12, 2471, 1–22. https://doi.org/10.3390/su12062471
40. Mckillop J., and Brown A., (1999), Linking project appraisal and development: The performance of EIA in large scale mining projects, Journal of environmental assessment policy and management, Vol. 1, No. 4, pp. 407–428
41. Kinnunen P., and Kaksonen A., (2019), Towards circular economy in mining: Opportunities and bottlenecks for tailings valorisation, Journal of Cleaner Production, 228, 153–160. https://doi.org/10.1016/j.jclepro.2019.04.171
42. Pell R., Tijselling L., Goodenough K., Wall F., Dehaine Q., Grant A., Deak D., Yan S., and Whattoff P., (2021), Towards sustainable extraction of technology materials through integrated approachs, Nature reviews, Earth and Environment, Vol. 2, 665-679pp.
43. Gonzalez R. M., Macias F., Meyer A., Schneider P., Nieto J. M., Olias M., Canovas C. R., (2023), Life Cycle Assessment of management/valorisation practices for metal sludge from treatment of acid mine drainage, Environmental Impact Assessment Review, 99, 107038, https://doi.org/10.1016/j.eiar.2023.107038
44. Pomykala R., and Tora B., (2017), Circular Economy in mineral processing, E3S Web of Conferences 18, 01024, pp. 1–6
45. Govindan K. and Hasanagic M., (2018): A systematic review on drivers, barriers, and practices towards circular economy: a supply chain perspective, International Journal of Production Research, https://doi.org/10.1080/00207543.2017.1402141
46. Lemiere B., A circular economy classification of mine waste. Resources for Future Generations - RFG2018, Jun 2018, Vancouver, Canada. ffhal-01740918
47. Zinck J., Tisch B., Cheng T., Cameron R., (2018), Mining value from Waste initiative: Towards a low carbon and circular economy, G. Gaustad et al. (eds.), REWAS 2019, The Minerals, Metals & Materials Series, https://doi.org/10.1007/978-3-030-10386-6_39

48. Korhonen J., Nuur C., Feldmann A., Birkie SE., (2018), Circular Economy as an essentially contested concept, Journal of Cleaner Production, Vol.175, 544-552pp.
49. Ioannis E. Nikolaou, Konstantinos P. Tsagarakis, An introduction to circular economy and sustainability: Some existing lessons and future directions, Sustainable Production and Consumption, Volume 28, 2021, Pages 600–609. ISSN 2352-5509. https://doi.org/10.1016/j.spc.2021.06.017
50. Falagan C., Grail B., and Johnson D. B., (2016), New approaches for extracting and recovering metals from mine tailings, Mineral Engineering, https://doi.org/10.1016/j.mineng.2016.10.008
51. Lebre E., Corder G., and Golev A., (2017), The role of mining industry in a circular economy, Journal of Industrial Ecology. https://doi.org/10.1111/jiec.12596
52. Adrianto L. R., Niu H., and Pfister S., (2021), Life cycle assessment of emerging processes to valorise mine waste, 7th International Slag Valorisation Symposium at Belgium, pp. 46–50

Online Real Time Elemental Analysis of Minerals and Raw Materials Through PFTNA Technique

Rajendra Mishra

Abstract With its high penetrating power, neutron activation is particularly well suited for online real-time elemental measurement. Unlike the conventional surface techniques viz. NIR phase analysis, LIBS laser micro-spot sampling or the hazardous radioactive isotopic sources, the D-T PFTNA based CNA (Controlled Neutron Activation) online analyzer—uses an electrical neutron source that can be switched ON and OFF any time. The CNA analyzer comprises a measuring chamber, with attached electrical box which uses digital signal processing technology and an operator interface. The CNA's measuring chamber, equipped with a robust BGO detector to detect gamma is installed around the conveyor belt and analyses the composition of material moving on it. Interfaced with a high-performance automatic system, the CNA not only makes it possible to measure and control the chemical composition of the raw materials but also adapt the manufacturing process. The CNA is widely used to control and optimize various manufacturing processes viz. in thermal power plants, cement manufacturing, mining & metals industry, ore-sorting, battery recycling and waste treatment plants.

Keywords Bulk analyzers · Cross belt analyzers · Neutron based · PFTNA · Gamma

1 Introduction

While the demand for raw materials has been growing to keep pace with production, scarcity owing to over-exploitation and geopolitical reasons is a stark reality. Hence, it is prudent to not only optimize usage but also minimize waste and unwanted rejections. This is achievable only with robust continuous analysis and monitoring of quality and other vital process parameters. This paper intends to highlight and introduce some novel innovative applications of the D-T PFTNA neutrons.

R. Mishra (✉)
Malvern Panalytical BV, Almelo, Netherlands
e-mail: rajendra.mishra@malvernpanalytical.com

K. Randive et al. (eds.), *Current Trends in Mineral-Based Products and Utilization of Wastes: Recent Studies from India*, Springer Proceedings in Earth and Environmental Sciences, https://doi.org/10.1007/978-3-031-50262-0_25

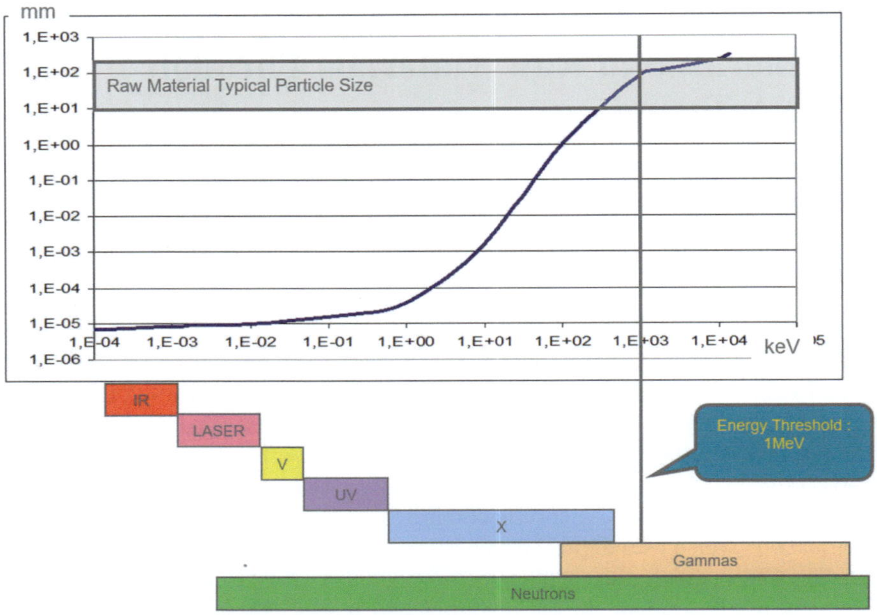

Fig. 1 Depth of penetration

Unlike laboratory sampling techniques, neutron radiation penetrates the whole material giving comprehensive, precise, and consistent values for measurements (Fig. 1). Furthermore, the technology does not need any sampling or sample preparation. Instead, continuous measurement of bulk material moving on a conveyor belt has been made possible.

Conventionally, radioactive isotopes viz. Cf-252 has been the prevalent source of neutrons. However, considering the operational hazards of unsafe inherent radiation, D-T PFTNA is fast emerging as a safe and most reliable alternative for online elemental analysis. Unfortunately, neither of the much-touted 'safe alternatives' viz. NIR (Fig. 2) or the LIBS (Laser Induced Breakdown Spectroscopy) get any closer when it comes to real-time analysis of minerals and raw materials in bulk (Fig. 3).

2 The Technology

PFTNA or the Pulsed Fast Thermal Neutron Activation (Fig. 4) is a unique technique of generating neutrons in a vacuum-sealed tube (akin to a particle accelerator) based on the deuterium–tritium (D-T) fusion reaction, resulting in the formation of a He-4 ion and a neutron with kinetic energy of 14 meV (Fig. 5).

Contrary to Cf-252, where the neutron emission is continuous, permanent and diminishes with time (drops to half of original value in 2.6 years), neutrons generated

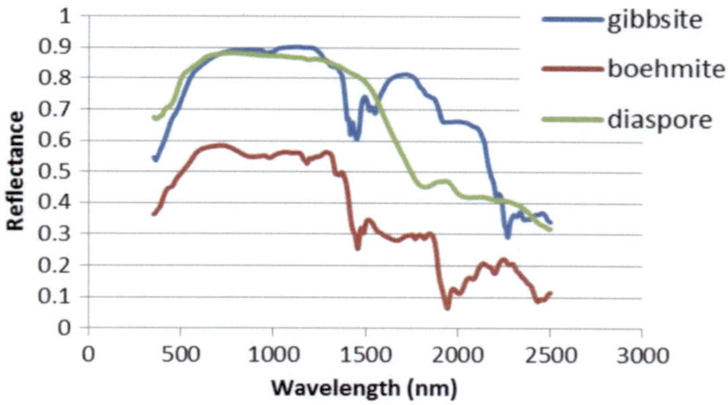

Fig. 2 NIR is only surface and not elemental

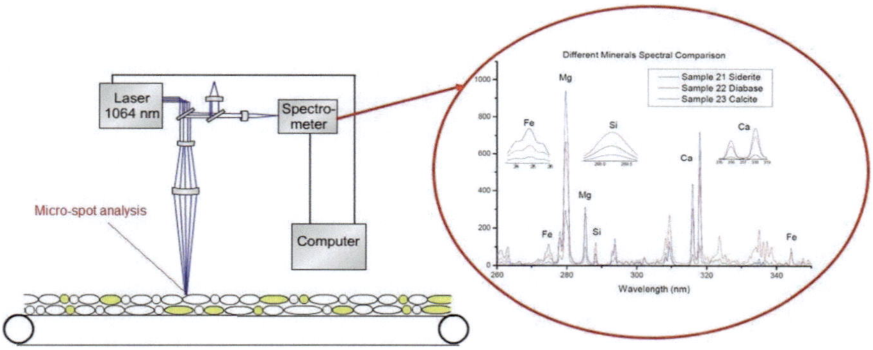

Fig. 3 LASER only surface based micro-spot analysis

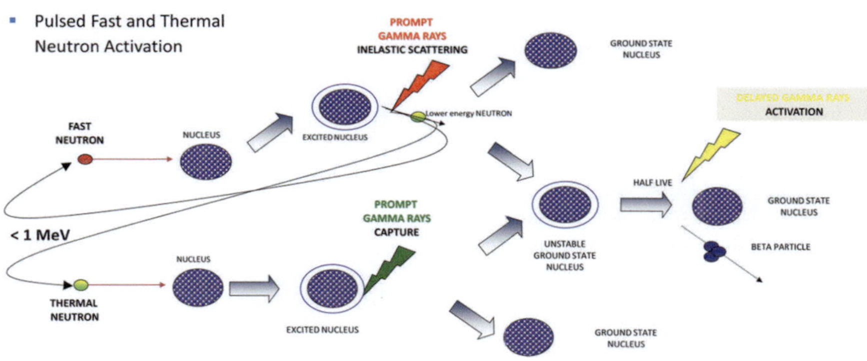

Fig. 4 PFTNA

Fig. 5 D-T PFTNA

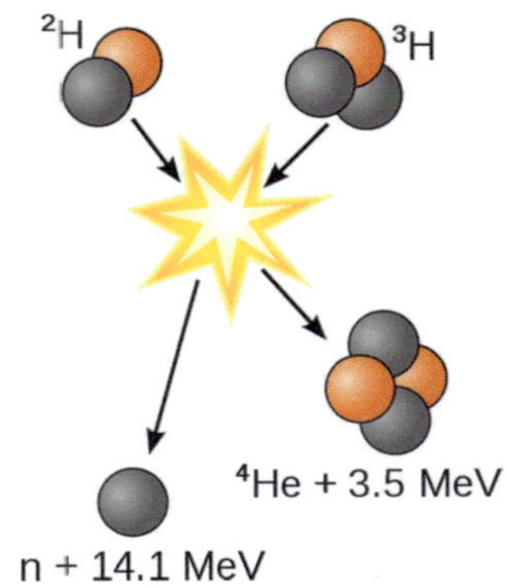

electrically are emitted on demand (on power-up) and the output is constant and stable over time (Fig. 6).

In addition to the possibility of switching off the neutron emission when not required, the flux level is maintained same for the lifetime of the instrument. It renders regular, laborious on-site recalibration unnecessary, leading to cost savings, such as maintenance and daily operations without unexpected downtime. The specific advantages and abilities of the system come from the electronic monitoring of these "controlled" neutrons, whose output never changes. The neutron flux is monitored to provide lasting stable calibration. Precision does not degrade over time and can be fully guaranteed. Applications of PFTNA technique in analysis of minerals viz. Iron ore, nickel, copper, limestone, dolomite, bauxite etc. are a few to mention. The CNA in conjunction with the NIR sensor doubles up as elemental cum phase analyzer, facilitating real-time analysis of bauxite with direct reporting of Al_2O_3, SiO_2, CaO, TiO_2, Fe_2O_3 and TAA (gibbsite + boehmite).

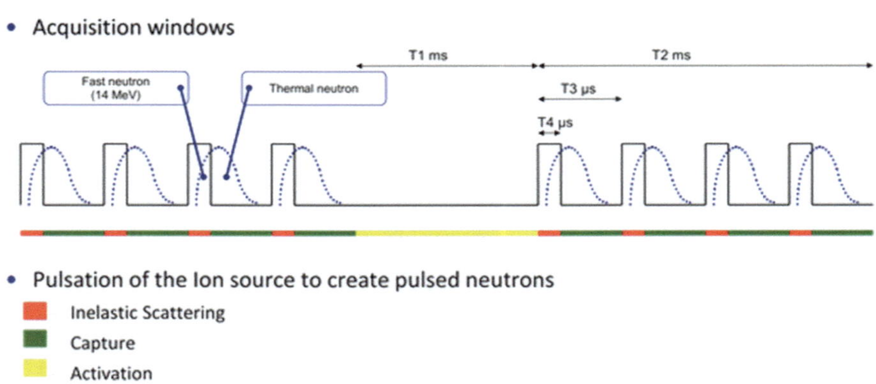

Fig. 6 Three stage activation

3 Summary

In the mining industry, elemental analysis of raw materials is a key step in the manufacturing process, for which quality optimization and profitability are primordial requirements. PFTNA technology makes it possible to do away with artificial isotopic sources (Cf-252 or Am-Be), which are still widely used even though they have been classified as "highly toxic" radionuclide by European regulations, and which pose an additional hazard as they are impossible to extinguish. The D-T PFTNA technology presents numerous advantages in terms of analytical performance, as well as environmental and operator safety.

The use of an electrical neutron source guarantees the stability of the system's analytical performance, through the constant production of neutrons throughout its life cycle. This characteristic eliminates periodic calibration operations, which are often difficult to carry out on-site and are always detrimental to production.

Moreover, with its on/off function, the electrical neutron source can be handled without risk and can easily be transported by air as an accepted package, stored as a spare part and replaced whenever necessary without interrupting the production process.

MIX
Papier aus verantwortungsvollen Quellen
Paper from responsible sources
FSC® C105338

If you have any concerns about our products,
you can contact us on
ProductSafety@springernature.com

In case Publisher is established outside the EU,
the EU authorized representative is:
Springer Nature Customer Service Center GmbH
Europaplatz 3, 69115 Heidelberg, Germany

Printed by Libri Plureos GmbH
in Hamburg, Germany